Computer-aided Prod

CAPE 2003

Conference Chairman

Professor J A McGeough
Regius Professor of Engineering
School of Engineering and Electronics
The University of Edinburgh, UK

Scientific Committee

Professor A Balendra	University of Strathclyde, UK
Professor A Bramley	University of Bath, UK
Professor G Byrne	University College, Dublin, Ireland
Professor K Cheng	Leeds Metropolitan University, UK
Professor A Day	University of Bradford, UK
Dr A de Silva	Glasgow Caledonian University, UK
Professor D Harrison	Glasgow Caledonian University, UK
Professor S Hinduja	UMIST, UK
Professor T Honda	Tokyo University of Agriculture and Technology, Japan
Professor B Hongzan	Huazong University of Science and Technology, China
Professor A Kochhar	Aston University, UK
Professor J Kubie	Napier University, UK
Professor G Levy	University of Applied Sciences, Switzerland
Dr M Lucas	University of Glasgow, UK
Dr E Lundy	University of Tennessee, USA
Professor A Nee	National University of Singapore, Singapore
Dr Pankaj	University of Edinburgh, UK
Professor M Player	University of Aberdeen, UK
Professor K P Rajurkar	University of Nebraska, USA
Dr A E W Rennie	Lancaster University, UK
Professor A Ruszaj	Institute of Metal Cutting, Poland
Dr D Seager	University of Edinburgh, UK
Professor J Simmons	Heriot-Watt University, UK
Professor B Sinha	University of Edinburgh, UK
Dr G Smith	University of the West of England, UK
Dr J Szpytko	University of Mining and Metallurgy, Poland
Professor R Teti	University of Naples Federico II, Italy
Dr A Usmani	University of Edinburgh, UK
Dr G Wollenberg	Institute Allgemeine Elektrotechnik und Leistungs Elektronic (IELE), Germany

18th International Conference on

Computer-aided Production Engineering

(CAPE 2003)

18–19 March 2003

The University of Edinburgh, UK

Edited by

Regius Professor Joseph McGeough
The University of Edinburgh, UK

Professional Engineering Publishing Limited
London and Bury St Edmunds, UK

First Published 2003

Related Titles of Interest

Title	Editor/Author	ISBN
Advances in Manufacturing Technology XVI	K Cheng	1 86058 378 4
Advances in Manufacturing Technology XV	D T Pham, S S Dimov, and V O'Hagan	1 86058 325 3
Computer-aided Production Engineering CAPE 2001	H Bin and co-edited by J A McGeough and H Wu	1 86058 367 9
Computer-aided Production Engineering CAPE 2000	J A McGeough	1 86058 263 X
Design and Manufacture for Sustainable Development	B Hon	1 86058 396 2
Engineering Design in the Multi-discipline Era	P Wiese and P John	1 86058 347 4
Manufacturing Automation ICMA 2002	S T Tan, I Gibson, and Y H Chen	1 86058 376 8
Software Solutions for Rapid Prototyping	I Gibson	1 86058 360 1

For the full range of titles published by Professional Engineering Publishing contact:

Marketing Department
Professional Engineering Publishing Limited
Northgate Avenue
Bury St Edmunds
Suffolk
IP32 6BW
UK

Tel: +44 (0)1284 724384
Fax: +44 (0)1284 718692
Email: Orders@pepublishing.com
Website: www.pepublishing.com

Foreword

18th International Conference on Computer-aided Production Engineering (CAPE)

These proceedings contain papers presented at the 18th CAPE Conference, held in the University of Edinburgh, between 18–19 March 2003. This series of conferences provides a forum for researchers from both industry and academia to discuss progress in research and development in this major branch of Engineering that affects all industrialized countries. Manufacturing globally has witnessed major transitions, since these CAPE Conferences began in 1986. The rapid advancement in electronics and communication technologies enables design and manufacture to proceed on an international dimension that hitherto could not be anticipated. The advent of these technologies has affected large, medium, and small -sized enterprises throughout the world.

The CAPE conferences reflect these changes. When the 14th CAPE was held in Japan in 1998, papers were presented there that dealt with Telecommunications and the Internet, as well as issues on Multi-media and significantly the Human Interface and Safety. The Durham 15th CAPE gave special attention to Global and Agile Manufacturing operations, and to new aspects of Supply Chains and Virtual Reality. The 16th (Edinburgh) and 17th (Wuhan) conferences continued these themes with papers dealing with the Internet, and included comparisons of supply chains in Europe and the Far East. In CAPE 17 there was further emphasis on the changes and opportunities for digital enterprises for product design and manufacture in the twenty-first century (e-manufacture). In this climate it was interesting to find creative design also featuring in the Wuhan CAPE conference in 2001.

Papers on these and more established topics for CAPE are included in the present 18th CAPE. In addition the 2003 CAPE has seen a rise in the number of papers that include innovative and new uses of computer-aided production engineering techniques as an aid to the manufacture of prostheses, by well-known methods of computer-aided design and manufacture, and the use of agile manufacture and robotics in medicine. It is a reflection of global awareness of an ageing world population, and how to meet its needs. (It may be recalled that the Tokyo CAPE was among the first to recognize the social requirements on computer-aided production engineering with papers on the human interface and safety.)

All offers of papers in the form of abstracts were fully refereed by at least two independent referees. Their reports were passed to contributing authors. The full papers were further refereed by local members of the organizing committee, to check scientific/engineering quality and that referees' comments had been met, and adherence followed detailed instructions for paper preparation. Papers are organized in the following main sections.

1. Computer-aided design and manufacture
2. Computer-aided production planning
3. Quality in manufacturing and design
4. Concurrent engineering and design for manufacture
5. Production and control

6. AI-applications in manufacturing
7. Manufacturing and supply chain management
8. Manufacturing processes
9. Rapid prototyping
10. Robotics in medicine
11. Computer-aided orthopaedic surgery

The international committee was drawn from nine countries, whose members are thanked for their help in publicising the meeting and reviewing the papers. The UK Institution of Mechanical Engineers has again been the main sponsor of this meeting, and has published the Proceedings. Its President, Mr John McDougall, is thanked for coming to the conference and addressing our delegates. The Engineering and Physical Sciences Research Council and the Royal Society of Edinburgh are also thanked for their interest and support. BAE Systems Limited in Edinburgh kindly arranged for delegates to visit their company in order to see the industrial applications of CAPE. As always we have arranged a social programme for guests accompanying delegates, showing them the sights of the City of Edinburgh.

Finally, I wish to thank my colleagues Ms Amy Middlemass and Elise MacDonald in the Office of Lifelong Learning at the University of Edinburgh on whom I have relied for much of the day-to-day administration of the conference, and to our University Principal, Professor Timothy O'Shea for formally opening the 18th CAPE.

I hope that all the delegates will enjoy the 18th International Conference on Computer-aided Production Engineering

J A McGeough
Regius Professor of Engineering
University of Edinburgh, UK
Editor and Chairman of CAPE 18, 2003

Contents

Computer-aided Design and Manufacture

Computer-aided Process Planning

Quality in Manufacturing and Design

Concurrent Engineering and Design for Manufacture

Production and Control

AI Applications in Manufacturing

Manufacturing and Supply Chain Management

Manufacturing Processes

Computer-aided Design
and Manufacture

Population seeding for genetic algorithms

G PASSANNANTI and **N SCALZO**
Dip. Tecnologia Meccanica, Produzione e Ingegneria Gestionale, University of Palermo, Italy

SYNOPSIS

Genetic algorithm (GA) success depends on its operator formulation, which must be specific for the particular problem under investigation. Among these operators, the greatest attention is generally turned to crossover, but mutation, coding/decoding, population dimension and its quality are also very important issues.

The initial population quality probably plays an important role in order to speed up the whole evolution process, and the present paper deals with such an issue. A particular stochastic seeding technique is introduced and it is applied to problems proposed in the literature to create some populations that are used as input for two different formulations of genetic algorithm. The results are compared with the ones obtained using random populations and they confirm the goodness of the seeding technique, which often has allowed a high reduction of the residual error or of the computational time required to yield the same final result.

1 INTRODUCTION

Initial population selection is an issue concerning the developing of GAs that probably has not attracted the attention it deserves. The initial population is usually formed by randomly generated chromosomes, provided of course that their phenotypes represent feasible solutions to the specific problem to be solved. Some researchers (1) maintain that although this is not the only way to get an initial population, nevertheless, considering that it is a very simple procedure to apply and that it yields good results, it is not worth trying more sophisticated methods.

From some of the rare studies carried out about this issue it seems possible to conclude that seeding the initial population with high quality solutions obtained with other heuristic methods could help the GA to find better solutions using less computational time, but, on the other hand, there could be the risk that the GA converges prematurely on solutions that are still quite far from the optimum.

As regard this issue (that is: the premature convergence of an inseminated GA) it is possible to make the following remarks. The population dimension should depend on the problem complexity and it should be ample enough so that it can contain different initial chromosomes dispersed over the whole solution space in order not to prevent any potential part of the solution space to be explored by the GA. But finding heuristic algorithms that provide a suitable number of initial solutions is very difficult, even for a flow-shop problem. It is possible to overcome such a problem if the initial population is formed by using solutions calculated with the few available heuristic algorithms in literature ("few" if compared with the population size that generally are used to implement a GA), and then integrating them with other randomly generated solutions. But this procedure will yield an initial population made up by two subsets: one with relatively high fitness values, and the other with poor quality ones. The rare high quality solutions will have much better chance of being selected if compared to the poor quality ones, and this will cause that they will get through the selection operator with higher probability to crossover their own genetic code each other, and this may cause the unwanted effect that the GA converges prematurely on a suboptimal solution, which can be a local minimum corresponding to a solution used to inseminate the GA initial population, or a solution of its neighbourhood.

To judge from the small number of papers produced about it (2), such an issue has not received much attention, amazingly enough. Besides, a new approach is needed to deal with this issue in order to surmount the objections that it is possible to raise against it.

The basic idea is that the whole initial population, and not only a part of it, should be made up by "seeds", which must not be local optima, anyway. So, rather than using good solutions calculated by applying other algorithms, it could be useful choosing random solutions instead, provided that they have been appropriately "adjusted" before they are inserted in the initial population. So, if a criterion was found that could provide some indications as to how the optimum solution structure should be, then this bit of information could be used to form proper initial chromosomes. For example: if it is known that when some particular jobs are placed in a particular position (locus) of the chromosome sequence then the algorithm would generally produce better results, an ample set of solutions could be assembled by inserting these jobs in these particular loci and then randomly adding the remaining jobs in the loci still empty. In this case there would not be a population made up by good solutions combined with completely random solutions, which may cause all the disadvantages previously discussed, but the whole set of initial population solutions will be made up by individuals with similar fitness levels, not necessarily good solutions, but generally better solutions if compared with randomly chosen ones.

As for finding indications about how to build appropriately adjusted solutions to be inserted in the initial population, it is possible to apply a particular technique, available in literature (3). As a matter of fact such a technique is a scheduling algorithm, but it could be easily adapted to this case. A brief description of this algorithm is in the next paragraph.

2 THE SCHEDULING ALGORITHM

For the sake of convenience a 9 job / 7 machine flow-shop problem is addressed. This problem has been solved encompassing all the possible solutions. Solutions have been grouped together according to the job in the first locus of the sequence ($J_{<1>}$), and for each of these groups the related mean makespan value (**MK)** has been calculated. The results are shown in figure 1.

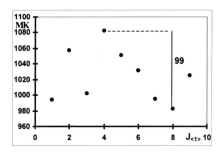

Figure 1. Mean MK values for the groups $J_{<1>}$

It can be pointed out that the $8_{<1>}$ solution set (i.e.: the solution subset with job number 8 in the first position of the sequence) has, if compared with $4_{<1>}$, a 99 unit lower mean makespan, that is about 10% lower. Of course there is no guarantee that the optimal solution will be within this solution set (incidentally, the optimum correspond to the sequence: 8, 9, 6, 3, 5, 1, 2, 4, 7, with MK = 843, which happens to be in $8_{<1>}$ subset), yet, if one wants to limit the search to a subset of the solution space, then this one should be taken into consideration. Similar results are obtained when it comes to $J_{<9>}$ groups, which are the subsets that fixe the job J at the other end of the sequence (Figure 2), whereas as for the central position of the sequence, that is the $J_{<5>}$ solution subsets, the differences between extreme mean MK values are much narrower (Figure 3).

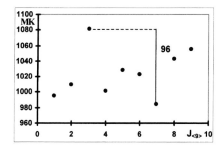

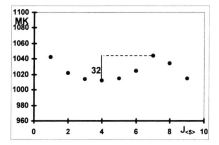

Figure 2. Mean MK values

for the groups $J_{<9>}$

Figure 3. Mean MK values

for the groups $J_{<5>}$

The current problem has been solved exploring the entire solution space, but usually it is not possible to do so, of course; but in these cases the search for the groups with the smallest mean MK values can be based upon random samples extracted from the solutions space.

Considering these results, the algorithm described in (3) is developed as follows.

Given a problem with N jobs, for each of the $N(N-1)$ $J'_{<1>}J''_{<9>}$ possible solution subsets (with $J' \neq J''$), **n** sequences are generated by randomly inserting in the **(N-2)** remaining loci the left behind jobs ($n \geq 3$). For each of these groups, MK values, mean MK value and standard deviation are calculated. Then the lowest mean MK value is compared with all the other mean values by using the Student's test. The result of this test indicates whether the two mean values are significantly different or not. Then all the groups with significantly higher mean value of makespan are left out of further analyses. As for the remaining groups, further sequences are generated for each of them by using the procedure described before, thus increasing the selective capability of the Student's test (in fact the degrees of freedom have increased), in this way detecting as significant even smaller differences (of course the standard deviation plays a crucial role in this procedure). This procedure will be repeated, by increasing sample sizes, until the number of surviving groups is equal to a predetermined value, **a(1)**, or when the sample size of each group has reached a predetermined value, n_{max}. In the latter case the number of surviving groups is greater than **a(1)**. Then a new parameter, d_i, has been introduced in order to reduce the number of samples to **a(1)**. d_i is defined as follows:

$$d_i = MK_{mi} - xs_i$$

where MK_{mi} is the mean makespan of group number **i**, s_i is the standard deviation of the same group, and **x** is a parameter greater than 0 (typically x=1 or 3). The surviving **a(1)** groups are selected on the basis of the smallest d_i values.

The next step of the procedure is to divide up each of the **a(1)** surviving groups into other subsets. Each of these subsets is formed by taking into consideration the positions adjacent to the ones that have already been assigned to two jobs (which will be considered "fixed" from now on), and applying to the (N-2) remaining jobs the previous step of the procedure. In this way other subsets will be eliminated from further explorations, just as it was done before. At the end of this second step, for each of the first two and the last two loci of the partial sequence characteristic of a group there will be a determined job (of course the **a(2)** survived partial sequences will differ for at least one job in the four fixed positions).

Other steps will be appropriately repeated so building even more longer partial sequences, until a predetermined stop condition is met.

It is very important to point out that the process that builds partial sequences is random, therefore different runs of the procedure produce different results.

3 THE SEEDING ALGORITHM

The procedure described in the previous paragraph, which will be called P1, is a step by step procedure. Partial sequences produced at each step have different lengths. The greatest number of different partial sequences available at step **k** ($1 \leq k \leq INT(N/2)$) is equal to:

$$\text{SEQ(k)} = \text{a(k-1)} \text{ [N-2(k-1)]} \text{ [N-2(k-1)-1]}$$

whit $\mathbf{a(0)}=1$

Let the population size that is to be used in the GA be **P**. Then, at every step **k** of the procedure, it is always possible to yield **P** partial sequences by just setting $\mathbf{a(k)} = \mathbf{P}$. Each of these sequences can be integrated with the left behind jobs by inserting them randomly in the remaining loci. Such **P** sequences will be called "oriented" solutions (even if a consistent part of every sequence was generated randomly).

The previous procedure can be called "standard", and it may raise some problems about the diversity of the genetic code of the initial population.

Let q_{in} be the gene in the position **n** of the chromosome number **i** of the GA population. Let define the quantity $\mathcal{D}_k$ as the "diversity degree" of a population at stage **k** of the procedure. $\mathcal{D}_k$ is calculated as follows:

$$\mathcal{D}_k = 100 \ (\ \Sigma_{i,j,n} \ x_{ijn} + \Sigma_{i,j,n} \ x_{ij(N+1-n)} \ ) \ / \ (k \ P \ (P-1))$$

with $i = 1 \div P\text{-}1$ $j = i+1 \div P$ $n = 1 \div k$

       $x_{ijn}=1$ if $q_{in} \neq q_{jn}$

       $x_{ijn} = 0$ if $q_{in} = q_{jn}$

The mean $\mathcal{D}_k$ values calculated for 40 different populations (P=20) obtained by the P1 procedure are shown in figure 4. The proposed problem is a 20/5 one.

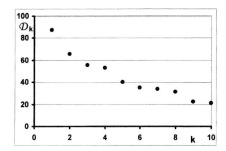 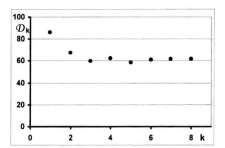

Figure 4. $\mathcal{D}_k$ mean values   **Figure 5. $\mathcal{D}_k$ mean values for grouped populations**

Since the $\mathcal{D}_k$ values seem to be too small, especially in association with higher values of **k**, a maximum value for **k** was introduced, $\mathbf{k_{max}} = \text{INT(N/2)-2}$. Besides, the value of $\mathbf{a(k)}$ was set equal to **P/4**, therefore, to build an initial population with size **P** the procedure P1 was repeated four times. The results are shown in figure 5.

4 TESTING THE SEEDING PROCEDURE

It may seem appropriate using small values of k in order to get better values of $\mathcal{D}_k$, but this is not the case when the objective function is also taken into consideration. In fact: the higher the value of **k** is, the better the mean fitness of the population is. So, in this case, the conclusions one can draw are opposed if compared to the former case. Figure 6 shows the mean per cent increase, with respect to the optimum, of the MK values obtained for different values of **k** (problem dimension is 30/15).

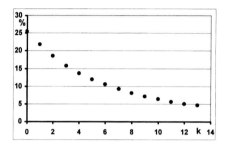
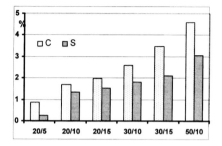

Figure 6. Mean MK per cent increase for different values of k.

Figure 7. Comparison between the two GAs based on randomly generated populations.

Finally, since the diversity degree of the population and its mean fitness level are conflicting, it is not possible to fix an optimal value of **k** *a priori*, so the subsequent analyses have been carried out considering all possible values of **k**, ranging from 0 (that is: the initial population is chosen completely randomly, with no seeds) up to k_{max}.

The tests have been carried out on problems proposed in literature (4) whose optimal solutions are known. Their implementation requires applying a GA. So, two different GAs have been developed using two different crossover operators: the first one, indicated as C, is the well known one-cut-point crossover, the other one, indicated as S, is Siswerda's uniform crossover. Both these techniques are extensively studied in the available literature. Just in order to verify the behaviour of the two GAs in a flow shop environment the two approaches have been compared. Figure 7 shows the results of this comparison expressed as per cent increase of the mean MK values with respect to the optimal values. Horizontal axis reports the problem dimensions. Each test is made up by 60 runs of the algorithm.

Considering that these problems are quite difficult to be solved, both algorithms provide good solutions, although the performance of S is much better than C's.

The tests have been carried out on ten different populations for each problem and each value of k, applying, in each case, the two different GAs six times.Considering the final results of the considered problems, figures 8 show the values of the following parameter:

$$R = (MK_r - MK_k) / (MK_r - MK_{opt}) * 100$$

where **MK$_{opt}$** is the optimal makespan, **MK$_r$** is the mean makespan value obtained with random generated initial populations and **MK$_k$** with inseminated initial populations instead. Then **R** measures the relative improvement yielded by using an inseminated population with respect to a random generated one, over the maximum improvement that is possible to obtain.

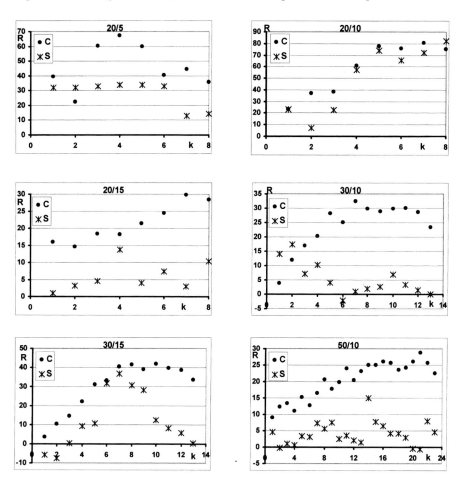

Figures 8. Values of R for the considered problems.

The analyses of the diagrams show quite clearly that using the proposed seeding procedure can assure a GA performance improvement. It is possible to point out that, excepting some sporadic points obtained by S, the use of inseminated initial populations determines consistent positive values of R. Whichever complexity the problem presents, the greater advantages are gained by using C, besides C has a much more regular behaviour with varying values of k. It seems that the best results are get when k ≈ N/3.

One objection can be raised at this point: it is meaningless to refer to mean makespan values. In fact: there could be an oriented population that is able to provide better mean results, but, on the other hand, just because it is an oriented population, the exploration of certain parts of the solution space could be much more difficult to be done, and if the optimal solution is within one of these areas, the GA will have poorer chances to get it. Besides, a GA is a stochastic process, and not a deterministic one, for this practical reason it is useful carry out multiple runs of a GA, but then again, evaluating which algorithm provide the best mean solutions it could be meaningless too: after all, what counts is the best value of makespan that a GA can provide, not the mean value of its multiple runs.

For these reasons, for every value of **k** the best result of each run has been selected and figures 9 show their per cent increases as to the optimal values. Each point must be compared with the first one on the left, which is the minimum increase obtained with 60 runs by using a randomly generated initial population (k=0).

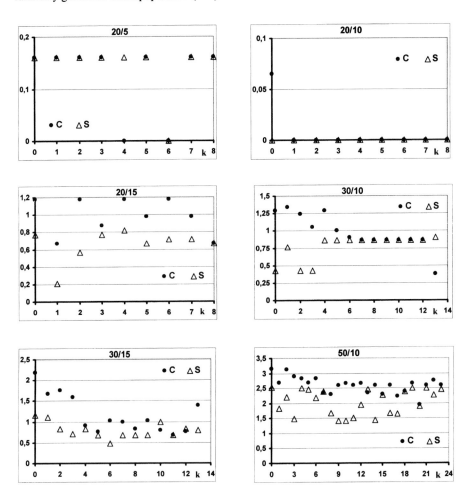

Figures 9. Best result per cent increases as to the optima

Aside from the 30/10 problem which shows some irregularity, particularly when solved with S, and a single point in the 20/15 problem, in all the other cases the results obtained by using inseminated populations are better than the ones obtained with a completely random approach, and just in a few cases they are the same. As for the 20/10 problem the optimal solutions are always found (with the exception of the case with C algorithm and k=0) and therefore the per cent frequency of the optimal solutions have been analyzed, as shown in figure 10.

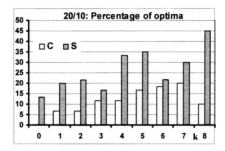

Fig 10: Optimal solution percentage for the 20/10 problem.

It can be readily seen that the number of optimal solutions found by using inseminated populations is greater: it can be concluded that a smaller number of runs is necessary in order to get the optimum (that is: the algorithm has gained more computational power).

5 EXECUTION TIME

As for the whole procedure execution time, this is made up by two parts: one for building the initial population, and the other to carry out the GA. As long as small values of k are concerned, even for problems with very large dimensions, the execution time for building the initial population can be readily determined: one order of magnitude less compared with the one required to execute the GA. But as soon as k values become greater, the execution time of the first part of the procedure increases so much that it eventually equals the time required to execute the GA (this is due to the recursive nature of the P1 procedure). The exceeding execution time is certainly a disadvantage but it usually is compensated by better results that the GA provides (and this means a more efficient scheduling of a productive system that certainly pays back the higher computational time used to find the solution).

6 CONCLUSION

The paper presents an initial population seeding procedure for GAs intended to solve flowshop environment problems. This procedure has been tested and it shows excellent results if compared with ones obtained using GAs with randomly chosen initial population. GAs with inseminated initial population allows to get better results or, with same results, to obtain the optimum using less computational time.

The goodness of these results encourages to apply seeding procedures in jobshop environment problems too. Some attempts are currently being made.

Another interesting issue to address is the complexity of the problem that seems to have a profound impact on the behaviour of GAs. New analyses seem to indicate that with particularly complex problem, using inseminated initial populations is the only practical way to increase the computational power of the GA; whereas when the complexity of the problem exceeds certain limits, only conventional heuristic methods seem to provide acceptable solutions. This issue will be addressed in an upcoming paper.

REFERENCES

1. **Bartlett, G.** (1995): "Genie: A First GA" in *Practical Handbook of Genetic Algorithms, Applications, Volume I*, Ed.Lance Chambers.
2. **Fichera, S., Grasso, V., Passannanti, G.** (1995): "Genetic algorithms for production scheduling", Proceeding of the Workshop on Integrated production system management, Padova, pp.47-65. (in italian)
3. **Passannanti, G.** (1999): "A new approach to the flow shop sequencing problem", Proceeding of the 15th International Conference on Computer-Aided Production Engineering, Cape'99, Durham, pp.756-761.
4. **OR-Library** : ftp://mscmga.ms.ic.ac.uk/pub/flowshop1.txt

Acknowledgments
This work has been supported by MIUR founds.

Application of reverse engineering to end-of-life cycle product disassembly

L M GALANTUCCI, G PERCOCO, and **R SPINA**
Dipartimento di Ingegneria Meccanica e Gestionale (DIMeG), Politecnico di Bari, Italy

ABSTRACT

The authors propose the implementation of a Reverse Engineering methodology to plan the automatic disassembly sequence of end-of-life sub-assemblies. Reverse Engineering is used to recognize geometrical characteristics and to reconstruct their 3D CAD representations. The approach uses Reverse Engineering devices characterized by a high scanning speed and the possibility to examine complex objects. Through the RE approach, an accurate reconstruction of part geometry is performed. Appropriate algorithms have also been developed to deal with products characterized by missing components. The optimal disassembly sequence of the product was thus identified for real product conditions. This methodology represents a powerful approach to resolve some critical aspects of real product disassembly.

1 INTRODUCTION

Recycling represents an important issue in manufacturing research due to its beneficial effects connected to the reduction in the environmental impact of the product. Several researches have highlighted the need to incorporate recycling considerations into new product designs using approaches such as Design for Disassembly (DFD) and Design for Recycling (DFR) 12. The evaluation of the function, manufacture, assembly, disassembly and other life-cycle aspects of a product can be performed in the early stage of its development cycle, thus making it more environmentally friendly. Automatic disassembly of a product can be planned based on information from its CAD models. The DFD and DFR approaches, which are very efficient during development of new products, can be not applicable during recycling of end-of-life components if CAD models are not available. In this case, disassembly planning is performed by human operators. The aim of this research is the formalization of procedures for end-of-life product disassembly by applying Reverse Engineering.

2 FRAMEWORK OVERVIEW

An operative framework for the definition, development and control of the disassembly planning of end-of-life products has been proposed to reach the objective of this research. The framework consists of several modules (Figure 1), necessary to: (i) reconstruct CAD models of the product by acquiring data directly from its physical representation using Reverse Engineering; (ii) transform the CAD model representation into a feature-based one to define functions for disassembly operations in a CAD system; (iii) plan the disassembly sequences, check their feasibility and identify the optimal sequence through the Disassembly Planner; (iv) verify the product state and adapt operations of the disassembly sequence to that state using an on-line control system. Particular attention has been paid to critical aspects related to the definition of information flows and to the search for the optimal disassembly sequence. The information flow was defined in terms of received and transferred data between framework modules in order to implement a concurrent engineering approach. For this reason, the level of details of each module was accurately chosen to support successive activities and allow data evaluation. All tasks linked to the evaluation of the disassembly sequences were carried-out by implementing software tools based on Genetic Algorithms and Fuzzy Logic. The advantages obtained by using these approaches are speeding-up the solution search and creating an explicit detailed knowledge of the process.

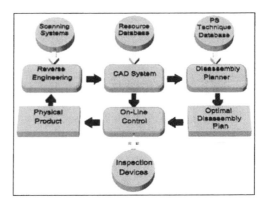

Figure 1: Proposed Framework.

In this paper, Reverse Engineering and its link with other framework modules are described in more detail. The acquired CAD model of the end-of-life product is used for efficient planning of its disassembly sequence. The authors report the analyses performed on a mechanical product to demonstrate the approach efficiency, also specifying procedures to realize a prototype of a disassembly station equipped with a vision system.

   A003/047/2003 © With Authors 2003

2.1 Reverse Engineering

Reverse Engineering is divided into several operations necessary for: the acquisition of the point cloud and data reduction, the triangulation for approximating product surfaces, the identification of part contours, the creation of surfaces and their connection 2. The RE tool is initialized selecting one system from the Scanning Database, taking into account both product and acquisition characteristics. The main selection factors of the scanning device are: the product shape complexity, scanning speed, surface conditions and precision required for the application. At present, all tasks of RE necessary to rebuild the product CAD model are carried-out in a manual way by skilled operators, due to their complexity. At the end of the RE phase, the reconstruction of component geometries and their CAD solid models are obtained.

2.2 CAD System

The role of the CAD system is essentially connected to data collection and analysis of the disassembly process. During data collection, CAD models of the different components are re-assembled into one model to restore the original configuration of the product. The aim of the following analysis phase is feature mapping between the CAD model and its disassembly representation to obtain a feature-based model. Disassembly Features, relations (contact, attachment and blocking between components) and directions are identified. Related resources (grippers and fixtures) linked to disassembly features are then selected to prepare the planning phase. At the end of the analysis phase, only entities that have a function in disassembly operations are considered. A preliminary review of the literature has pointed out that the graph-based representation is the most used to describe a mechanical product in terms of disassembly features and relations between them 45. Graph data, in the form of matrices, are transferred to the disassembly tool in order to plan the disassembly sequence.

2.3 Disassembly Planning

Disassembly Planning represents a very important link between the end-user and final product recycling 6. Its function is mainly connected to the identification of the optimal sequence for efficient separation of end-of-life product components and materials. At this level, the optimization task is realized through an off-line procedure. Data on disassembly features and related resources are retrieved from the CAD and Resource Database in order to plan operation sequences. The creation of these sequences, checks on their feasibility and the identification of the optimal sequence are only a few of the principal tasks performed by the disassembly planner. The efficiency of these tasks is strictly dependent on the problem solving techniques used. The application of an efficient problem-solving technique allows the reduction of the computational effort without affecting the process model. Moreover, an objective criterion must be used to evaluate disassembly sequences. This function is normally built using some parameters such as: the number of product orientations, the number and type of grippers used. Additional constraints and precedence between components can be added to face the real industrial environment.

2.4 On-Line Control

Once the optimal disassembly sequence of the end-of-life product has been identified, the proposed solution must be adapted to real product conditions. Defective parts, upgrading/downgrading of the product during consumer use and missing components are only a few possible uncertainties 7. The presence of uncertainties can influence the previous disassembly planning phase, making the proposed sequence not optimal or even unfeasible. To avoid this situation, an on-line control phase must be implemented to verify the product state and give useful feedback information to the disassembly planner. The on-line control is activated after the completion of two phases: inspection system selection and process control training. The first phase is performed in the same way as the scanning system choice, taking into account the different characteristics required by the control phase. The control training is realized through a semi-automatic process, creating a verification model from CAD models to compare with the real product. The control system then verifies the product state of maintenance, transferring recognized differences between the model and the physical part and requiring a new optimal sequence generation. As a result, some critical aspects of the real product disassembly are solved.

3 CASE STUDY

The oil-pressure sub-group of a car pump braking system was chosen as a case study to describe the Reverse Engineering approach and test the planning sequence of disassembly operations. The Reverse Engineering was performed by a specialized CAD system that manipulated point clouds acquired using a 3D laser scanner device. The optimization of the disassembly sequence of this product was carried-out with a Fuzzy-Genetic Algorithm.

3.1 Reconstruction of solid objects

As mentioned in the previous paragraph, RE is performed in a manual way by skilled operators and the efficiency of RE activities strictly relies on operator experience. Moreover the characteristics of the reconstructed CAD model are closely related to application requirements. For these reasons, RE phases were carefully carried-out to achieve well-defined CAD models oriented to the disassembly planning and on-line control. Disassembly Features and their relations were accurately examined during the segmentation, classification and fitting steps. Through segmentation, part edges are detected and the original scanning cloud is divided into sub-sets to simplify the reconstruction of component surfaces. During classification, the selection of the proper surface type to associate to each point sub-set is carried out. Fitting between surfaces and sub-sets is then performed, so identifying parameters for balancing surface quality and accuracy. The scanning device chosen for this application was a non-contact 3D laser scanner. The main device specifications are: scanning speed up to 20000 points/s; triangulation angle equal to 30°; measurement resolution equal to 10 μm; distance between two following points about 40 μm. This choice was justified by the speed and robustness offered by this device. The 3D scanner was mounted on a 6-DOF articulated arm with a workspace diameter equal to 2.5 m. Due to the presence of multiple components, the pump was manually disassembled and point clouds of each component were acquired. A registration procedure was also necessary to join different scanned point clouds of a complex shaped part into a single one. The different shapes of components and the presence of cavities required a two step procedure for the Reverse Engineering phase.

In the first step, the reconstruction of the external shapes of parts was performed while internal cavities of components were reconstructed during the second step. The reconstruction of the external surfaces of a part was carried-out by using some geometrical conditions.

A003/047/2003 © With Authors 2003

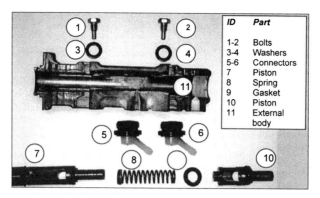

ID	Part
1-2	Bolts
3-4	Washers
5-6	Connectors
7	Piston
8	Spring
9	Gasket
10	Piston
11	External body

Figure 2: Exploded view of the product (Case A).

For components with approximate external axial-symmetry (e.g. part 1 in Figure 2), the revolution axis and main revolution curve were identified by selecting points belonging to only one section of the point cloud. The cloud was then fitted with surfaces connected to the approach direction for part disassembly. The advantages of this procedure were the reduced computational time and effort required for fitting a planar curve compared to those required for a surface. Components consisting largely of complex external surfaces required a different technique, due to their shape (e.g. part 11 in Figure 2). During the segmentation phase, the original cloud was initially divided into sub-clouds, taking into account the complexity of the part and the number of connections with other parts. For axial symmetric sub-clouds, the segmentation task was performed in an identical way to that used for axial symmetric parts, paying particular attention to the identification of disassembly directions of the connected component. For complex sub-clouds, the segmentation was carried-out by identifying the main contours of sub-clouds. The successive classification phase led to the use of curve network fitting because of the fairly large smooth area involved.

The external surfaces of the components were obtained by the above reconstruction step. The accurate representation of these surfaces is very important for the identification of blocking surfaces, selection of fixtures and layout design of the disassembly station. Information on part surfaces was directly linked to point clouds of the part itself. In presence of cavities, the internal surfaces of a part were obtained by manipulating geometries of other components in contact with it. The surface representation was initially transformed into a solid one. The solid objects were then subtracted from the recognized component in order to obtain its cavities, through a procedure similar to the Destructive Solid Geometry (DSG) approach. In addition, the accuracy of contact and interference areas was assumed as the main requirement to be satisfied rather than surface quality. Some modifications were necessary to consider deformations of non-metallic components (part 5 and 6 in Figure 2). In this case, the solid object associated to part 5 was subjected to a scale operation between its original geometry and some reference points taken on part 11 (Figure 3). The advantages of the DSG procedure are the elimination of the section cutting necessary to inspect the internal geometry of a part and of deformations of the part. Moreover the cavities of a part are represented in more details than obtained using fragment point data because of limitations related to the scanner device accessibility.

The final step in the reconstruction phase was checking the topological constraints of the model to assure the correctness of solid objects.

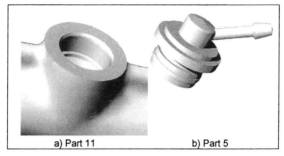

| a) Part 11 | b) Part 5 |

Figure 3: An external cavity of Part 11.

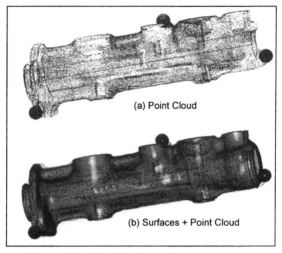

(a) Point Cloud

(b) Surfaces + Point Cloud

Figure 4: External shape of Part 11.

The point cloud of part 11 and its reconstructed CAD model are shown respectively in Figure 4-a) and 4-b). The disassembly graph-based representation of each component was created and transferred to the planner in the form of matrices 8.

3.2 Fuzzy-Genetic Algorithm approach

In the literature, several problem-solving techniques are applied to the optimization of the product disassembly sequence. Among them, Genetic Algorithms (GAs) are very efficient during the search for optimal solutions. GAs are more robust than existing direct search methods (hill-climbing, simulated annealing, etc) because they present a multi-directional search in the solution space (set of individuals) and encourage information formation and exchange between these directions. The chromosome consisted of three sections that included the component identifier, disassembly direction and selected gripper for each part, similar to one representation used in

A003/047/2003 © With Authors 2003

literature 89. The possible disassembly directions and available grippers of each part are reported in Table 1. The number of disassembly direction was reduced from 6 to 4, thanks to the existing symmetry condition along x-z plane for the complete product. Moreover, the *b* disassembly direction and the *G0* gripper were associated to part 11 in order to represent the blocking condition (part cannot be disassembly).

Part ID	Possible Direction	Available Grippers
1	± z	G1 G6
2	± z	G1 G6
3	± x, ± z	G2 G7
4	± x, ± z	G2 G7
5	± z	G3 G6
6	± z	G3 G6
7	-x	G3
8	± x	G4
9	± x	G2
10	± x	G3
11	b	G0

Table 1: Disassembly directions and grippers.

The evaluation of the disassembly sequences can be realized using the fitness function 9:

$$f = w_1 \cdot l + w_2 \cdot (N-1-o) + w_3 \cdot (N-g-1) + w_4 \cdot s \qquad (1)$$

where constant N is the component number while parameters l, o, g, s are respectively the maximum length of the feasible sequence, the orientation change number, gripper change number and maximum number of similar grouped disassembly components. The weights w_i with i=1,4 are associated to each parameter. In the proposed approach, GA has been modified, substituting the algebraic fitness function with a Fuzzy one to obtain a closer control of the technological knowledge of the disassembly process.

The aim of the proposed Fuzzy-GA approach was to avoid convergence problems, sub-optimal and/or non-feasible solutions, dependency from the disassembly product characteristics due to the weight balancing. For each parameter, a fuzzy set with 3 triangular membership functions was created to specify the process condition (bad, medium or good). The fuzzy sets of each parameter were then used as input of the Mamdani Inference system, obtaining a single de-fuzzified output, used to evaluate the quality of chromosomes generated by the GA.

4 DISCUSSION OF RESULTS

Two product configurations have been used to test the efficiency of the Fuzzy-GA approach. In the first test, the end-of-life product consisted of all the 11 components (Figure 2) while 3 less components were considered in the second test (Figure 5). In both cases, the initial population of the Fuzzy-GA approach was set at 150 chromosomes. The best chromosomes (30% of the entire population) were kept for generating off-springs at each iteration, with a mutation probability equal to 80%. The convergence was verified for 100 runs.

Figure 5: Product with missing components (Case B).

As a result, Fuzzy-GA always converged to a feasible solution. In particular for case A, 27% and 73% of runs respectively led to a sub-optimal and an optimal solution, with an average CPU time equal to 104 s (Pentium III 1GHz). Two of these solutions are reported in Table 2.

Case A - Optimal sequence: fitness value=0.8055										
10	→9	→8	→7	→5	→6	→1	→2	→3	→4	→11
+x	→+x	→+x	→+x	→-z	→-z	→+z	→+z	→+z	→+z	→b
G3	→G2	→G4	→G3	→G3	→G3	→G6	→G6	→G2	→G2	→G0
Case A - Sub-optimal sequence: fitness value=0.6650										
10	→9	→8	→1	→2	→4	→3	→6	→5	→7	→11
+x	→+x	→+x	→+z	→+z	→+z	→+z	→-z	→-z	→-x	→b
G3	→G2	→G4	→G1	→G1	→G7	→G7	→G6	→G6	→G3	→G0
Case B - Optimal sequence: fitness value=0.6700										
1	→2	→4	→10	→7	→8	→9	→11			
+z	→+z	→+z	→+x	→-x	→-x	→-x	→b			
G1	→G1	→G2	→G3	→G3	→G4	→G2	→G0			

Table 2: Disassembly sequence of Case A

One advantage of the Fuzzy-GA approach is the identification of several equivalent solutions from the disassembly point of view.

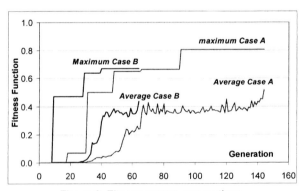

Figure 6: Fitness versus generations.

A003/047/2003 © With Authors 2003

Comparing case A and B, the number of generations of case A is greater than case B because of the reduced search space (Figure 6). The fitness value of the optimal sequence of case B is lower than that of case A because the number of orientation changes was considered more important than the number of gripper changes. The maximum value of the fitness function can be realized only if the orientation and gripper changes are both equal to 0. This condition cannot be achieved for these two tests because it would lead to unfeasible solutions.

5 CONCLUSIONS

Considering the results obtained using RE and Fuzzy-GA, the proposed framework can be enhanced by: (i) promoting more extensive data integration between RE, CAD and disassembly systems, (ii) improving the feature extraction phase between CAD and disassembly planner systems, (iii) standardizing the RE procedures in order to support non-skilled operators in the reconstruction of the product surfaces, and (iv) training artificial intelligence systems (e.g. neural networks) that embed planning activities in a real-time system for product control. Moreover, thanks to the accurate representation of component surfaces, further research could lead to the design of specialized fixtures and grippers and the setting-up of a vision system to detect missing components. In addition, the stability and flexibility of the Fuzzy-GA approach for product disassembly planning can be further enhanced by studying a greater variety of product types and configurations.

6 REFERENCES

1. **Boothroyd G., Alting L.**, 1992, Design for assembly and disassembly, Annals of the CIRP, Vol.41/2, pp.625-636
2. **Sodhi M., Knight W.A., 1998**, Product design for disassembly and bulk recycling, Annals of the CIRP, Vol.47/1, pp.115-118
3. **Varady T., Martin R.R. & Cox J.**, 1997, Reverse engineering of geometric models – An introduction, Computer Aided Design, Vol.29, pp.255-268
4. **Subramani A.K., Dewhurst P.**, 1991, Automatic Generation of product disassembly sequences, Annals of the CIRP, Vol.40/1, pp.115-118
5. **Laperrière L., ElMaraghy H.A.**, 1992, Planning of products assembly and disassembly, Annals of the CIRP, Vol.41/2, pp.5-9
6. **Kaebernick H., O'Shea B., Grewal S.S.**, 2000, A method for sequencing the disassembly of products, Annals of the CIRP, Vol.49/1, pp.13-15
7. **Gungor A., Gupta S.M.**, 1998, Disassembly sequence planning for products with defective parts in product recovery, Computers and Industrial Engineering, Vol.35, pp.161-1
8. **G. Dini G., Santochi M.**, 1992, Automated sequencing and subassembly detection in assembly planning, Annals of the CIRP, Vol.41/1, pp.1-4
9. **Dini G. et all, 1999**, Generation of optimized assembly sequences using genetic algorithms, Annals of the CIRP, Vol.48/1, pp.17-21

Authors
Prof. Luigi M. GALANTUCCI,
Full Professor at DIMeG – Politecnico di Bari; Viale Japigia 182 – 70126 – Bari (Italy);

Email: galantucci@poliba.it; Phone:+39 080 5962764; Fax:+39 080 5962788.
Mech. Eng. Gianluca PERCOCO

D.d.R. Student at DIMeG– Politecnico di Bari; Viale Japigia 182 – 70126 – Bari (Italy); Email: g.percoco@dimeg.poliba.it; Phone:+39 080 5962782; Fax:+39 080 5962788.

D.d.R. Mech. Eng. ROBERTO SPINA
Lecturer and Researcher at DIMeG– Politecnico di Bari; Viale Japigia 182 – 70126 – Bari (Italy); Email: r.spina@poliba.it; Phone:+39 080 5962768; Fax:+39 080 5962788.

A003/047/2003 © With Authors 2003

Identifying three-dimensional object features using shape distributions

H J REA, R SUNG, and **J CORNEY**
Department of Mechanical Engineering, Heriot-Watt University, Edinburgh, UK
D CLARK
Department of Mathematics, Heriot-Watt University, Edinburgh, UK

SYNOPSIS

The collaborative re-use of design and manufacturing data is one way that e-commerce can significantly reduce development costs and lead times of new products. However, the proliferation of web based catalogues for standard components (e.g. brochure-ware for nuts and washers etc.) only hints at the productivity gains that might be possible if bespoke components could be located electronically. Consequently the research reported here is motivated by the belief that shape matching technology is key to enabling a much deeper form of Internet-based collaborative commerce.

This paper starts by describing the architecture, geometric meta-data and user interface of a 3D search engine called ShapeSifter. Later sections describe how a novel form of shape distribution could be used to enhance the performance of the current shape matching algorithms. This is explained by showing how shape distributions can be used to separate information about the objects internal features from its gross shape, or convex hull. The paper ends by assessing how the use of this new shape distribution would change the system's performance.

1 INTRODUCTION

Crucial to shape retrieval systems (i.e. 3D search engines) is the means to characterise the topology and geometry of 3D models in a way that allows the *similarity* of shapes to be assessed. The need for such tools is growing as the number of models held in CAD/CAM databases is increasing rapidly. Indeed, already even small manufacturers frequently have over 10,000 models in their systems. Although the need for retrieval systems for CAD data has long been recognised (1) the widespread use of the Internet technologies both within and between manufacturing enterprises has created new opportunities for this technology.

The search engine architecture developed in this project is shown in Figure 1. There are two distinct activities 1) indexing or filling the database and 2) searching the database. The database is indexed by analysing the models to extract various shape metrics (i.e. geometric meta-data). Models can either be uploaded individually or in bulk (however longer term it is hoped that a web-crawler, which identifies manufacturer's Web pages containing shape models, could be used to fill the database automatically). To search the database, a target model is analysed to generate its global shape metrics or meta-data (e.g. Table 1). The database is then searched, via an SQL query, to identify models with metric (i.e. geometric meta-data) values within a stated tolerance (e.g. ±10%).

In this paper we describe a novel shape distribution and assess its effectiveness at extracting characteristic information about the geometry and topology of the object. After describing the search engine's user interface, architecture and the geometric meta-data it uses, we use a previously published shape function (D2) to generate the distributions of a basic shape and discuss its interpretation. The distributions of a more complex shape is then considered, together with the distributions of its convex hull, and a new distribution, which emphasises an object's depressions, is introduced. The paper ends by presenting the results of using these distributions to assess the similarity of components in an experimental database.

2 THE SHAPESIFTER 3D SEARCH ENGINE

2.1 Geometric Meta-Data

For every 3D model loaded into the system, a number of geometric parameters are calculated (2) and stored in the system's database. Thus meta-data (summarised in Table 1) is numerical and mostly independent of an object's orientation (e.g. how it is aligned with the coordinated axis). It should also be noted that the metrics based on ratios are also independent of scale.

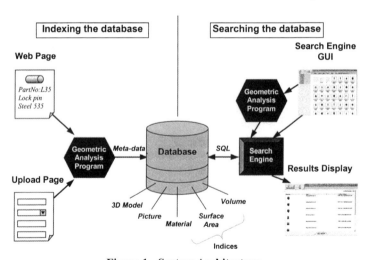

Figure 1 - System Architecture

Table 1 – 3D Model Meta-Data Example

Model		Model's Convex Hull	
Surface Area	1202.06	Surface Area	868.41
Volume	251.52	Volume	1157.66
Number of Facets	2164	Number of Facets	46
Crinkliness	190.43	Crinkliness	82.70
Compactness	27455.38	Compactness	488.67
Surface Area to Volume	4.78	Surface Area to Volume	0.75
Number of Symmetry Axes	3	Number of Symmetry Axes	3
Number of Holes	8	Hull Crumpliness	0.72
Bounding Box Length	27.44	Hull Packing	0.78
Bounding Box Max	24.00	Hull Heel	1.10
Bounding Box Med	12.32		
Bounding Box Min	5.00		
Bounding Box Ratio	4.79		
Center of Gravity X	-0.43		
Center of Gravity Y	0.00		
Center of Gravity Z	-1.17		

2.2 ShapeSifter User Interface

To carry out a search for a similar 3D model, the user has to login to the system (www.ShapeSearch.net). The search engine is an application of Computer Associate's Cleverpath ECM software. Collection folders containing over 800 3D models (in STL format) are available and detailed information regarding the collection, model or search results is presented. Buttons to selected various options related to a particular collection or individual model exist and provide the following functions:

- **Summary** – gives a description (e.g. ownership, origin etc) of the selected collection or 3D model.
- **Properties** – show the values of shape meta-data (e.g. volume, surface area etc) for a selected model.
- **3D View** – displays a 3D view of the selected model that can be manipulated by the user. Note that although STL files are used to support meta-data extraction, VRML representations are used for visualisation.
- **Thumbnails** – displays a thumbnail of an entire collection or an individual model.
- **History** – shows the version of the current model and its edit history.
- **Security** – allows the owner of the model to control the user type that can view the model. In other words access to particular collections and models can be restricted and associated with particular passwords.

- **Search** help – displays a step-by-step description of how the user can perform a "simple" or "advanced" search.
- **Submit** help – displays a step-by-step description of how the user can submit (i.e. upload) a model to the system.

To view all the models in a particular collection, the user first selects a particular collection, in the left section, and then the 'thumbnail' button, on the top section. This displays small images of each of the models in that collection (Figure 2). To search for a 3D model that is geometrically similar to a particular model, the user can then click the "Search for similar part" link under each model. This search uses a predefined filter (i.e. tolerance and meta-data selection) that has been found by experiments to give good results (3).

An advanced search, that allows the search tolerance and meta-data (i.e. geometric properties) to be manually selected, is also possible. To do this, the user right clicks on a particular model (in the left hand section) and then selects "Advanced Search". This presents the user with a list of shape properties and their values together with a drop down list of possible search tolerance percentages.

2.3 Search Engine Performance

In previous studies (3, 4), we have shown that the most effective filter of the geometric-metrics has been to use a search based on a hull compactness[1], hull packing[2] and hull crumpliness[3] with tolerances of 25%, 50% and 10% respectfully.

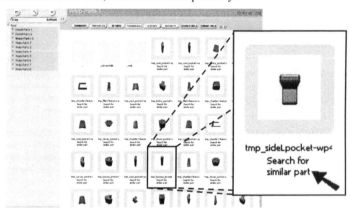

Figure 2 - Thumbnail View

---

[1] **Hull Compactness** (*Hc*): The non-dimensional ratio of the surface-area of the convex hull cubed over the volume of the convex hull squared (5): $Hc = \dfrac{A_{Hull}^{3}}{V_{Hull}^{2}}$.

[2] **Hull Packing** (*Hp*): The percent of the convex hull volume not occupied by the original object: $Hp = 1 - \dfrac{V_{mod}}{V_{Hull}}$.

[3] **Hull Crumpliness** (*Hcp*): The ratio of the surface area of the object to the surface area of its convex hull: $Hcp = \dfrac{A_{mod}}{A_{Hull}}$.

      A003/054/2003 © IMechE 2003

The false positive/false negative rates (or performance curve) of these three metrics at various tolerances were calculated for a benchmark database of 450 parts. This data set contains known families of "similar" components generated by a number of scaling, warping and editing functions. In this context a false negative result would be to wrongly classify one member of a family as dissimilar from its known relations. These values are reproduced here in Table 2 and Figure 3.

The ideal filter has a performance curve that has the two axes as asymptotes and a point of inflection near the origin. Figure 3 clearly shows that while the three types of meta-data considered approximate this behaviour there is still considerable scope for improvement. Because of this the project started to investigate the use of the geometric probability distributions (described in the next sections) as filters.

3 SHAPE DISTRIBUTIONS AS A METRIC

3.1 Geometric Probability

The work discussed here has its roots in a branch of mathematics known as geometric probability. Classical geometric probability is concerned with the probability of an event related to the relative location of geometric figures placed at random on a plane or in a space. The results of these analyses often take the form of a distance (or shape) function that is a mathematical expression that describes the probability of a system taking a specific value or set of values. A typical result is that two randomly chosen points in an n-dimensional unit cube have an average distance of $\sqrt{n/6}$ (6).

So, essentially, shape distributions are a means of representing an object as a probability histogram. For example Figure 4 plots the frequency of lengths between random points on the surface of a sphere. The graph clearly shows that the distance between any two points is more likely to be large (i.e. close to the diameter) than small. Note that ideally the region under the graph in Figure 4, and all the probability curves, has unit area, implying all possible outcomes are represented. Beyond unit spheres and cubes the analytical study of geometric probability of 3D shapes is extremely complex. However, computers allow probability distributions to be generated empirically via explicit calculations. Various forms of probability distributions have been widely used in image analysis and recognition.

Table 2 - Geometric Filter Performance Data

Search Metric	Tolerance %	False positives %	False negative %
hull	2	26.77	21.87
crumpliness	10	54.08	10.38
	25	80.69	5.06
hull packing	10	3.97	50.90
	25	10.24	43.73
	50	22.35	30.76
hull	2	3.23	64.73
compactness	10	15.63	37.55
	25	39.32	16.55

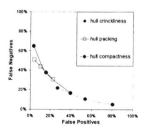

Figure 3 - Geometric Filter Performance Curves

Figure 4 - D2 distribution of a sphere

3.2 Shape Distributions

Recently several researchers (7-9) have studied probability distributions for three-dimensional objects that are generated by explicit computation. These methods have two distinct steps:

1. Generation of random points over the surface of the object, and
2. Application of a shape function.

Points are generated from facetted representations of objects by a procedure described in detail in (10) but summarised here as:

- The area of each triangle is calculated and the triangle is stored in an array along with a cumulative area.
- A triangle is selected with probability proportional to its area, by generating a random number between 0 and the total surface area, and performing a search on the array of cumulative areas.
- For each selected triangle a random point, P, is generated on it surface using the vertices of the triangle (A, B, C), two random numbers, r_1 and r_2, between 0 and 1 and the equation:

$$P = (1 - \sqrt{r_1})A + \sqrt{r_1}(1 - r_2)B + \sqrt{r_1}r_2C .$$

3.3 The Behaviour of the D2 Shape Function

On first encounter it is not obvious why the probability distributions that arise from the different shape functions are of the form they are. However, familiarity with how form and shape functions interact soon enables one to anticipate the results. Figure 5 shows, for example, how the probability distribution for a cube generated by the D2 function is a combination of the probability distributions for individual faces, pairs of orthogonal faces and pairs of parallel faces. After noting that the distances/lengths of the probability distributions were scaled by the diagonal length of their bounding box and that the same number of points were used to generate each curve the following observations can be made.

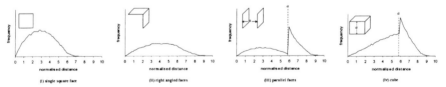

Figure 5 - The build up of D2 distribution of a cube

A003/054/2003 © IMechE 2003

- Both the single face (Figure 5 (i)) and the orthogonal faces (Figure 5 (ii)) produce distributions that progress smoothly from zero to the maximum distance.
- However, the distribution produced by *parallel* faces in Figure 5(iii) has a sharp increase at the distance of separation between the faces. This is due to the fact that there is a high probability of the distance between two points closely approximating the distance between the faces.
- For the entire cube in Figure 5 (iv), the smooth curve from 0 to the point of sharp increase is due to the distribution of distances on each of the faces.

4 DISTRIBUTIONS FOR COMPLEX SHAPES

Osada et al (9) have successfully used the D2 shape function to differentiate between grossly dissimilar shaped objects (e.g. aeroplanes and animals). However Ip et al (7) have shown that the difference between mechanical parts cannot be distinguished using this simple distribution alone, and have accordingly developed the function further to identify the measured distances as within the solid shape, across a void, or through both solid and void. In other words the D2 shape distributions of objects with even moderate re-entrant features prove difficult to analyse as they have surfaces with very similar separations. The gross shape of the complex object drowns information about individual features, and, conversely, the effect of internal features on the distribution obscures information about the gross shape.

In response to these problems Ip et al describe three new distributions that separate the distance measures generated based on geometric properties of the line. The authors demonstrate that their approach improves the ability of the D2 shape function to distinguish between mechanical parts. However, the generation of these distributions is computationally intensive, requiring repeated intersection calculations and it is reasonable to ask if simpler functions can produce comparable performance.

4.1 Isolating the Contribution of Concaved Features

Using the point insertion algorithm introduced by O'Rourke (11) and implemented by Pudney (12), we are able to generate the convex hull of the object and hence its D2 shape distribution. Typically this distribution is a relatively simple one as it is devoid of all concaved features. Indeed, the resulting distributions frequently reflect the "stock" from which the part is manufactured. We postulated that subtracting this distribution from the overall shape distribution will provide a distribution reflecting the nature of the concaved features alone (see Figure 6), we call this the CHD (Convex Hull Difference) curve. In Figure 6 note that 1) negative values are ignored as this indicates that these are derived from values of length more frequently encountered on the convex hull than the original D2 distribution and 2) the CHD curve has been enlarged for clarity of illustration.

While the resulting curve is no longer a probability density curve derived from an explicit shape function, it is a reproducible characteristic signature of the object's form. Our premise was that the CHD curve would emphasis the internal features, as can be seen in the thin-walled object in Figure 7.

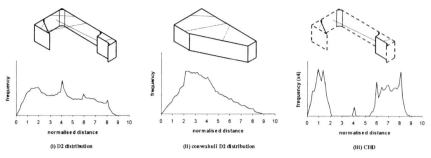

Figure 6 - The make up of the CHD distribution

| (i) | (ii) | (iii) | (iv) |

Figure 7 - CHD of a thin walled object

We reasoned as follows:

i)   The D2 distribution for the original part contains a higher frequency of small lengths than the convex hull's distribution due to the thin walls of the part (see Figure 7 (i)).

ii) The convex hull distribution contains a higher frequency of lengths "around" the depth of the part's pocket depth than the D2 distribution of the original part. (see Figure 7 (ii)).

iii) The original part's distribution contains a higher frequency than the convex hull D2 of the lengths around the width and breadth of the pockets (see Figure 7 (iii)).

iv) Distributions of lengths between convex hull faces that are coincident with faces on the original body are approximately the same on both the convex hull and the D2 curves (see Figure 7 (iv)).

When the difference between the average length arising from the types of D2 lengths (i.e. Figure 7 (i)-(iv)) are significantly distinct, such as with thin-walled objects, the contribution each of the characteristics makes to the CHD curve is easily identified, as can be seen in Figure 6 and for the "simple Boeing" model in Table 3. The effect is not so evident when the contributions made by the walls and convex hull etc are similar as in the other models in Table 3 and the resulting signatures are less distinct.

5 DISCUSSION

As a means of comparing the CHD distributions of a number of objects, the Minkowski L_1 value (13) can be calculated, which is effectively the area of the region under the CHD curve. This measure can be used as an initial filter to reduce the number of comparisons required, prior to calculating the Minkowski L_1 norm between the target object's D2, convex hull D2 and CHD curve and those of each object in the reduced data set.

   A003/054/2003 © IMechE 2003

The false positive/false negative rates (or performance curve) of the CHD Minkowski L_1 value at various tolerances were calculated for the same benchmark database as used in Section 2.3. These values are shown in Table 4 and are plotted in Figure 9 alongside the three filters based on the ratios of geometric parameters described earlier (i.e. hull crumpliness, packing and compactness).

The ideal filter curve is one that has the two axes as asymptotes and a point of inflection near the origin. Of the four measures considered, the CHD L_1 curve is closest to this ideal, indicating that it would acts as a more efficient filter. Choosing a tolerance of ±0.002 means that it is possible to reduce the data set by more than 50% while only eliminating 10-15% of the possible good matches.

Table 3 - Shape Distributions of Complex Objects

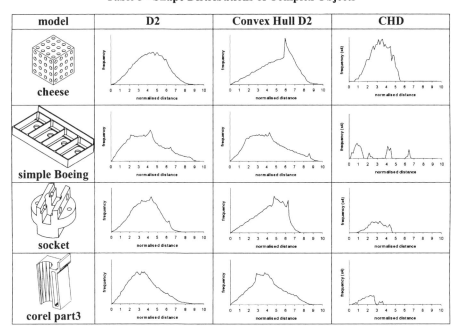

Table 4 - CHD Filter Performance Data

Search Metric	Tolerance ±	False positives %	False negative %
CHD	0.0001	3.30	82.89
	0.0002	6.43	69.05
	0.0005	13.51	39.41
	0.001	20.71	24.46
	0.001	27.14	17.05
	0.002	32.96	12.85
	0.003	43.95	9.76
	0.005	64.77	5.56
	0.01	89.80	0.74

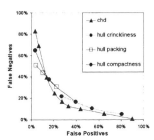

Figure 8 - CHD Filter Performance Curve

6 CONCLUSIONS AND FUTURE WORK

The work on the CHD filter described here has been implemented on a stand-alone experimental system. The CHD performance curve is plotted alongside the performance curve of the filters previously developed by the authors (3) and currently available on our publicly accessible search engine at http://www.ShapeSearch.net. This clearly shows that the CHD acts as a more efficient filter and therefore it should be incorporated into the search engine. Hence future work will involve:

- incorporating these methods into our search engine, http://www.ShapeSearch.net;
- trials on larger and more diverse test data;
- looking at the effectiveness of applying Minkowski L_1 norms to rank the results.

ACKNOWLEDGMENTS

The authors gratefully acknowledge the contributions to this research from the following organizations: The UK Engineering and Physical Science Research Council; Pathtrace plc; First Index Ltd; Computer Associates and the Scottish Polymer Technology Network. We also express our appreciation to: Alistair Thomas for his help in producing the model images; Athol A Korabinski for his advice regarding the similarity matrices used in the computation of the filter performance; Bill Regli for his assistance with the Design Repository.

REFERENCES

1. **Cybenko, G., Bhasin, A., and Cohan, K.D.**, *Pattern Recognition of 3D CAD Objects: Towards an Electronic Yellow Pages of Mechnical Parts.* Smart Engineering Systems Design, 1996. **1**: p. 1-13.
2. **Rea, H., Corney, J., Clark, D., Pritchard, J., Breaks, M., and MacLeod, R.** *Part sourcing in a global market.* in *International Conference on eCommerce Engineering: New Challenges for Global Manufacturing in the 21st Century.* 2001. Xian, China.
3. **Sung, R., Rea, H.J., Corney, J.C., Clark, D.E., Pritchard, J., Breaks, M.L., and MacLeod, R.A.** *Assessing the effectiveness of filters for shape matching.* in *2002 ASME International Mechanical Engineering Congress & Exposition.* 2002. New Orleans, Louisiana.
4. **Corney, J.C., Rea, H.J., Clark, D., Pritchard, J., Breaks, M., and MacLeod, R.**, *Coarse Filters for Shape Matching.* IEEE Computer Graphics and Applications, 2002. **22**(3): p. 65-74.
5. **Bribiesca, E.**, *A Measure of Compactness for 3D Shapes.* Computers & Mathematics with Applications, 2000. **40**: p. 1275-1284.
6. **Santalo, L.**, *Integral Geometry and Geometric Probability.* Encyclopaedia of Mathematics and its Applications, ed. G.-C. Rota. Vol. 1. 1976, Reading, Massachusetts: Addison-Wesley Publishing Company.
7. **Ip, C.Y., Lapadat, D., Sieger, L., and Regli, W.C.** *Using Shape Distributions to Compare Solid Models.* in *Solid Modeling.* 2002. Saarbrucken, Germany: ACM.
8. **Ankerst, M., Kastenmuller, G., Kriegal, H.-P., and Seidl, T.** *3D Shape Histograms for Similarity Search and Classification in Spatial Databases.* in *SSD 99 - Advances in Spatial Databases.* 1999. Hong Kong: Springer-Verlag Berlin Heidelberg.
9. **Osada, R., Funkhouser, T., Chazelle, B., and Dobkin, D.**, *Shape Distributions.* ACM Transactions on Graphics, 2002. **21**(4).
10. **Osada, R., Funkhouser, T., Chazelle, B., and Dobkin, D.**, *Matching 3D Models with Shape Distributions.* Shape Modelling International, 2001.
11. **O'Rourke, J.**, *Computational Geometry in C.* 1998: Cambridge University Press.
12. **Pudney, C.**, *Java 3D Convex Hull Algorithm.* 1998, The Stony Brook Algorithm Repository, Department of Computer Science, State University of New York, www.cs.sunysb.edu/~algorith/implement/CONVEX-HULL-JAVA/implement.shtml: New York.
13. **Kullback**, *Information Theory and Statistics.* 1968: Dover.

Integrated facilities planning system for automated factory

K MORI and **S SHINDO**
Advanced Technology R&D Center, Mitsubishi Electric Corporation, Amagasaki, Japan

SYNOPSIS

This paper presents a facilities planning method with an integrated software system consisting of several software systems. The system can support each of planning and design processes seamlessly by using a common factory model. This paper shows effectiveness of the system through a facilities planning of virtual liquid crystal display panel fabrication line.

1 INTRODUCTION

As semiconductor wafer fabrication and liquid crystal display panel fabrication facilities (fab) become more automated to handle large material, planning of the facilities including automated material handling systems (AMHS) becomes more difficult. All aspects of facilities planning and operation must be applied to save new semiconductor fab investment cost beyond the billion U.S. dollar and increase efficiency of the fab. Many methodologies have been exploited in industrial engineering and operations research. However, since each methodology has been developed for each planning process, it is difficult for a planner to apply them to all planning processes of actual fab.

In this paper, we propose a facilities planning method by using a facilities planning software called FPSS (facilities planning and simulation system) which consists of a decision making system, modelling system of manufacturing systems, facilities layout system and simulation system. This system enables facility planners to generate the best possible designs in the shortest possible time.

The decision-making system based on analytic hierarchal process (AHP) is used to determine weighted rankings over all criteria of facilities planning and select alternative plans. The modelling system generates a common model represented on spreadsheet to use in a facilities layout system and a simulation system. The layout system is main system consisting of three functions. The first function is capacity planning function that induces optimal number of each facility. The second function is evaluation function of generated plans from several viewpoints. The third function is automatic generation function of layout plan based on an optimisation method. The simulation system based on timed petri nets is used for dynamic evaluation of plans and determination of dispatching rule in practical operations.

In this paper we present these important functions and show usefulness of the systems through a facilities planning of a virtual liquid crystal display panel fabrication line.

2 FACILITES PLANNING FOR FACTORY

Object of facilities planning is not only production factory but also office, hospital, hotel, warehouse and department store. A handbook [1] tells that *facilities planning determine how an activity's tangible fixed assets best support achieving the activity's objective.* For production factory, facilities planning implies the determination of how the manufacturing facility best supports production. Facilities planning is subdivided into subject of facilities location and facilities design. Moreover facilities design is subdivided into facility systems design, layout design and material handling systems design. There are many researches and support systems to support these design processes in areas of industrial engineering and operations research. Each of support systems is independent of others, so planning engineer must use several support systems in each design process. Moreover, it needs long design period and big efforts to achieve overall objectives of facilities planning because of repeated design processes.

Sly presented a systematic approach to factory layout and design combining with Factory PLAN/OPT and FLOW software systems that work in AutoCAD. FactoryFLOW can generate a variety of diagrams from product data [2,3]. These software systems are applied to several layout plans, and it is shown that computer support to layout design process is effective.

However, these systems cannot generate and evaluate alternative plans from dynamical viewpoint. Therefore, after planning with these systems, simulation process is needed to evaluate the plans from dynamical viewpoint. It will take a long time to complete planning processes because static planning process and dynamic planning process with another simulation system are iterated. Moreover, plans generated by traditional support systems have not been taken account of dynamical behaviour of AMHS.

In this paper, we present planning system that support facility systems design process, layout design process and material handling systems design process using both static method and a discrete event simulation method to solve these problems. As this system has static evaluation function of AMHS route, it can reduce iterations of

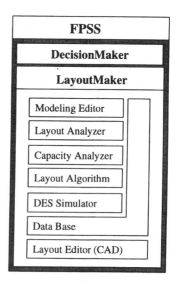

Fig. 1 System configuration

   A003/055/2003 © With Authors 2003

design process with discrete event simulation of AMHS.

3 PROPOSED SYSTEMS

3.1 System configuration

Figure 1 shows a system configuration of proposed systems called "FPSS (facilities planning and simulation systems)". This system consists of DecisionMaker, LayoutMaker and a discrete event simulator for AHMS dynamical evaluation.

3.2 DecisionMaker

DecisionMaker based on analytic hierarchal process method has been developed to determinate weight of criteria and evaluate alternative plans. First of all, human planner classifies a hierarchal structure of several criteria (objectives) shown in figure 2. The planner determines each weight of criteria in each hierarchy by comparisons of each pair of criteria. Finally human planner determines priorities of generated alternative plans as same way as determination of each weight of criteria. For example, the evaluation window shows that the criterion "simple layout" (the weight is 0.419) is higher than others in layout criteria.

3.3 LayoutMaker

LayoutMaker has been developed to support capacity planning and layout design. The capacity planning process plays an important role to ensure desired factory production plan. The layout design is important to reduce load and cost of material handling systems.

Factory model of LayoutMaker can be modelled on table expression such as spreadsheet shown in figure 3.

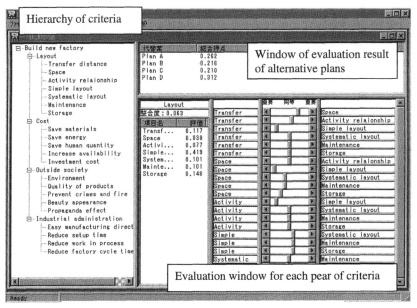

Fig. 2 Screen of DecisionMaker

3.3.1 Modelling editor

3.3.1.1 Factory modelling

Factory model consists of equipment model, AMHS model, preventative maintenance model, tool model, operator model, AMHS From-To model, setup matrix model and obstacle model. For example, the equipment model includes equipment group, equipment, equipment cost, batch type, batch longest wait time, dimension of equipment, shape of equipment, and layout position shown in figure 3(a). The obstacle model includes safety area or other unusable facilities.

3.3.1.2 Process flow modelling

Process flow model consists of part model and process route model as shown in figure 3(b). The part model defines parts of products and their process flow, which is defined by the process route model. The process route model includes process sequences, relation between process step and processed equipment group, processing time, processing unit type, sampling rate, setup tool, operator, and setup time.

3.3.1.3 Production plan modelling

Production plan model includes order name, lot name, start time, due time, lot size, lot quantity, and lot priority as shown in figure 3(c).

These models can be edited on spreadsheet editor.

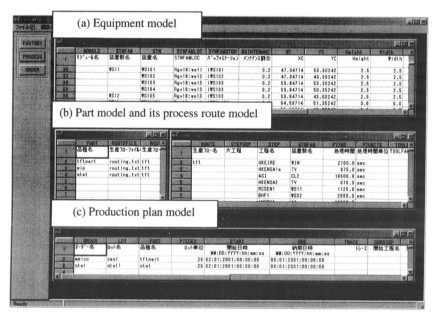

Fig. 3 Production model in LayoutMaker

A003/055/2003 © With Authors 2003

3.3.2 Capacity planning

3.3.2.1 Estimation of number of equipment

Capacity planner can calculate least equipment number of each equipment group from given process flow model and desired production plan model. Calculation results can be applied to equipment data of factory model.

3.3.2.2 Cost of equipment ownership

Estimator of cost of equipment ownership (CEO) can generate a CEO template sheet from model data. If human planner set additional cost data such as building cost, utilities cost, energy cost (electricity, gas, oil), material cost, and labor cost to the template sheet, the estimator can calculate each manufacturing cost of products and portion of the cost.

3.3.2.3 Process simulator

LayoutMaker has an original discrete event simulator based on timed petri nets [4]. This simulator can simulate factory process operations from both directions (forward and backward). The forward simulation can investigate how dispatching rule is effective and when lots will be completed, and the backward simulation can investigate when lots have to release into factory to meet with required due date.

3.3.3 Layout design

Layout design is a key of facilities planning and most difficult process, so it is needed to support this process by computer system.

3.3.3.1 Flow analysis

Flow analyser can generate several charts such from-to as equipment/module matrix chart and activity relationship chart. The activities relationship chart based on Muther's systematic layout planning procedure (SLP) enables us to set each relationship between activities to closeness values such as A, E, I, O, U, X [1, 5].

    A: Absolutely necessary, E: Especially important, I: Important, O: Ordinary closeness,
    U: unimportant, X: Undesirable

Figure 4 shows a relationship chart. Figure 5 shows a relationship diagram of the relationship

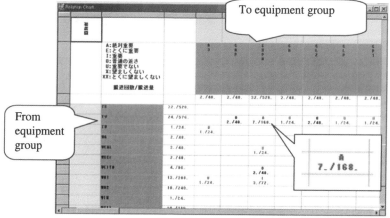

Fig. 4 Relationship chart

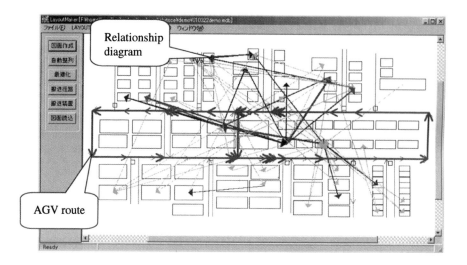

Fig. 5 Relationship diagram on layout CAD

chart. A cell in the relationship chart presents an information move in process flow/transfer volume of each from-to among equipment group in default value. In figure 4, value 7 and 168 mean from-to count in all flow and from-to volume corresponding to a production plan respectively. Human planner can define the relationship in each cell. For example, figure 4 shows that a character "A" is set. These relations of relationship chart are automatically translated to relationship diagram on layout CAD as shown in figure 5.

3.3.3.2 Layout editor
Layout editor can move each activity (equipment) freely by using edit function of CAD system. And the layout editor can make routes of AMHS with direction. Editing result is reflected to layout evaluator immediately.

3.3.3.3 Layout optimiser
Layout optimiser can generate several alternative plans automatically by using optimisation method. In real world, layout planning has too many criteria considered in practical use so that a generated plan by the layout optimiser, which is lacking for some criteria might not be satisfied requirement of human planner. However it is useful to generate an initial plan.

3.3.3.4 Layout evaluator
Layout evaluator has some criteria from several viewpoints that are total Euclid distance from equipment to equipment, its total moving time, total/average moving distance of AMHS along the AMHS routes and its average moving time per trip of AMHS, average utilization rate of the AMHS, and least vehicle number of the AMHS. These criteria from the AMHS view are calculated by shortest path based on Dijkstra's method.

3.4 AMHS simulator
The discrete event simulator of LayoutMaker is used to evaluate dispatching rule and factory cycle time, but the simulator cannot simulate detail AMHS vehicle model. As detail AMHS

     A003/055/2003 © With Authors 2003

simulator, AutoMod/AutoSched of Brooks-PRI Automation Inc. is popular and widely used in semiconductor industry. The model of LayoutMaker is subset model of AutoSched, so the model constructed on LayoutMaker can be translated to a model in AutoSched easily. By using translation from LayoutMaker model to AutoSched model, it becomes easy to move to simulation process step of detail AMHS. Layout generated by LayoutMaker can be evaluated by using detail AMHS simulator. Comparisons of static evaluation result by LayoutMaker and dynamic evaluation result by detail AMHS simulator are described in chapter 4.

4 FACILITIES PLANNING PROCESSES

4.1 Planning process cycle
Figure 6 shows the planning process cycle [1]. There are three phases of the planning process; (I) definition of the problem, (II) generation of layout plan, and (III) implementation of the plan. The definition of the problem is to define objectives of the facility and processes of each product. The generation of layout plan consists of six process steps; specify the primary and support activities to be performed in accomplishing the objective, determine the interrelationship among all activities, determine the space requirements for all activities, generate alternative facilities plans, evaluate alternative facilities plan, select a facilities plan. The implementation phase consists of three process steps; implement the facilities plan, maintain and adapt the facilities plan, redefine the objective of the facility. These phases are iterated until objectives of each phase are satisfied.

4.2 Comparison of static evaluation and dynamic evaluation of AMHS
To evaluate static evaluation method of AMHS, we compare the result of static evaluation with that of dynamic evaluation.
A virtual factory liquid crystal display panel is selected to compare these methods. The production of the panel is characterized by the repeated resorting to some process classes such

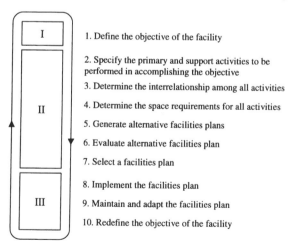

Fig. 6 Planning process steps

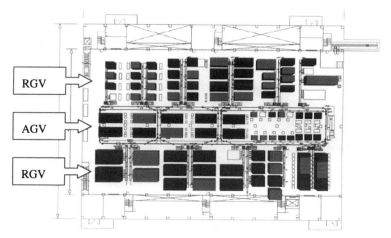

Fig. 7 Layout on AMHS simulator

as semiconductor wafer fabrication [6]. There are more than 90 of operation process steps to fabricate a panel. There are about 100 equipments consisted of 23 types of equipment in the factory. The equipments in the factory are allocated to job-shop type layout as shown in figure 7. The AMHSs consist of one AGV (automated guided vehicle) system in central area and 20 RGVs (rail guided vehicle) in upper and under areas. Each RGV has one vehicle, but AGV system has multiple vehicles. Direction of the AGV is unidirectional, but that of the RGV is bi-directional.

Table 1 is the comparison of both methods of 3 cases in which number of vehicle is varied from 7 to 9. Average moving time per trip consists of retrieval time and delivery time in the simulator. In LayoutMaker, it is supposed that delivery time is equal to retrieval time, because retrieval time depends on vehicle number.

Table 1 Comparison of estimated vehicle number in each method

Vehicle Number	Vehicle Utilization Rate (%)		Average Moving Time/Trip (sec)	
	LayoutMaker	Simulator	LayoutMaker	Simulator
7	81.8	70.9	267	220
8	71.5	65.8	267	233
9	63.6	62.7	267	250

The table 1 shows that vehicle utilization rate estimated by LayoutMaker (static method) is not so different from that by AMHS simulator (dynamic method) in the case of 9 vehicles. Average moving time of LayoutMaker is longer than that of simulator, because delivery time is longer than retrieval time in general. From our experiments, static method is useful for human planners to evaluate and improve alternatives in initial planning step, because design process by using static method is easier and faster than that by using dynamic method.

     A003/055/2003 © With Authors 2003

5 Conclusions

We presented a facilities planning method with integrated software system consisting of several software system. The system can support each of planning and design processes seamlessly using a common factory model. We showed that proposed system is useful through a facilities planning of virtual liquid crystal display panel fabrication line.

Future work is to apply the strict criteria to evaluation of AMHS by using additional information such as load port size, its direction type, batch size, internal buffer size, retrieval time, delivery time, processing time, preventive maintenance time, and so on.

REFERENCES

1 **Tompkins, J. A., White, J. A., Bozer, Y. A., Frazelle, E. H., Tanchoco, J. M. A.** and **Trevino, J.** (1996) *Facilities Planning Second Edition*, John Wiley & Sons, Inc.
2 **Sly, D. P.** (1996) A Systematic Approach to Factory Layout and Design with FactoryPLAN, FactoryOPT, and FactoryFLOW. Proceedings of the 1996 Winter Simulation Conference, pp. 584-587.
3 **Plata, J. J.** (1994) Enhancing the Semiconductor Fab Layout Process. Proceedings of 1994 IEEE/SEMI Advanced Semiconductor Manufacturing Conference, pp. 11-15.
4 **Mori, K., Tsukiyama, M.** and **Fukuda, T.** (1991) Dynamic Manufacturing Scheduling with Petri-Net Modeling and Constraint-based Editing. Proceedings of 7th International Conference on Computer Aided Production Engineering, pp. 477-486.
5 **Muther, R.** (1973) *Systematic Layout Planning Second Edition*, Cahners Books.
6 **Mori, K., Tsukiyama, M.** and **Fukuda, T.** (1998) Petri Net Modeling of an Artificial Immunity Based Production Control System. Proceedings of 14th International Conference on Computer Aided Production Engineering, pp. 89-94.

From concept design to CAD models using reverse engineering

G SMITH and **T CLAUSTRE**
Faculty of Computing, Engineering, and Mathematical Sciences, University of the West of England, Bristol, UK

ABSTRACT

This paper is concerned with reverse engineering strategies to build valid geometric models of prototype parts with both analytic and free-form features. The problems associated with the construction of CAD representations from multiple sets of point-cloud data are discussed. Techniques for scanning parts from single and multiple directions, with fixed and variable orientations, will be investigated and a comparison of data quality from parts scanned using various strategies will be presented.

1 CONCEPT DESIGN

Concept design is the first step in the process of creating a new product. Early stages in this process generally involve the divergent development of many alternatives, focussing on innovation, structure and function. Information gathering and brainstorming are typical activities at this stage, and there is considerable emphasis on the designers creative skills (1). With very few exceptions, notably aero and ship design, initial concept design is more the domain of the artist than the engineer.

Concept designers often prefer to work with traditional design tools, rather than computer-based ones. They perceive that traditional methods are less restrictive of the creative design process, and that sketches and physical models complement development and design evaluation activities. One of the major tools in concept design is the prototype model and one consequence of this is that engineers are faced with the challenge of converting prototype concept models into CAD models at the appropriate stage in the product development/manufacturing cycle.

Design prototypes may be very complex, incorporating both analytic and free-form geometric entities on fully three-dimensional models. For example, Baranek (2) presents a case study based on the re-design of Ford's GT40 super car. The process started with design and concept drawings, based on the original GT40 produced in 1966, then the production of sculptured clay models. The clay models were full scale but represented just one half of the car, that is one complete side up to the centre-line. These were modified and developed to represent current trends in proportion and aerodynamics. The resulting design was a combination of aerodynamic and aesthetic features incorporating many complex, free-form, surfaces. The challenge then was to generate a CAD representation of the geometry of the sculptured half-model, and to use this to develop CNC cutter paths to mill the matching half in order to produce a complete model of the concept car. The CAD model could also be used for subsequent engineering activities.

Producing geometric representations of free-form, or sculptured, parts from scratch is a time consuming and expensive process. CAD workstations with advanced curve and surface modelling capabilities, such as B-Spline and NURBS facilities, are used to develop three-dimensional geometric representations of sculptured parts, but even with these tools ab-initio modelling of complex parts is a lengthy process. It is not unusual for engineering designers to spend weeks, or months, developing CAD models of such parts. It would probably have taken many months to develop a geometric model of the GT40 concept car from scratch. Consequently, Ford used reverse engineering tools to develop the CAD model.

2 REVERSE ENGINEERING

In the most general sense, reverse engineering entails the prediction of what an existing product should do, followed by dissection, modelling and analysis of its actual characteristics. Redesign follows reverse engineering, where a product is developed to its next level. Within the context of this paper, reverse engineering is limited to the process of generating CAD models of existing products. This is in contrast to conventional engineering design which starts with a CAD model and ends with a product.

2.1 Digitisation
The reverse engineering process starts with the capture of three-dimensional data from existing parts. Traditional methods would involve the physical measurement of a limited number of key features on the part and interpolating a model based on these. More recent techniques employ digitising systems to generate masses of three-dimensional position vectors, known as point-cloud data, which are then used to aid the development of CAD models. A point-cloud data set for a relatively small component may consist of tens of thousands of individual data points (figure 1). Whereas a large complex part may require many millions of points. The density of points depends on the level of detail to be captured; plain flat areas will require few points, highly curved areas and fine detail will require many points.

A003/059/2003 © IMechE 2003

The method of digitisation used for a particular application will largely depend on the size of the part to be digitised and the accuracy required. An overview of three different methods is given below.

2.1.1 Digitising with co-ordinate measuring machines

Co-ordinate measuring machines (CMM) have been used to measure analytic and free-form features for several decades. The initial approach was to use touch-trigger probes to gather discrete position data. This data could be very accurate, systems with sub-micron resolution are available, but the process of collecting data with such probes is relatively slow.

Many factors are involved in limiting the speed of data capture, significantly the need to back-off and reset touch trigger probes and the inertia of moving parts of the CMM limit the rate of data collection. A typical CMM with touch trigger probe could gather data at about the rate of 200 points per minute. Recent developments in CMM technology and contact scanning probes have improved this situation. Inertia has been reduced considerably and there is no requirement to back-off and reset scanning probes. Consequently, scanning rates may be increased to the region of about 800 points per minute.

The high accuracy, but low data collection rate, tends to limit the application of CMM digitisation to geometric inspection rather than reverse engineering.

2.1.2 Digitising with scanning machines

The requirement to capture geometric surface data led to the development of dedicated three-dimensional scanning machines. In principle these are similar to CMM but the controlling software and scanning heads are purpose designed for the collection of point-cloud data. Systems range from those based on analogue contact scanning, such as the Renishaw Cyclone (3) machine, with resolutions in the order of 10 microns collecting data at the rate of 400 points per second, to those based on laser digitisation with higher scanning rates but lower resolution, such as the Lectra Digilast (4) with a resolution of about 100 microns.

Scanning machines typically have a working volume in the range of a 400 mm cube. This makes them well suited to scanning small to medium sized parts.

2.1.3 Digitising with optical systems

Optical systems for the digitisation of three-dimensional parts generally employ some form of stereo-imaging. This involves capturing two images of the part and using these to generate a set of three-dimensional vectors representing the surface geometry. Such systems may use either ambient or structured light. Optical systems are capable of digitising large areas very rapidly. Commercial systems are available with the capacity to capture areas up to a square meter in a few seconds.

One of the more successful optical digitising tools is ATOS from GOM (5). This employs two cameras and a fringe projection system to capture three-dimensional data. It is capable of measuring up to 1,300,000 points in seven seconds. Accuracy depends on the measuring area and varies between about 30 to 250 microns. Areas up to 1200 x 960 mm can be captured in one set up, and the system is free-standing so very large parts can be digitised as a series of sections. Baranek (2) describes the digitisation of the GT40 concept car using ATOS.

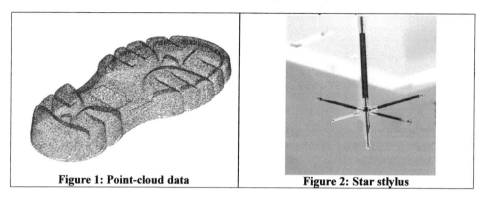

| Figure 1: Point-cloud data | Figure 2: Star stlylus |

2.2 Multiple Sets of Point-Cloud Data

Digitising parts which can be viewed, scanned or probed from one direction is a relatively routine activity. Such parts, subject to limitations on physical size, may be represented by a single point-cloud data set. Large parts and parts with re-entrant features pose more of a challenge. Re-entrant features will necessitate scanning from more than one direction, large parts will need multiple scans to cover the entire area. Large parts with re-entrant features will need many scans from several directions. This will necessitate either changing the part orientation during scanning or changing the orientation of the scanning device.

The general principle adopted in machine shops for decades is that it is better to keep changes in part orientation to a minimum. Changes in orientation will require additional fixturing and result in some loss of accuracy. For parts with fixed orientation, accuracy will be dictated by the machine's motion control systems. Changing part orientation introduces another set of variables relating to the efficiency and accuracy of jigs and fixtures. It is also apparent that some parts, for example the clay models used in concept car design, are inherently difficult to move. These factors mitigate against changing part orientation when digitising.

Scanning parts from different orientations is also problematic because it necessitates the collection of multiple sets of point-cloud data then assembling these into one three-dimensional model. Miss-alignment of point-cloud data sets may result in ambiguities and unwanted holes in the assembled model. These can cause serious problems for subsequent activities such as the generation of machining data, STL files or the development of surface models.

Many sculptured parts have no obvious datum points, and points of reference between different scans can be difficult to determine. Gom (5) address this problem by employing a two-stage process for the digitisation of large parts. This involves photogrammetry to determine the co-ordinates of a set of reference targets placed at strategic positions on the part. These targets define the part co-ordinate system. Digitising is then carried out from as many positions as necessary to cover the scanned area. Each scan must include at least three reference targets. The position of these targets is transformed onto the global co-ordinate system, and the point-cloud data is then transformed onto the part co-ordinate system. Hence, the reference targets provide a common co-ordinate system for all sets of point-cloud data, facilitating their assembly into a single model.

Renishaw (3) adopted a different strategy involving the use of a fixture to hold the part vertically on the Cyclone Scanning machine, then scanning from multiple directions using a star stylus (figure 2). The process starts with the calibration of each arm of the stylus against a reference sphere so that the precise geometry of the star is known. The part is then sectioned and each section is scanned using whichever arm of the stylus is most appropriate. There is a common reference between each scan because of the fixed co-ordinate system and the known geometry of the star stylus. The common co-ordinate system facilitates the assembly of point-cloud data sets into a single model.

The authors have carried out extensive tests with the Cyclone and star styli. Full three-dimensional models have been created by assembling multiple sets of data (figures 3 and 4). It has been established that this approach facilitates the assembly of point-cloud data sets with sufficient accuracy to facilitate subsequent machining, the generation of STL files and the development of valid surface models without unwanted holes or ambiguities. The main restrictions of this system are the limited scanning volume (600 x 500 x 360 mm), the need for long styli to scan deep overhangs, and the inaccessibility of the area held by the fixture.

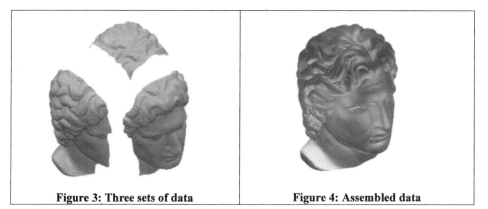

| Figure 3: Three sets of data | Figure 4: Assembled data |

3 GEOMETRIC MODELLING

Point-cloud data is the most basic form of three-dimensional geometric information. It gives an incomplete representation, but it is relatively easy to generate and it can be very accurate. Any subsequent modelling activities based on point-cloud data may increase the completeness but will almost certainly reduce the accuracy. For example, fitting splines and surfaces to point-cloud data can give a complete representation of the surface of a part, but the process of creating these will add further approximations to the model.

The process of constructing surface models based on point-cloud data consists of: trimming to remove excess data, tolerancing to remove unnecessary data, triangulation to create a tessellated model, segmentation to split the data into logical subsets, boundary creation and finally surface fitting. Boundary curves and fitted surfaces may take many forms, but Bezier and B-Splines are by far the most common for free-form modelling.

The use of reverse engineering to create models may appear lengthy, but it is generally much quicker than ab-initio modelling for parts with complex geometry. For example, the triangulated surface model shown in figure 4 was scanned and created in less than 8 hours. It would have taken weeks to produce such a model from scratch. Baranek (2) describes the digitisation and modelling of the GT40 concept car. This process was completed in three days, compared to three weeks without the aid of point-cloud data.

3.1 Trimming, tolerancing and triangulation
Trimming of point-cloud data consists of removing any points lying outside the boundary of the intended model. Such points are often generated as a direct result of the data capture process; for example when a tactile probe tracks off the scanned part, or when a laser scanner scans over the edge of a part. Trimming may be done manually, by selecting and deleting individual points or groups of points. Alternatively trimming may involve removing all points lying outside some plane or boundary curve. The latter approach is generally preferred when large point-clouds are involved.

Tolerancing is employed to reduce the amount of point-cloud data in a model to facilitate subsequent manipulation and modelling activities. The process is generally performed automatically based on some form of chordal deviation tolerance. It results in a high density of points in areas of the model with high curvature and lower densities in areas of low curvature.

Triangulation converts the trimmed and toleranced point-cloud data set to a tessellated surface consisting of triangular facets, with the points forming vertices for adjacent triangles. The effect of tolerancing will result in a high density of small triangles in areas of high curvature, and a low density of larger triangles in flatter areas. The resulting tessellated surface forms the basis for curve and surface fitting processes. It is also well suited to the generation of STL files for rapid prototyping, and as a basis for CNC tool-path planning.

The major limitation with both point-cloud data and triangulated surfaces is that they only support global modifications; such as scaling, translation and mirroring. Localised modifications are not a realistic proposition because they would necessitate the selection and grouping of each individual element (point or triangle) in the region to be altered.

3.2 Segmentation and boundary creation
The major challenge in model construction is the detection of logical boundary curves for the segmentation process, and the establishment of appropriate continuity conditions at these boundaries. Logical, or natural, boundaries are created when one type of surface intersects

another (for example a cylinder intersecting a plane) or when some functional characteristic is required (for example the creation of a parting or joining line). Varady (6) discussed the problems associated with automatic surface segmentation and the creation of boundary curves. He observed that it is relatively easy for humans to recognise natural boundaries on complex objects, but it is generally very difficult for computers to perform this task without some form of human guidance. Consequently, the automatic segmentation of point-cloud data usually results in arbitrary boundary curves bearing little resemblance to the natural topology or functional requirements of the model.

It is, therefore, common practice to have some form of human intervention in the process of boundary curve generation. This may be done at the data capture stage by manually tracing along functional boundaries. Alternatively it may be done at the modelling stage, by selecting points which lie on the chosen boundary and then creating a spline through these. However, both of these approaches have inherent problems; manual tracing of complex curves can be both tedious and error prone, the selection of points along a boundary will also pose problems when large point-clouds are involved.

3.3 Surface Fitting
The processes of fitting surfaces to data points and boundary curves are well established. They generally involve some form of least squares approximation to give the best fit surface to a set of data points. The resulting network of curves and patchwork of surfaces are combined to give a smooth, or fair, representation of the modelled part. Such representations are achieved by applying tangent plane (G^1) or curvature (G^2) continuity conditions over the entire surface. Complex parts are thus represented by a set of surface patches, rather than as individual points or triangles. Sarkar (7) and Milroy (8) give consideration to the development of smooth surfaces in the reverse engineering process.

Segmented surface models support localised modifications, such as the alteration of individual patches or boundary curves, and the control of continuity between adjacent surfaces. From a design perspective, this makes them far more useful than basic point-cloud data or triangulated surface representations.

The authors have attempted to evaluate the accuracy of curve and surface fitting techniques based on the triangulation, segmentation and surfacing of a series of moulding tools. A mould would typically be represented by between six and twenty surface patches with G^1 continuity. Two approaches were considered, one based on the segmentation of point-cloud data along arbitrary boundary curves, the other based on segmentation along curves corresponding to areas of very high curvature. In practice, areas of very high curvature may correspond to functional features, such as parting lines on moulding tools. Any nominal sharp-edge will be represented as a region of high curvature on triangulated point-cloud data. Software routines were developed to identify points in areas of high curvature, these points were sorted into sets and splines were fitted through them. The splines were used as boundary curves in the segmentation process. An arbitrary approach was adopted for segmentation in areas of lower curvature.

Error maps were produced by projecting surface models onto the original point-cloud data. The results of segmentation along arbitrary curves are shown in figure 5, results for segmentation along curves corresponding to regions of high curvature are shown in figure 6. There is clear evidence that segmentation along boundaries corresponding to high curvature gives more accurate results than arbitrary segmentation. It is also more logical to segment models along functional features, such as parting lines, because this will facilitate subsequent design revisions.

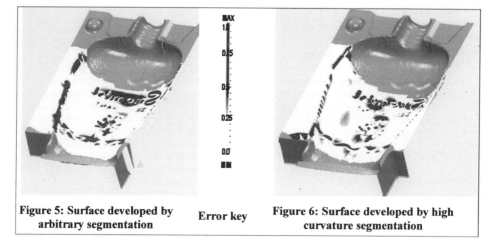

| Figure 5: Surface developed by arbitrary segmentation | Error key | Figure 6: Surface developed by high curvature segmentation |

4 CONCLUSIONS

Case studies have been presented to support the case that the reverse engineering approach to geometric modelling has some advantages over ab-initio modelling. Digitised data, from concept models, can considerably reduce the time taken to develop CAD representations.

The difficulties associated with digitising large parts and parts with re-entrant features have been discusses. The conclusion that it is generally advantageous to fix part orientation and scan/digitise from different positions was presented. Strategies for combining point-cloud data sets obtained from different scan positions have been discussed. It has been shown that multiple point-cloud data sets can be assembled to give full three-dimensional representations of complex parts. These representations have sufficient accuracy/validity to support subsequent activities such as rapid prototyping, CNC machining and surface modelling.

Alternative strategies for the segmentation of point-cloud data sets have been presented. Arbitrary segmentation and segmentation along lines of high curvature were considered. Projecting constructed surfaces back onto the original point-cloud data tested the relative accuracy of each approach. It was shown that segmentation along lines of high curvature can give significantly more accurate results.

 A003/059/2003 © IMechE 2003

REFERENCES

1. **Otto, K. and Wood, K.** Product Design, Techniques in Reverse Engineering and New Product Development. Prentice Hall, UK, 2001, pp 411-432.
2. **Baranek S.L.** Designing the Great American Supercar. Time Compression Technologies Magazine. Communication Technologies Inc, USA, September 2002.
3. **www.Renishaw.com.** November 2002.
4. **www.Lectra.com.** November 2002.
5. **www.Gom.com.** November 2002.
6. **Varady,T. Martin, R.R. and Cox, J.** Reverse Engineering of Geometric Models – an Introduction. Computer-Aided Design, Vol 29, No 4, 1997.
7. **Sarkar, B. and Menq, C.H.** Smooth-Surface Approximation and Reverse Engineering. Computer-Aided Design, Vol 23, No 9, 1991.
8. **Milroy, M.J. Bradley, C. Vickers, G.W. and Weir, D.J.** G^1 Continuity of B-Spline Surface Patches in Reverse Engineering. Computer-Aided Design, Vol 27, No 6, 1995.

Development of an experimental container for the fluid science laboratory of the international space station

C STEIN
Space Infrastructure Division, Astrium GmbH, Friedrichshafen, Germany
D K HARRISON and **A K M DE SILVA**
School of Engineering, Science, and Design, Glasgow Caledonian University, UK

ABSTRACT

"GeoFlow" [1] is an experiment designed for the "Fluid Science Laboratory" on board the International Space Station. The experiment focuses on the simulation of fluid flow conditions in the gap between two concentric rotating glass spheres driven by imposed electrohydrodynamics and thermal fields. This paper describes the design of an experimental container for the GeoFlow experiment that meets the specific challenges of modularisation and miniaturisation of the assembly. The modular approach facilitates varying experimental parameters to be used at a later stage.

1 INTRODUCTION

The International Space Station (ISS), which is developed and built by 18 nations, represents a new and unique platform allowing research under permanent microgravity conditions [2]. From 2001, an increasing number of experiments from different disciplines have been performed on the ISS in the absence of normal gravity. The Fluid Science Laboratory (FSL) will be accommodated in the so-called COLUMBUS module, the major European contribution to the ISS, which will be launched in 2004. The FSL will allow to fluid science research in the absence of gravitational influence. The gravitation is prevalently a disturbing factor of e.g. crystallisation processes or convection in liquid media.

The FSL is a multi-user facility for scientific experiments, built up of different modular sub-units for the control and monitoring of the individual exchangeable experimental modules called Experimental Containers. There is a common envelope for all experiments with standard interface options to the facility itself in order to be accommodated within FSL. This entails that each experiment has a specific set of requirements with regard to shape, optical field of view, observation direction, and electrical, thermal and functional interfaces. For the observation of experiments, glass windows with a diameter of up to 115 mm are mounted on the top, bottom, front and rear side of the housing. The volume of the housing is about 30 l with outer dimensions of H x W x D = 280mm x 400mm x 270mm. The housing constitutes a hermetically sealed container with a maximal design pressure of about 2 bar, which has to be taken into account for the experiment.

The FSL shown in figure 1 is built up of the following major sub-units:
- Master control unit (MCU), for data handling and control of the entire facility
- Power control unit (PCU), which performs power switching and protection function
- Lap-Top unit (LTU) for man/machine interface
- Video management unit (VMU), for the control of different cameras, recording and compression of video data
- Optical diagnostics module (ODM), which contains different interferometers, digital CCD cameras, holography cameras and different illumination
- Secondary water loop assembly (SWLA) as thermal control for the Experimental Container

For each experiment a separate Experimental Container is sequentially placed in the Lower Central Experiment Module (CEM 1), a drawer-like retainer, which hosts parts of the optical equipment. The Upper Central Experiment Module (CEM 2) contains a laser for light scattering and the interface for the external camera. The individual experiment is accommodated within the Experimental Container assembly representing a small independent experimental facility.

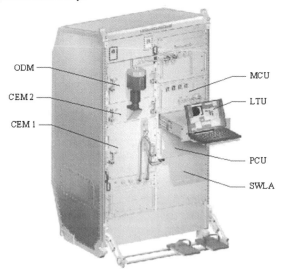

Figure 1 The FSL and its units

The project "GeoFlow" was started in early 2002 under the ownership of the ESA and has successfully passed the state of a feasibility analysis in October 2002. The fundamental

     A003/070/2003 © With Authors 2003

principles described are based on the results of this state and further detailed design will be established during the next months.

2 THE SCIENTIFIC BACKGROUND OF "GEOFLOW"

The "GeoFlow" contains an experiment that allows the investigation of fluid flows within a spherical gap, driven by imposed electrohydrodynamics and thermal field. The research focuses on so-called "geophysical fluid flows" which are of particular importance for a large number of problems such as the explanation of mantle convection of the earth, the flow in a planet's interior or the behaviour of planet's atmospheres. The comprehension of these spherical Couette flows will also help improving engineering tasks such as spherical bearings or centrifugal pumps [2].

The core of the experiment is a fluid cell with a diameter of about 80 mm that consists of a polished hard metal sphere which is surrounded by two concentric glass spheres and rotated by a maximal rate of 120 rev/min. The glass spheres form two separated gaps, filled with silicone oil, whereat the inner gap is the observed area and the outer gap is used for temperature gradient control.

The electrohydrodynamic field is a central force field, similar to the gravity field acting on planets, established through the dielectrophoretic effect that is caused by applying high voltage to the cell. It is considered that an adjustable multi-purpose high voltage power supply with U = 0 - 12 kV, f = 0 – 500 Hz will be implemented either within the FSL or within the Experimental Container. The thermal gradient is induced by heating the inner sphere and cooling with a fluid flow across the outer gap of the cell [3].

The expected flow patterns strongly depend on the temperature gradient, the rotating velocity, the fluid viscosity and the gap width. The different parameters will be varied during the experimental time. It is designated to accomplish about 120 parameter variations with each of 9 experiment cells, all having the same outer diameter. With these experiment cells 3 gap widths and 3 different viscosities are processed sequentially. The gap widths are varied by different diameters of the inner sphere and the viscosities are changed by filling these cells with different silicone oils.

2.1 The dielectrophoretic effect
The dielectrophoretic effect is an effect of higher order and is based on the motion of media due to a non-uniform electrical field. The uncharged molecules are polarised by the electrical field and interact with variations of this field. These dipoles are pulled towards the smaller electrode due to the force-difference acting on the opposite charges as the field intensity increases. In a spherical system the force acts as a central body force towards the centre. An alternating field is used to compensate for the effect of electro-convection. The central force field is proportional to r^{-5} and depends quadratically on the potential U across the electrodes [4].

2.2 The principle of interferometry

Interferometry is based on the wavelike nature of light and its property, that superposition of wave fronts result in an irradiance that can be derived from the sum of the individual irradiances. Generally, the technique of interferometry is to emit a wave front towards a surface or through a media being measured and superpose the reflected wave front, with its impressed structure, with an unchanged reference wave front. The resulting image therefore contains the structure of the surface or the media as a fringe picture.

The Wollaston-Shearing Interferometer (figure 2), which is used in this experiment, is based on a single planar light sheet which is emitted to a reflective surface. The returned wave front is split by a Wollaston prism into two shifted wave fronts, which superpose with each other. This means, that one picture is superposed with itself under a defined shift and thus produces a derivation of the wave front. Because the temperature in the observed fluid correlates with the refraction index, the picture of the interferometer represents the temperature distribution within the fluid [5].

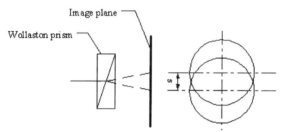

Figure 2 The principle of the Wollaston Shearing Interferometer

3 THE "GEOFLOW" DESIGN

The design of the experiment is primarily determined by the requirements of the visualisation technique. To observe the spherical system by the use of interferometry, the inner metal sphere acting as a mirror, a non-rotating adaptation optics is needed to form the planar wave front used within the FSL to a spherical wave front used within the experiment and vice versa. In addition to this, the optical system must have the property to view both the equatorial as well as the polar region of the cell at the same time due to predicted non-symmetric flow patterns, occurring in the case of the transition from laminar to turbulent flow. This property will be realised by an optics unit that has a flare angle of more than 90° and that is mounted inclined to the optical axis of the experimental cell, the focal point being the centre of the inner sphere. It has been shown, that an optics system with these properties is about 127 mm long with a diameter of about 126 mm. To save space within the Experimental Container, the adaptation optics is also used as hermetically sealed window mounted with a metal diaphragm to the container to allow heat expansion without influencing the alignment [6].

Due to the numerous exchanges demanded of the experimental cell that are required for parameter variations, the possibility must be given to carry out this task in an on-orbit operation. To allow this exchangeability without opening any fluid circuits, a so-called "Orbital Replaceable Unit (ORU)", a cohesive unit concept that contains the fluid cell and its

A003/070/2003 © With Authors 2003

corresponding infrastructure, has been developed. The space limitation of the housing entails that the exchange can not take place within the Experimental Container. For this purpose coherent components are mounted on a common base plate that is dismounted in the case of a cell exchange. The plate is also the reference surface for the precise alignment between the cell and the adaptation optics. To circumvent inhomogeneous heat distribution and resultant different thermal expansions, the base plate is mounted on top of a water-cooled cold plate.

3.1 The fluid cell

Each of the two glass spheres consists of two hemispheres, which are mounted mechanically by a metal ring and glued. The inner sphere has a spark eroded thread and is mounted on top of a shaft composed of two nested tubes which carry the fluid for heating the sphere. To induce the central force field, the inner and the middle sphere constitute the electrodes of a spherical capacitor in which the silicone oil is used as dielectric fluid. The inner side of the middle sphere is coated with transparent Indium Tin Oxide (ITO), to ensure reliable conductivity for the high voltage. The inner sphere is connected via the shaft to the high voltage contact and the coated glass sphere is connected to ground potential via a slip ring device.

3.2 The fluid cell thermal control

The fluid cell thermal control unit consists of a inner sphere fluid heating loop and a outer sphere fluid cooling loop with an analogous structure. Each thermal control is composed of a Thermo-Electrical-Cooler (TEC) coupled to a heat exchanger for heating or cooling the fluid which is circulated by a pump. Two temperature sensors are placed at the inlet and outlet of each heat exchanger. The sensor at the heat exchanger outlet is used for temperature control. The temperature difference between inlet and outlet of the heat exchanger is, at known fluid flow rates, a measure of the heat input respectively the heat removal of the thermal control unit.

The maximum temperature gradient to be achieved is about $\Delta T = 10$ K and is adjustable with an accuracy of ± 0.1 K, whereat the absolute temperature of the media is of minor importance for the experiment result and thus will be chosen close to the ambient temperature that is about $20 - 30°C$.

3.3 The Orbital Replaceable Unit (ORU)

Figure 3 shows a cross section through the Orbital Replaceable Unit (ORU). The ORU consists of a support structure for the experimental cell, which contains the tubing of the fluid to the cell, the fluid cell itself, the inner and outer sphere thermal control circuits and the electrical interface to the rotating tray via Sub-D connectors, placed on a heat spreader plate. Embedded in the support structure of the cell is the high voltage bushing to the inner sphere, immersed in the working fluid, to circumvent partial discharge as the dielectric strength of silicone oils is about twice as high as of air. The high voltage transmission is designed as a central contact, consisting of a spring-loaded pin with a ball-shaped tip that turns in a fixed bowl with the same radius. The advantage of this design is the reliable transmission, as the electrical charge can even be transmitted by capacitive coupling in the improbable case that the pin lifts off to a little distance to the bowl. A drawback is, that additional sealing must be assigned between the fixed bowl and the rotating support structure and that a distance of about 10 mm must be kept to the support structure to circumvent flashover.

Embedded in the heat spreader plate is a centring ring with centring balls, which are designated to position the ORU on a rotating retainer with repeatable accuracy. The centring

ring is screwed separately onto the heat spreader plate as provision is made for the accurate centring of the entire ORU by micrometer in assembled state during the pre-launch integration. Each of the 9 experimental cells is aligned to the optical axis of the adaptation optics in the same way on ground. Because of the inclined alignment of the rotation axis to the optical axis of the optics, the accurate alignment is vital for the experiment, as errors of concentricity affect the interferometrical results by a factor of 4. It has to be taken into account, that a necessary contact between the heat spreader plate of the ORU and the rotating tray is vital to ensure reliable heat conduction.

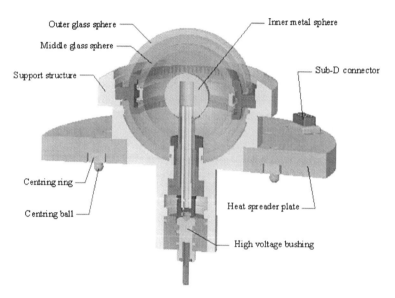

Figure 3 A section through the ORU, shown without infrastructure

3.4 Additional parts of the assembly

The assembly of the parts is determined by the requirements of the cell and the optics. Figure 4 and 5 show the complete assembly. The ORU is inserted in a rotating tray, that is pivoted in a bracket which is mounted on the base plate. The structure provides the rotational drive, the essential high precision alignment versus the adaptation optics and the heat removal from the heat spreader plate. The heat removal is performed by fins, attached to the tray, that are cooled by an air flow circuit generated by fans, coupled to a heat exchanger with TECs and water cooled cold plate. The precise guide of the tray is provided by high precision pre-stressed angular contact ball bearings. Embedded in the tray are the self-mating Sub-D connectors to the ORU and the precise intakes for the centring balls. The signal and power provision is done by a slip ring device that is connected to the Sub-D connectors on the rotating side and to the appropriate interfaces on the fixed side. The rotation drive for the tray is a belt drive system, the rotor pulley being embedded in the tray, driven by a stepper motor that is controlled by the FSL stepper motor interface.

    A003/070/2003 © With Authors 2003

To achieve the autarkic property of the Experimental Container, data editing is done by an integrated controller board. The FSL provides standard interfaces for the most common tasks, such as stepper motor interface, TEC interface or data lines but the specific experimental features has to be adopted to the options, that can be handled within the FSL.

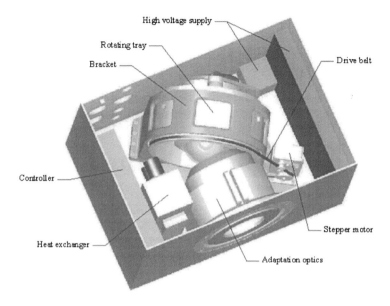

Figure 4 Schematic overall assembly, shown without hoses and cables

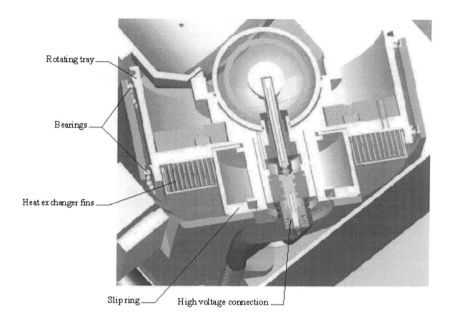

Figure 5 Sectional schematic drawing of the assembly

CONCLUSION AND OUTLOOK

The results of the feasibility analysis show that the accommodation of the experiment within an envelope of an Experimental Container is in principle possible. At present, investigations are continuing on the accuracy analysis of the entire assembly and the optimum ways to manufacture the parts. Some critical design points will be analysed and tested by breadboarding. It is envisaged that by the end of 2003 the experimental container would achieve a state, which can be used for the flight model.

REFERENCE

1   **ESA-ESTEC, Alenia Aerospacio** (1999) Fluid Science Laboratory Preliminary Design Review (FSL-PDR), FSL-PB-AI-0008

2   **ESA** (2001) Simulation of Geophysical Fluid Flows under Microgravity. EC-SRD Phase A/B, FSL-ESA-RQ-003

3   **Egbers C., Brasch W., Sitte B., Immohr J. and Schmidt J-R.** (1999), Estimates on diagnostic methods for investigations of thermal convection between spherical shells in space. Meas. Sci. Technol. 10 866-877

4   **Sitte B., Brasch W., Junk M., Travnikov V. and Egbers C.** (2001), Thermal flow in a rotating spherical gap with a dielectrophoretic central force field. Dynamo and dynamics, a mathematical challenge (Eds. P. Chossat et al.), Kluwer Academic Publishers, 93 – 100.

5   **Eugene Hecht** (1987), Optics Second Edition. ISBN 0-201-11611-1, Addison-Wesley Publishing Company, Inc., p 333 ff.

6   **John Hegseth, Laudelino Garcia and M. Kamel Amara** (1998), A compressible geophysical flow experiment (CGFE). Fourth Microgravity Fluid Physics & Transport Phenomena Conference, August 1998, Cleveland, Ohio

Computer aided programming for NC-electrochemical contour evolution machining (NC-ECCEM) of integral impellers

J XU, Y ZHU, P HU, and **N YUN**
College of Mechanical and Electronic Engineering, Nanjing University of Aeronautics and Astronautics, China

SYNOPSIS

An algorithm to calculate the evolution movement trace and a method to design a NC-ECCEM program was presented in this paper. The above-mentioned Computer Aided Programming (CAP) for NC-Electrochemical Contour Evolution Machining (NC-ECCEM) of integral impellers was fulfilled by means of C+ + language. The machined samples of integral impellers show us that the CAP for NC-ECCEM of integral impellers is successful and referable.

1 INTRUDUCTION

Electrochemical Machining (ECM) is a non-traditional machining technology based on electrochemical anodic dissolution and has some outstanding advantages, including being able to machine difficult-to-cut metals and to produce complex shaped parts with high machining rate and low surface roughness, and the used tool-cathode without wear, which is different from the electrode wear happening in the EDM process. For the above advantages ECM had been quickly developed and widely applied. But in the last 20 years ECM has not developed so much and its application is also limited. The reasons are mainly as follows: the ECM process is not so steady, it is hard to reach high machining accuracy, and it is also necessary to design and make a complex profiled tool-cathode, which is skilled and costly work. In order to make ECM applicable for a new field, such as either micro-precision machining or producing special parts, the researchers try to find some new methods not only to overcome the above-mentioned disadvantages of ECM but also to carry forward its particular merits [1-4]. With the NC and CNC techniques rapidly developing and constantly perfected, like most of the mechanical cutting technologies, the ECM combined with NC and CNC may reach this goal. Some brilliant and attractive results have been produced in the experimental research on NC-ECCEM.

Like NC milling technology, in the NC-ECCEM process a workpiece is machined by a simple, small and non-profiled tool-cathode moving along a contour trace. The movement of the tool-cathode is controlled by a NC system (Fig. 1) to evolve into a desirable shape of surface. The "machining edge" of a tool-cathode can be designed as a certain shape, such as a point, one linear edge or two linear edges, or a curved edge. Obviously NC-ECCEM has good machining flexibility, the designing-making work of a complex profiled tool-cathode is replaced by programming the moving trace of a simple tool-cathode, thus enhancing its machinability. For example, it is very difficult, costly, and time-consuming to machine an integral impeller with twisted blades by NC mechanical cutting, but this type of impeller can be easily machined by NC-ECCEM, not only with the production cost and time reducing but also with the machinability increasing. In this paper, the key technique of NC-ECCEM, Computer Aided Programming (CAP), including the method to process the data of the coordinates, the machining shaping law, the movement distribution law and the scheme to program the trace of machining movement, are discussed in detail. Finally the machining result will be presented to show the success of the programming.

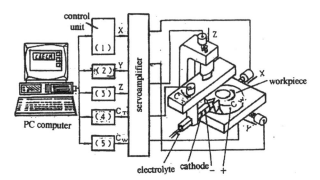

Figure 1. Schema of the NC-ECCEM process system

2   THE MACHINING MOVEMENTS AND THE SHAPING LAW OF NC-ECCEM

The blades of an impeller commonly have the complex surfaces, and so the movement of the NC-ECCEM machine tool must have multi-axis in line. A 5-axes linkage machine tool is used in the experimental research. In accordance with the experiments the optimized movement distribution was schemed (seeing Fig.1). In the machining process the tool-cathode moves along the z-direction, as well as the workpiece rotates around the z-direction and moves along the x-direction and the y-direction. So the shaped surface is machined by means of the 4-axes combined movement of the tool-cathode relative to the workpiece. The trapezoidal plane cathode (referring to Fig. 2) is used in the machining process and the electrolyte is spurted out from the interior of the cathode through a slot (seeing Fig. 3). It should be pointed out that the NC-ECCEM is a non-contact machining process different from the contact machining process such as NC-milling. So the machining gap between the tool-cathode and the workpeice, and the gap distribution along the moving trace of tool-cathode relative to the workpiece (referring Fig.

2 and Fig. 3), in general we call them as shaping law of ECM, will be emphatically considered in the following paragraph to discuss the moving trace and the machining accuracy.

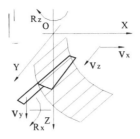

Figure 2. Illustration of NC-ECCEM a shaped surface

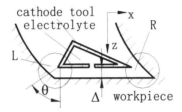

Figure 3. The diagram of movement of the cathode relative to the workpiece

From Fig. 2 and Fig. 3 it is clear that the machining accuracy of NC-ECCEM is depended on the gap between the "machining edge" of cathode and the machined surface of anode (workpiece). The edges of the cathode can be supposed as round and the radius as r. In the Fig. 3 the feed rates of the z-direction and the x-direction are v_z, v_x separately. The gap in the θ-direction is δ, but at the z-direction is Δ. The feed rate in the z-direction keeps constant, but v_x will be alerted to meet the needed shape of the workpiece, so producing δ. Assuming $tg\beta$ as v_x/v_z and $A=\eta\omega\kappa U_R$ as a constant (here η-current efficiency of anodic dissolution, ω-volume equivalent of anodic dissolution, κ-electrolyte conductivity, $U_R=U-\delta_E$, U-applied voltage between anode and cathode, δ_E-decomposition voltage), the shaping law of the L region (referring to Fig. 3) will be expressed by the following equation:

$$\frac{A}{\delta} - (v_z \cos\theta - v_x \sin\theta) - (v_z \sin\theta + v_x \cos\theta)\frac{d\delta}{(r+\delta)d\theta} = 0 \qquad (1)$$

Noting $A/v_z=\eta\omega\kappa U_R/v_z=\Delta$, then the shaping law of the L region will be deduced as the following equation:

$$\frac{\Delta}{\delta}\cos\beta - \cos(\theta + \beta) - \sin(\theta + \beta)\frac{d\delta}{(r+\delta)d\theta} = 0 \qquad (2)$$

Noting v_x is much less than v_z, then β will be approximate to 0. If $r \to 0$ and $\theta = \pi/2$, the Eq. (3) will be deduced:

$$\delta\big|_{\theta=\pi/2} = \Delta\pi/2 \qquad (3)$$

The similar result could be obtained in the R region. Based on the analysis of the movement and the NC-ECCEM shaping law, the method to process the feature coordinates could be determined.

3   PROCESSING THE DATA OF SHAPED SURFACE

3.1 Reconstructing the designed blade surface into a quasi-parallel straight-generatrix surface

The NC codes are produced from the feature coordinate points of a blade surface. The total $n \times m$ feature points at the different sections on a surface will be used respectively to depict either the basin or the back surfaces of an impeller blade. There are n sections along the y-direction paralleled to the x-o-z plane and m points at each section to depict a curve (see Fig.4). If taking m sections along the z-direction paralleled to the x-o-y plane, then n points will appear at the each section and approximately on a straight line (referring Fig. 4). The following job is how to make the m quasi-straight lines to construct a quasi-parallel straight-generatrix surface (referring Fig. 4) by which the designed blade surface is described in a necessary accuracy.

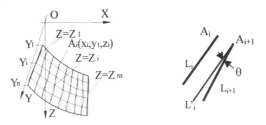

Figure 4. The illustration to construct a quasi-parallel straight-generatrix surface

Assuming the equation of the line at the section is

$$\begin{cases} y = a_i x + b_i \\ z = z_i \end{cases} \quad (1 \le i \le m) \qquad (4)$$

Then a_i and b_i can be calculated by the least-square method, therefore, m straight lines are obtained. Each line has one point intersecting with each plane $y = y_i (1 \le i \le n)$. The m straight lines intersect the y = y₁ plane to get m points, which could be made into one curve with the cubic B-spline and could be interpolated according to the necessary machining accuracy. In the same way another curve on the plane $y = y_n$ would also be obtained. Both of the two curves are the baselines of the straight-generatrix surface in request. The boundaries of the surface were determined by the lines or the curves that the surface intersects with the plane $z = 1, z = z_m$ and the cylinder $x^2 + y^2 = R^2$, $x^2 + y^2 = r^2$. R and r respectively equal y_n, y_1, those are the radius of the impeller crest and the radius of the impeller root.

In accordance with above-mentioned analysis, the surface composed by the separate straight generatrixes was obtained. The segment L_i and L_{i+1} were supposed to be the adjacent two generatrixes (see Fig.4). The cathode would move from the position L_i to the position L_{i+1} relative to workpiece in the t_i time, then the shifts of every axis could be calculated as follows:

$$P(x, y, z) = A_{i+1}(x, y, z) - A_i(x, y, z) \qquad (5)$$
$$\theta = \arg(\vec{L}_{i+1}) - \arg(\vec{L}_i) \qquad (6)$$

$P(x,y,z)$ in the Eq. (5) expresses the shifts at the x-direction, the y-direction and the z-direction. θ in Eq. (6) stands for the angle shifts rotated around the z-direction. It must be emphasized that the segment L_i', which paralleled to L_i, was also on the $i+1$ section but with an intersecting angle to the line L_{i+1} (referring Fig. 4), namely L_i' was also on the plane $z = z_{i+1}$, so the method only suits the impeller with parallel straight-generatrix surface.

3.2 The unparallel straight-generatrix surface matched with a blade surface
A blade surface matched with the unparallel-generatrix surface is also described by a set of feature points, but at the different sections paralleled to neither the xoy plane nor the xoz plane. The ith generatrix of the surface was a space line and could be expressed as the follows:

$$x = \frac{y + d_i}{a_i} = \frac{z + e_i}{c_i} \qquad (7)$$

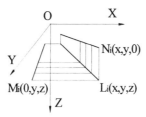

Figure 5. A space line and its projections

The space line $L_i(x, y, z)$ was supposed as one of generatrixes of the surface. It is projected on the xoy plane to form the line $N_i(x, y, 0)$ and on the yoz plane to form the line $M_i(0, y, z)$ (Fig.5), therefore, the parameters of the Eq.(7) could also be calculated by the least-square method.The two baselines and the boundary were determined as same as mentioned in the paragraph 3.1. It is clear that five-axes linkage movement in line was needed to complete machining the blade with unparalleled straight-generatrix surface; the bottom plane of cathode should rotate around the x-direction.

The segments L_i and L_{i+1} were also supposed as the adjacent two generatrixes (referring Fig.4), and the segments N_i and N_{i+1} were supposed as their projections on the xoy plane separately. Then in the time t_i, a vector $\vec{n}_i$ normal to $\vec{L}_i$ and $\vec{x}$ could be calculated as follows:

$$\vec{n}_i = \vec{L}_i \times \vec{x} \tag{8}$$

$\vec{x}$ in the Eq.(8) was the unit vector of the x-axis. So the following equations could be obtained to calculate the shifts of every axis.

$$P(x, y, z) = A_{i+1}(x, y, z) - A_i(x, y, z) \tag{9}$$
$$\theta = \arg(\vec{N}_{i+1}) - \arg(\vec{N}_i) \tag{10}$$
$$\phi = \arg(\vec{n}_{i+1}) - \arg(\vec{n}_i) \tag{11}$$

$P(x, y, z)$ and θ were the same as mentioned in the paragraph 3.1. The ϕ stands for the angle shift rotated around the x-direction by the cathode.

4 COMPUTER AIDED PROGRAMMING BY C++ LANGUAGE

Using Microsoft Visual C++ 5.0 as the programming tool, according the overall scheme shown in Fig. 6, a computer aided programming software was developed. By means of this software, inputting the coordinate points of a designed blade surface, a NC program to control the linkage machining movement of every axis could be produced automatically, which was applied to machine several kinds of impellers in the experimental research.

5 THE MACHINED SAMPLES

The NC program produced from the Computer Aided Programming Software was inputted into a 5-axes linkage NC system, and then to control a 5-axes NC-ECM machine tool. The relative NC-ECCEM experimental result shows us its success. The machined samples are shown in Fig. 7. The error of machined blade surface for a middle size of impeller (left of Fig. 7) is less than 0.15mm, and for a small size of impeller (right of Fig. 7) less than 0.10mm.

                                                    A003/079/2003 © IMechE 2003

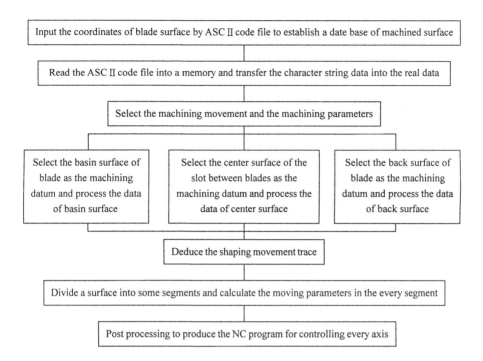

Figure 6. The overall scheme of Computer Aided Programming

Figure 7. The impeller samples machined by a 5-axes NC-ECM machine tool
(left-ϕ190mm, right-ϕ120mm)
Machining conditions: electrolyte-15%NaCl, $T_0=35^0C$, $p_0=0.5MPa$; U=15v;
feedrate $v_z=0.8mm/min$; the others v_x, v_y, and C_w are relatively fitted with v_z;
workpiece and tool-cathode are made of stainless steel.

6   CONCLUSION

In summing up the technological merits, The NC-ECCEM process synthesizes the advantages of both ECM and NC techniques, overcomes their own different disadvantages, and then reaches very good machining steadiness and high accuracy. Therefore NC-ECCEM will be applied in machining some complex shaped parts, such as an integral impeller. Especially it will pay an important role in the aviation and aerospace manufacturing engineering.

The above-mentioned Computer Aided Programming is a key technique of the NC-ECCEM technology. The experimental research results, including the machined impeller samples, showed us that both the analysis and design method for Computer Aided Programming and the programmed software are successful and referable.

REFERENCES

1 **Rajurkar, K. P., Zhu, D. etc.** (1999) New Development of Electrochemical Machining. Annals of CIRP, Vol.48/2, pp 48-56.

2 **Masuzawa,T. and Takawashi, T.** (1998) Recent Trends in EDM/ECM Technologies in Japan. Proceedings of the ISEM-12, pp1-15.

3 **Xu, Jiawen.** (1992) The non-traditional machining method of integral impellers, Aviation Precision Manufacturing Technology. Vol. 28, No. 4, pp19-21. (In Chinese)

4 **Sermatech-Lehr.** (2000) Leading the Way in Electro-Chemical Machining Technology For Turbomachinery. Turbomachinery International, January/February, pp37-39

5 **Zhu, Yongwei. Xu, Jiawen.** and **Hu, Pingwang.** (2001) Study and application of NC-Electrochemical Contour Evolution Machining integral impeller. ACTA Aeronautica et Astronautica Sinica, Vol.22, No.4, pp376-378. (In Chinese)

6 **Xu, Jiawen. Tang, Yaxin. Yun, Naizhang.** (1993) The investigation on the shaping law of NC-ECCEM with a rotary cathode. Proceedings of IMCC, pp97-100.

7 **Yun, N. Z., et al.** (1989) Investigation on Application of Electrochemical Contour Evolution Machining. Proceedings of the ISEM-9, pp143-145.

8 **Wei, B., and Rajurka, K. P.** (1990) Accuracy and Dynamics of 3-Dimensional Numerical Control Electrochemical Machining (NC-ECM). Proceedings of the Winter Annual Meeting of the ASME. PED-Vol. 45, pp33-45.

9 **Kozak, J., Rajurka, K. P., et al.** (1990) Sculptured Surface Finishing by NC-Electrochemical Machining with Ball-End Electrode. Advances in Technology of Machines and Equipments, Vol. 22, No. 1, pp53-74.

10 **Belogorsky, a., et al.** (1995) Flexible Automation ECM Technology Based on 3-Coordinate CNC-Machines. Proceedings of the ISEM-11, pp585-592.

Optimization of NC-point distribution for five-axis high-speed milling

H FENKL
Clever Engineering, Stuttgart, Germany
D K HARRISON and **A K M DE SILVA**
School of Engineering, Science, and Design, Glasgow Caledonian University, UK

SYNOPSIS

This paper describes an algorithm for an optimised NC-point distribution from a given continuos three-dimensional NC-path on a free-form surface in five-axis high speed milling. High Speed Cutting (HSC) technology applied on NC-paths with high variations of curvature leads to enormous changes of velocity on each axis of the machine tool. The requirements of the NC-points and their distribution on the desired NC-path is important, because the CNC-controller has to compute the complete dynamic behaviour of each axis of the machine tool in real time. Current Computer Aided Manufacturing (CAM) systems which use very simple algorithms do not take into consideration the influence of feed rate, the block cycle time of the CNC-controller and the dynamics of each machine axis on the NC-point distribution. The "Adaptive Speed-up Control (ASC)" algorithm is designed to take into account the factors mentioned above and therefore to give an optimised NC-point distribution. The ASC algorithm is based on differential equations for the accelerations of each machine axis. A velocity profile for the cutter path is computed so that acceleration and velocity limit of a machine axis will be reached smoothly and with limited jerk. ASC has been successfully implemented for producing turbine blades on five-axis Machining Centres (MC) with a significant reduction of cutting times compared to conventional methods.

1 INTRODUCTION

Five axis milling has become the standard method for manufacturing 3-D complex shapes. It comprises three linear axes with two rotary degrees of freedom to give an optimum adaptation between the tool and the workpiece surface. Free-form surfaces can be milled more efficiently with cylindrical cutters with a corner radius then with ball end cutters, because the same surface quality is reached by considerably less cutter paths at osculating-optimal orientation of the tool axis [1].

The determination of such optimised cutter paths is very difficult, because of the need for optimal adaptation between tool and workpiece surface while simultaneous minimisation of the milling time by extending cutter path distances. This requires an algorithm to determine the direction of the tool axis and from this another one to calculate the contact curves between tool and workpiece.

A general solution for cutter path determination, which is optimal for all workpiece geometries does not exist at present. The automated determinations of five axis cutter paths offered in CAM systems confine themselves to the calculation of geometrically simple curves, along which the contact point of the tool is led on the free-form surface of the workpiece. The tool axis is determined at the contact point from the normal vector and additional angles related to this vector.

For practical reasons the continuous information about the tool trajectory is discretised to adjacent points along the contact curve. At each point, the direction of the tool axis is computed and from given tool diameter and tool corner radius the centre of tool tip is determined. The standard data format for description of tool trajectories in the workpiece coordinate system consists of a sequence of the following FICAPT-statement:

```
GOTO/x,y,z,i,j,k        x,y,z = coordinates of tool centre point
                        i,j,k = normalized vector of tool direction
```

This sequence of geometric information will be expanded further with dynamic and technological instructions such as spindle speed and cooling fluid information in order to create the part program. The part program contains all information necessary for manufacturing the component [2]. The following postprocessor step adapt the part program to the CNC-controller of the machine tool and transforms the coordinates of the tool centre point and tool direction into the positions of the five axis of the machine tool [3,4]. The tool direction vector i,j,k determines the positions of the two rotary axes, the tool centre point x,y,z and the positions of the rotary axes determines the positions of the linear axes of the machine tool.

The relative velocity of the contact point between the workpiece and the tool is defined by the feed rate. In order to achieve a uniform milling process, this value should be constant. In three-axis milling there is no difference between the velocity of the contact point and the tool centre point, because the direction of the tool axis is constant. In five-axis milling, however, due to great direction changes of the tool axis there are significant differences between the velocities of the contact point and the tool centre point. The given feed rate can be related only to the motion of tool centre point and not, as desired, to the motion of the contact point between tool and workpiece, because the information about the contact point was lost at the interface from the CAM-System to postprocessor.

The task of the CNC controller is to coordinate the movement of all five axes of the machine tool from the discrete NC-data points to the desired continuous relative motion between tool and workpiece. From the discrete NC-data points and the feed rate the positions of each machining axis is computed in real-time in the interpolator cycle of the position control circuit. The given feed rate is subjected to the dynamic constraints of the machine tool. Limitations of velocity, acceleration and jerk of each axis have to be considered during computation of the positions in the interpolator cycle of one or two milliseconds.

A003/071/2003 © With Authors 2003

The processing steps described above from CAM-system via postprocessor to CNC-controller is the established standard practice in industrial manufacturing. The significant progress made during the last decade on the dynamic behaviour of the machine tool and the cutting process can only be used if the data generation is considered in the context of the complex process chain. The geometric part of this chain starts with the surface quality which in turn effects the quality of the contact curve generation and the point distribution of tool trajectory. The dynamic part starts at this trajectory, to which the feed rate is related and hence influences the algorithm for determining the smooth velocity profile with regard to the dynamic limitations of the machining axis.

2 DETAILED PROBLEM DESCRIPTION

2.1 Quality of Free-Form Surface and Contact Curve
The quality of the finished surface and the production rate are influenced fundamentally by the quality of the free-form surface and the curve, on which the contact point of the tool is moved on the workpiece.

Within the CAM-system the mathematical description of free-form surfaces is made mostly with polynomial functions, which enables the mapping of the two independent parameters of a surface description to the three-dimensional space. Any polynomial function, whose parameter range is frequently chosen between 0 and 1, describes a part of the surface, a so-called "patch". A patch is defined as an area of a free-form surface that can be described by a single polynomial function. Complex free-form surfaces are a composite of concatenated patches [5].

The following formula shows the mathematical description of a patch with a polynomial degree of five for both parameters.

$$
\begin{aligned}
x(u,v) = {} & \alpha_{11} + \alpha_{12}u + \alpha_{13}u^2 + \alpha_{14}u^3 + \alpha_{15}u^4 + \alpha_{16}u^5 \\
& + \alpha_{21}v + \alpha_{22}uv + \alpha_{23}u^2v + \alpha_{24}u^3v + \alpha_{25}u^4v + \alpha_{26}u^5v \\
& + \alpha_{31}v^2 + \alpha_{32}uv^2 + \alpha_{33}u^2v^2 + \alpha_{34}u^3v^2 + \alpha_{35}u^4v^2 + \alpha_{36}u^5v^2 \\
& + \alpha_{41}v^3 + \alpha_{42}uv^3 + \alpha_{43}u^2v^3 + \alpha_{44}u^3v^3 + \alpha_{45}u^4v^3 + \alpha_{46}u^5v^3 \\
& + \alpha_{51}v^4 + \alpha_{52}uv^4 + \alpha_{53}u^2v^4 + \alpha_{54}u^3v^4 + \alpha_{55}u^4v^4 + \alpha_{56}u^5v^4 \\
& + \alpha_{61}v^5 + \alpha_{62}uv^5 + \alpha_{63}u^2v^5 + \alpha_{64}u^3v^5 + \alpha_{65}u^4v^5 + \alpha_{66}u^5v^5
\end{aligned}
$$

$$u,v \in [0..1]$$

analogous

$$y(u,v) = \beta_{11}...$$
$$z(u,v) = \gamma_{11}...$$

Together with the formation law, the algebraic coefficients $\alpha_{ij}, \beta_{ij}, \gamma_{ij}$ contain the information for mapping the two-dimensional parameter level into the three-dimensional space to create the shape of the surface. Since the contact curve lies on the surface of the workpiece, this curve can be described with an independent parameter which maps to the two-dimensional parameter area of the workpiece surface. A curve representation in a surface is called an edge, similarly to the surface being defined as a composite of patches an edge can be defined as a composite of arcs [6].

Compared to the parametric curve description in three-dimensional space, this representation has the advantage, that the normal vector of the surface at the contact point (which is necessary for determination of the direction of tool axis) can be computed directly from the algebraic coefficients of the surface. The following formula shows the mathematical description of an edge with a parameter of w and a polynomial degree of five.

$$u(w) = \alpha_0 + \alpha_1 w + \alpha_2 w^2 + \alpha_3 w^3 + \alpha_4 w^4 + \alpha_5 w^5$$
$$v(w) = \beta_0 + \beta_1 w + \beta_2 w^2 + \beta_3 w^3 + \beta_4 w^4 + \beta_5 w^5 \qquad w \in [0..1]$$

The xyz-coordinates of a point on the contact curve get the following dependency from the edge parameter w:

$$x = x(u(w), v(w))$$
$$y = y(u(w), v(w))$$
$$z = z(u(w), v(w))$$

By insertion of w into the edge function, both surface parameter u, v are computed, the following insertion of u, v into the surface function leads to the desired xyz-coordinates. The point sequence on the contact curve can be computed by an increasing sequence of edge parameter w. The direction of the normal vector $\bar{n}$ of each point at the surface can be calculated directly from the polynomial function of the surface with the following formula:

$$\bar{n} = \begin{pmatrix} x_u \\ y_u \\ z_u \end{pmatrix} \times \begin{pmatrix} x_v \\ y_v \\ z_v \end{pmatrix} \text{ with } x_u = \frac{dx(u,v)}{du}, \quad x_v = \frac{dx(u,v)}{dv}, \quad y_u = \frac{dy(u,v)}{du}, \dots$$

2.1.1 Influence of discontinuity on patch transition
While polynomial functions up to their polynomial degree are differentiating steadily within an arc or a patch by definition, the degree of differentiating on the transition between two arcs or two patches is often reduced to first or second order. The reasons for this lie in inadequate algorithms for patch concatenation or in the nonobservance of the continuity requests from manufacturing side during the construction of the free-form surface.

The continuity requirement on the surface for five-axis milling increases, since the normal vector of the surface directly influences the position of a rotary axis of the machine tool. Only very few CAM programs fulfil the continuity requirements, which are necessary for high speed cutting. Therefore the CNC-controllers have integrated algorithms to smooth these data. However this causes undesirable effects due to additional accelerations and decelerations of the machining axes.

2.1.2 Influence of inflection points of contact curve
Each inflection point on the contact curve leads to a lean back of the normal vector. This leads to an inversion of the direction sense of a rotary axis in which the velocity of the rotary axis has a zero-crossing. The consequence is an unavoidable marking by the tool on the finished surface of the workpiece.

                                   A003/071/2003 © With Authors 2003

2.1.3 Influence of fluctuations of curvature of contact curve

The directions of normal vector and tool axis on a point on the workpiece surface are determined by the curvature of the contact curve at that point. Every curvature fluctuation has a direct influence on the direction of the tool axis, which in turn affects the position of the rotary axes. Furthermore the curvature fluctuation about the tool radius influences the positions of the linear axes of the machine tool.

In five-axis case the positions of the linear axes are determined by the motion of the pivot of rotary axes and the motion of tool centre point. The centre point of tools with larger diameter is more sensitive against curvature fluctuations due to the lever arm effect between the contact point and the tool centre point. The tool centre point describes a circular arc around the contact point, if the direction of the normal vector of the surface changes.

2.2 Point distribution on contact curve

Since the contact curve lies in the workpiece surface a simple local coordinate system can be defined for every point of the contact curve. The axis consists of the tangent vector of the curve, the normal vector of the surface and the so-called binormal vector, which is formed by tangent and normal vectors from the cross product.

The theoretical solution space for the direction of the tool axis is formed by a hemisphere, which spreads itself out over the tangential plane in this point of the surface. The tool axis is determined by the handicap of a constant lead and/or tilt angles. After the direction of the tool axis is specified the tool centre point can be calculated for a given tool diameter and tool corner radius. Theoretically the tool axis and tool centre point can be expressed explicitly in terms of the parameters of the contact curve and therefore described continuously as a function. However, this leads to a very high polynomial degree of the curve parameter, which is not very useful in practise. Therefore tool axis and tool centre point are computed at discrete points along the contact curve and submitted to the postprocessor for transformation into machine coordinates.

2.2.1 Block cycle time

The block cycle time of a NC-controller determines the read-in velocity of NC-records. This time is > 10 ms for older CNC-controllers and is < 1 ms for modern CNC-controllers and the NC-data processing has to be achieved in real time. The number of processed NC records has to fit with the number of read-in NC-records. This has the consequence, that the point distribution on the contact curve cannot be chosen arbitrarily in the CAM system, because point distribution and block cycle time are connected over the desired feed rate. If the point density is too high, the CNC-controller has to reduce the specified feed rate to avoid a violation of the block cycle time.

2.3 Mode of Operation of CNC-Controller

For computation of the set values of the position control circuit it is necessary that the following three steps be carried out in real time [7]:
- path planning: to build a continuous trajectory in the five-dimensional space of the machine tool from the given discrete NC-data points means a interpolation process
- generation of velocity profile: to compute the velocity on the trajectory under consideration the given feed rate and the dynamic limitations of each machining axis

- path generation: to compute the positions of each machining axis in the interpolator tact from the planned path and the velocity profile

2.3.1 CNC-controller with linear interpolation of discrete NC-points

The simplest assumption for interpolation of the discrete NC-points is that in each interval between two NC-points the axis velocity is constant. The velocity is determined by assuming that the endpoint of the interval is reached by all axis at the same time. In order to calculate each axis velocity from the given feed rate, it is necessary, to compute for each interval a distance or space element, on which the feed rate can be related. Exactly, the given feed rate defines the velocity of the contact point between workpiece and tool.

Two formulae for the calculation of a distance between two NC-points can be built for 5-axis machine tools in a simple way:

$$\Delta s = \sqrt{\Delta x^2 + \Delta y^2 + \Delta z^2 + \Delta A^2 + \Delta B^2}$$

$$\Delta s = \sqrt{\Delta x^2 + \Delta y^2 + \Delta z^2}$$

The first equation calculates an interval distance under consideration of the rotary axes, in the second equation only the distance of the linear axes of the machine tool is considered.

Using the given feed rate v and the interval distance Δs in the following equation, for each interval i a period Δt_i can be calculated. This period is available for each single machining axis to reach the endpoint of the interval simultaneously. The constant axis velocity of each machine axis in the interval i is determined as follows:

$$v_{x_i} = \frac{\Delta x_i}{\Delta t_i} \quad v_{y_i} = \frac{\Delta y_i}{\Delta t_i} \quad v_{z_i} = \frac{\Delta z_i}{\Delta t_i} \quad v_{A_i} = \frac{\Delta A_i}{\Delta t_i} \quad v_{B_i} = \frac{\Delta B_i}{\Delta t_i}$$

Due to the adopted linear connection between two points and the constant axis velocity within an interval, jumps arise in velocity, acceleration and jerk, for every machine axis at the interval limits.

For every interval a maximal permitted profile velocity can be calculated, in which the single axis velocities are not exceeded. Furthermore the maximal permitted profile velocity at the end of every interval can be calculated under consideration of the acceleration of each machine axis in connection with the change of axis velocity at the interval change. Following a rough interpolation of the profile velocity course inclusive the ascent and descent flanks at the interval boundaries in the clock of about 5 ms is calculated.

The fine interpolation for the calculation of the set positions of the position control in the clock of about 1 ms uses the computed profile velocity of the rough interpolation. Ascent and descent flanks of the profile velocity can be modified by different acceleration profiles. All these methods lead to deceleration and following acceleration at significant direction changes at the interval boundaries.

A003/071/2003 © With Authors 2003

Different proposals and solutions exist to overcome these discontinuities in velocity, acceleration and jerk at the interval boundaries. In CNC-controllers the algorithms for integration of connecting pieces, like circle or polynomial curves between two successive intervals are available. For example, Prasetio [8] presents a connecting curve where the change of the direction angle is a sinus-quadratic function of time. This approach allows a constant profile velocity during the complete course of the curve.

2.3.2 NC-controller with spline or polynomial interpolation of discrete NC-Points

While only the point continuity in the discrete NC points is ensured at the linear interpolation, the continuity of tangency and curvature at the interval boundaries can be ensured with higher interpolation methods. To achieve this continuity in the CNC-controller an algorithm is necessary, which build a steady space curve from the discrete NC-points for all 5 axes of the machine tool. Those algorithms lead to a vector function, depending of one parameter. In principle, the parameter can be voted freely, but a parameterisation after the covered tool path is favourable. An ascending parameter value can be assigned to every discrete NC-point:

$$x_i = x_i(s_i), \quad y_i = y_i(s_i), \quad z_i = z_i(s_i), \quad A_i = A_i(s_i), \quad B_i = B_i(s_i)$$

Piecewise spline or general polynomial functions can be calculated by interpolation for every machine axis. The quality of the produced interpolated path understandably depends on the point distribution of the discrete NC-points. In unfavourable cases only very few NC-points can be interpolated by a spline function and tangential or curvature steady transition functions must be calculated between two sections.

The advantage of this time intensive interpolation lies in the fact, that the course of the tool path becomes steady and independent of the discrete NC-point distribution. The computation of the course of the profile velocity is simplified because the points of discontinuity are eliminated and a continuous course of axis velocity and acceleration can be achieved.

2.3.3 Look ahead functionality

In dependence of the processor power of the CNC-controller the so-called look ahead function computes from the current position the profile velocity course of the future tool path. This is necessary during strong changes in the direction of the tool path, because a transgression of the axis velocities and accelerations can only avoid by a punctual reduction of the profile velocity. This functionality becomes more and more important, because the application of the enhancements of the cutting process by higher-order cutting substances allows higher and higher profile velocities.

2.3.4 Following error compensation

The concept of following error describes the permanent deviation between the calculated set path and the produced workpiece contour. The following error has to be distinguished from an deviation between the current set points for the position control and the current values from the measuring system of the machine axis during the cutting process. This deviation during manufacturing time is caused by a proportional position control, which is used in conventional CNC-controller.

If all machine axes have the same control amplification $-K_v$-factors - the current deviation goes back to zero at the end of straight movements. Only at curvilinear tool path a lasting deviation surrenders, i.e. the superposition of the movement of the single machine axes with changing axis velocities leads to an permanent deviation. The following error depends on acceleration and velocity of each machining axis and increases with higher profile velocity. Higher profile velocities, which are usual in HSC, lead to an amount of the following error, which exceed the permitted tolerance band. Therefore suitable measure must be found for compensating.

3 Adaptive Speed-Up Control

The computation of a soft and regarding the time an optimised velocity profile under real time conditions and consideration of the dynamic limits of the machine tool is the target of the geometric kernel of a CNC-controller. This task exceeds the limits of processor power of many CNC-controllers at the market by far. Most of CNC-controllers, applied in the area of free-form surface milling, are not able, to make completely use of the progress in the cutting substance technology which allows an enormous increasing of the feed rate. In order to drive curvilinear tool paths with high profile velocities, often extreme accelerations and velocities are necessary in the single machine axes, which is connected with high effort of real-time computations.

3.1 Wording of Formulation:

The formulation consisted of the following two parts:

- To build up an offline geometric kernel of a CNC-controller without consideration of calculation time and actual position. This means a pre-computation of the set values of the position control of each machine axis. The calculated data set is directly submitted to the CNC-controller of the machine tool and carried away to the position control unit. The profile velocity has to be optimised in such a manner that under retention of the feed rate and the dynamics of the machine axes the complete proceeding time gets minimal. The set point values for the machine axes should be computed at each millisecond.
- The output should be a normal NC program written in the conventional interface format. The distance between two NC records is adapted to the block cycle time of the CNC-controller. The appropriate feed rate value for this interval is calculated and added to the NC program. The CNC-controller then processes a conventional NC program in real time with an adapted NC-point distribution.

3.2 Prerequisites: Surface Quality, Smooth Tool Axis Path

As already mentioned in the chapter of problem description, the quality of the free-form surface and a smooth tool path are important for a good quality of the finished surface and for a short manufacturing time.

The following formula shows exemplarily for the Y-axis the direct coherence between tool path and axis acceleration. If the tool path is parameterised over the arc length of the contact curve, then the acceleration of a machine axis is described as follows. The derivation after the time is marked with a dot, the derivation after the parameter s as a stroked quantity:

$$\ddot{y} = \frac{d^2 y}{dt^2} = y'' * \left(\frac{ds}{dt}\right)^2 + y' * \frac{d^2 s}{dt^2} \qquad \text{with } \ddot{y} = \text{acceleration} \left[\frac{length}{time^2}\right]$$

$$y', y'' = \text{first and second order derivative by}$$

A003/071/2003 © With Authors 2003

$$\left(\frac{ds}{dt}\right)^2 = \text{profile velocity in square}$$

$$\frac{d^2s}{dt^2} = \text{profile acceleration}$$

The direct influence of the derivations over the parameter of the arc length on the acceleration of the machine axis gets obviously. Those derivations are pure functions of the geometric tool path in space and are totally independent of the dynamics with which the tool path will drive through.

3.3 Presentation of the Optimisation Problem

As already mentioned, a spline interpolation from the given discrete NC data points generates a vector function of the five machine axes with the parameter s as a function of the arc length of the contact curve. The advantage of this specific parameter setting lies in the fact, that the derivation of the parameter s over the time is the velocity of the contact point between tool and workpiece. This velocity is defined as the feed rate. The velocities of the machine axes can be expressed respectively as a product from the derivation of the axis positions after the parameter s and the common profile velocity $v = \dfrac{ds}{dt}$. The second derivation of the axis position over the time leads to the acceleration of the machine axis:

$$
\left.
\begin{aligned}
\ddot{x} &= x'' * \left(\frac{ds}{dt}\right)^2 + x' * \frac{d^2s}{dt^2} = x'' * \dot{s}^2 + x' * \ddot{s} = x'' * v^2 + x' * \dot{v} \\
\ddot{y} &= y'' * \left(\frac{ds}{dt}\right)^2 + y' * \frac{d^2s}{dt^2} = y'' * \dot{s}^2 + y' * \ddot{s} = y'' * v^2 + y' * \dot{v} \\
\ddot{z} &= z'' * \left(\frac{ds}{dt}\right)^2 + z' * \frac{d^2s}{dt^2} = z'' * \dot{s}^2 + z' * \ddot{s} = z'' * v^2 + z' * \dot{v} \\
\ddot{A} &= A'' * \left(\frac{ds}{dt}\right)^2 + A' * \frac{d^2s}{dt^2} = A'' * \dot{s}^2 + A' * \ddot{s} = A'' * v^2 + A' * \dot{v} \\
\ddot{B} &= B'' * \left(\frac{ds}{dt}\right)^2 + B' * \frac{d^2s}{dt^2} = B'' * \dot{s}^2 + B' * \ddot{s} = B'' * v^2 + B' * \dot{v}
\end{aligned}
\right\} \ddot{X}
$$

These equations for the axes velocities and accelerations build the prerequisite to formulate the auxiliary conditions, which are given by the limits of the dynamic of the machine tool.

limitation of axis velocity : $\qquad -\dot{X}_{max} \le \dot{X} \le \dot{X}_{max}$

limitation of axis acceleration: $-\ddot{X}_{max} \le \ddot{X} \le \ddot{X}_{max}$

Two points of the representation of the differential equation system are meaningful:
- the coupling of the system of equations is carried out via the common profile velocity and its time derivative
- the single contributions of axis acceleration can be separated into a factor which depends only from the given geometric tool path in space (x'' and x') and into a factor, which depends only from profile velocity and its time derivative (v^2 and $\dot{v}$)

The task is to determine such a course of the profile velocity, which leads at any time to no transgression of the dynamic limitation and remains along the given tool path as long as possible with the predefined feed rate. The determination of the profile velocity course means to calculate the functional coherence of $v(s)$ under consideration of the given feed rate and the auxiliary conditions. The determination of the optimal solution can be represented in accordance with Johanni [9] as the following optimisation problem:

$$\text{Determine } s(t) \text{ with: } s(t=0)=0, \quad s(t=t_{end})=s_{end}$$
$$\dot{s}(t=0)=0, \quad \dot{s}(t=t_{end})=0$$

$$\text{so, that} \qquad \int_0^t f(s,\dot{s}^2,\ddot{s})dt \qquad \text{becomes a minimum.}$$

This is a variation problem of second order in s and t, with a free upper limit t_{end}, because the total time t_{end} for the complete movement is still unknown. It can be transformed into a problem of lower order and fixed upper limit.

Finally the task can be formulated: determine a maximal $v=v(s)$ which fulfil the following auxiliary conditions:

$$-\dot{X}_{max} \quad \leq \quad v * X' \quad \leq \dot{X}_{max} \qquad (1)\text{axis velocity}$$
$$-\ddot{X}_{max} \quad \leq \quad v^2 * X'' + \frac{1}{2}\left(v^2\right)' * X' \quad \leq \ddot{X}_{max} \qquad (2)\text{axis acceleration}$$
$$0 \quad \leq \quad v(s) \quad \leq v_{Pr\,max} \qquad (3)\text{given maximum profile velocity}$$

The first two auxiliary conditions relate to the dynamic characteristics of the machine tool, the third condition is determined from the technology of the cutting process.

3.4 Solution of the Optimisation Problem
The determination of the maximal allowed profile velocity $v=v(s)$ can be divided into several steps, arising from an analysis of the auxiliary conditions. Conditions (1) and (3) restrict only profile velocity, while (2) affect both, the profile velocity and its derivation. In a first step the calculation of the maximal possible profile velocity is carried out only under consideration of the limitation of the axes velocities. The result is the so called limit velocity $v_{lim} = v_{lim}(s)$.
In a second step the determined course of the limit velocity has to be checked by the auxiliary condition (2). A further reduction of the allowable profile velocity course could be forced, because (2) limits the combination of profile velocity and its derivation.

Interpreted clearly meant (2), that a change of the actual profile velocity is permitted only in a definite range. The velocity of the actual contact point may neither be accelerate too strongly nor braked too strongly. For this reason a look ahead capability from the actual contact point to the forward tool path course is necessary. If the current profile velocity was namely too high in front of a strong curve then it cannot be avoided that the axis acceleration get too high in the following curve. The only possibility is, to reduce the profile velocity punctually before a strong direction change of the contact point. The consideration of the auxiliary condition

(2), leads to the maximal profile velocity $v_{max}(s)$. A new limit velocity $v_{lim}(s)$ is determined by the minimum of $v_{max}(s)$ and the limit velocity calculated before from condition (1) and (3).

The general demand for a time-optimal tool path means clearly, that the profile velocity under compliance of the given restrictions have to be as great as possible in every point of the path. With the evaluation of the auxiliary conditions till now carried out, it can be say, that the profile velocity has to proceed either with maximal or minimal derivation or at $v_{max}(s)$. The limit curve can be subdivided into sections, each section has one of the following characteristics. The derivation of the limit curve:

- is greater than the maximal permitted v' - acceleration too high. This section of the limit curve cannot be part of the maximal $v(s)$. - is called trajectory source.
- is less than the minimal permitted v' - deceleration too high. This section of the limit curve cannot be part of the maximal $v(s)$ - is called trajectory drain.
- is in the range of the permitted v'. Only this section can be part of the maximal $v(s)$.

3.5 Construction of the optimal Profile Velocity
The subdivision of the limit curve can be used for the successive determination of a velocity profile, which completely fulfil the three auxiliary conditions. An iterative algorithm calculates the course of profile velocity $v = v(s)$ after determination of the velocity at the starting point and at the end point of the considered path. The start and the end velocities can arbitrary be chosen between zero and the given feed rate.

The result of the calculated course of profile velocity does not take into account any condition for smooth acceleration and deceleration. To achieve this, a further algorithm is implemented, which fits into the course of the profile velocity smoothing ramps. These ramps are calculated by a defined acceleration profile with limited jerk. The algorithm works iterative in both directions of the considered path and leads to the desired smooth course of the profile velocity.

3.6 Determination of NC-Positions

With the calculation of the profile velocity the relative motion between workpiece and tool is determined. The information is presented as a function $v(s)$ that can be used in different ways. If the CNC-controller is able to read-in the set positions of the position control in its own cycle time instead of the conventional NC-data positions, an offline calculation of these set positions can be obtained directly from the function $v(s)$ and a starting position. This calculation can be done with or without error compensation. In the case of error compensation the set positions are distorted so that the remaining path deviation is as low as possible. The distortion can be computed, because velocity and acceleration of each machine axis is known for each position of the tool path.

Another possibility is, to use the algorithm for an optimised NC-point distribution under consideration of the block cycle time. The NC-points are distributed in such a way, that for each clock of the block cycle time a point is generated. Between two points an appropriate feed rate value for the CNC-controller is calculated. This application of the algorithm generates a conventional NC-program as input to the CNC-controller. The third possibility is, to integrate the algorithm into the CAM system as a method for an optimised point distribution directly computed from the contact curve.

4 CONCLUDING REMARKS

It is evident that the quality of the free-form surface in a CAD system has a direct influence on the quality of the finished workpiece surface. This influence can be characterised by the partial derivative of the axis positions in a coupled differential equation system for the acceleration of the machine axes. The influence can be reduced using specific algorithms in the CNC-controller. However, these algorithms, carried out at the expense of the dynamics of the machine tool, lead to higher cutting times. If the surface quality in the CAD suffice the requirements on high speed cutting, the Newton differential equations system can be used for offline computation of an optimised course of the profile velocity.

The presented algorithm is implemented in a production chain for turbine blades in more than 60 machine tools. Without a change of the CNC-controller it leads to a time reduction of 30 % of the finish process for producing the airfoil of a turbine blade.

REFERENCES

1 **Schnider, M.** (1998) Berechnung schmiegeoptimaler kollisionsfreier Werkzeugverfahrwege für die fünfachsige Fräsbearbeitung von Freiformflächen mit Torusfräsern, [Computation of osculating-optimal collision-free tool trajectories for five-axis milling of free-form surfaces with toric cutters], Eidgenössische Technische Hochschule Zürich Diss. ETH Nr. 12522, Zürich, pp. 1-4

2 **DIN,** (1987), Taschenbuch 200: NC-Maschinen, Numerische Steuerungen, [DIN paper pocket 200: NC machines, Numerical Controller], German Institut of Standardization, DIN 66215 analogous ISO 3592, Beuth Verlag Berlin, Köln , pp. 49-60

3 **DIN,** (1987), Taschenbuch 200: NC-Maschinen, Numerische Steuerungen, [DIN paper pocket 200: NC machines, Numerical Controller], German Institut of Standardization, DIN 66025 analogous ISO 6983, Beuth Verlag Berlin, Köln , pp. 23-39

4 **Kief, H. B.,** (2001), NC/CNC Handbuch, '01/02, [NC/CNC Handbook, '01/02], Hanser Verlag, München, pp. 294-297

5 **Mortenson, M.E.** (1985), Geometric Modeling, John Wiley & Sons, New York, pp. 156

6 **CATIA,** (1998), Solutions Application Programming Interface Reference, Manual, Version 4, May 05, Edition, pp. 20-26

7 **Olomski, J.** (1989), Bahnplanung und Bahnführung von Industrierobotern, [Path planning and Path guiding of Industrial Roboters], Herausgegeben von W.Ameling, Fortschritte der Robotik 4, Vieweg Verlag, Braunschweig, pp. 38-41

8 **Prasetio, J.** (1994), Vorkorrektur der Sollbahn zur Erhöhung der Genauigkeit von CNC-Maschinen, [Pre-Correction of Set Path for Increasing of Accurancy of CNC-Machines],

Herausgeber: Prof. Dr. h.c .mult. Dr.-Ing. Günter Spur, Dissertation, Forschungsberichte für die Praxis 148, Produktionstechnik – Berlin, Carl Hanser Verlag, München, Wien

9 **Johanni, R.** (1988), Optimale Bahnplanung bei Industrierobotern, [Optimzed Path Planning for Industrial Roboters], VDI-Verlag, Düsseldorf

Computer-aided Process Planning

A method for the representation of assembly processes on mobile devices

S PARISI
Department of Production Engineering, University of Palermo, Italy
C MATYSCZOK
Computer Integrated Manufacturing, Heinz Nixdorf Institut, University of Paderborn, Germany
H KRUMM
Unity AG, Büren, Germany
A BARCELLONA
Department of Manufacturing Processes, University of Palermo, Italy

SYNOPSIS

In car industrial assembly today almost everything is done by machines and human direct intervention is only in operations like welding, assembly of doors or assembly of electrical parts; the present representation of assembly processes is done with slides or CAD drawings of the parts that have to be assembled. This way is a bi-dimensional representation of a three dimensional reality; furthermore visualization is static and with no assembly sequence. With a dynamic and solid representation, all the process could be easier to understand and faster to realize and this means obviously more productivity. In this work a prototype, based on devices like PDAs, has been realised in order to display video sequences in a three dimensional solid environment to assist the workers on the assembling operations.

1 INTRODUCTION

One of the problems of industrial assembly processes is that new and unskilled employees need much time to understand the operations they have to do and, normally, employees with more experience must teach them the most delicate steps. This of course means loss of productivity for the company, because of the time that specialised workers have to dedicate for teaching.

Besides, Market today requires great product differentiation, with different kinds of product to produce and to assemble as well and many problems for the workers, that need to adapt the new product configurations in the shortest possible time. This means high learning times and a reduction of the "learning effect".

The objective of this research is to give a support to the workers with a three dimensional representation of the process, through the use of a mobile device. By means of it, they can watch a solid animation that explains how the parts have to be assembled, as many times as they need. PDA (Personal Data Assistant) represents a good solution for the display of video assembly sequences because of its easy handling, while for example a laptop computer could be too cumbersome and heavy to use in an assembly division. An effective PDA, able to be employed in a Virtual Reality environment, requires to have the following features:

- Enough computing power: it is necessary to handle compressed and large video files;
- Good memory space: at first it will make the hypothesis that video files are stored in the device so we need enough free memory to store the video files, or part of them;
- Colours screen display: in a colours display it is easier to distinguish a higher number of particulars, for example coloured, than on a black and white screen;
- Screen display size as large as possible: on a too small screen it would be impossible to display small particulars;
- Resolution as high as possible: for the screen size, higher resolution supported means more possibilities to display comprehensible animations;
- Microphone and speakers integrated: microphone is used to record eventual voice suggestions and animations may contain also sounds;
- Compatible bar code scanner available: a bar code scanner generates the input variables of the process;
- Possibility of synchronization, thorough a docking station, with a computer, which to transfer files with;
- Possibility of other kinds of communication: video files can be better managed with a wireless data transfer, for example Bluetooth or WiFi.

The result of a executed Market analysis is that the most suitable device is the Pocket PC because it represents the device that satisfies better the technical requisites mentioned above. For these reasons this work was developed on this kind of device.

2 DEVELOPMENT OF A PROCESS CHAIN

2.1 Software concept

It has been found that there is no software available today that solves the above mentioned assembly processes problems, so it was necessary to develop a dedicated software. The software is developed through two main parts: the "preliminary operations" in figure 1 and the "process operations" in figure 2.

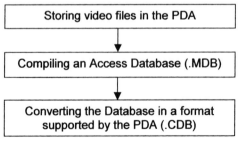

Figure 1. Preliminary operations

   A003/007/2003 © IMechE 2003

In this first part of the process, the actual "engine" of the software is created: the database contains all the information necessary to the software to select the correct operation and to display the right video sequences. A conversion of the database is necessary because a Pocket Access database is not the same created for desktop computers. Video files can be stored directly in the device or also transferred with a LAN but for now, to make the process easier, videos are stored in device's memory, such as RAM, memory cards or hard disk. The full video sequence of a generic operation is divided into smaller parts to simplify user's comprehension.

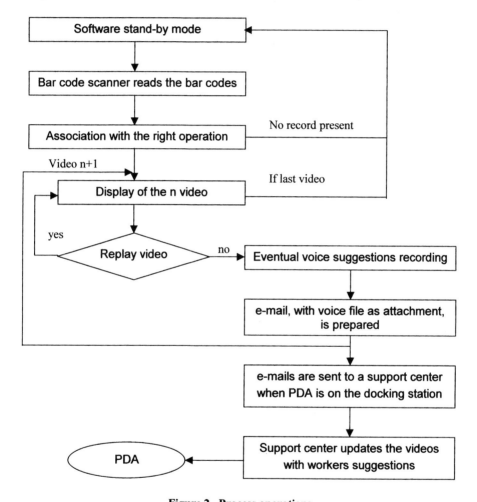

Figure 2. Process operations

The bar code, or the bar codes to give more flexibility, of the parts that have to be assembled are associated by the software, which looks for them into the database, to the correct operation. There is the possibility that no video files are present for some operations; in this case the software will print a "no record" message and it will return to the stand-by mode, waiting for another bar code input. If the bar codes correspond to a record in the database, the software prepares the sequence of the video files that have to be displayed.

After the first video, the user can decide to display it again, all the times he needs to comprehend that part of the operation; if he wants, user can also record his suggestions, regarding how to improve video quality, by means of a voice file that will be attached to an e-mail. The e-mail, whose body contains the information about what video is referred to and who is the person in charge between the designers, will be sent to a support center as soon as the PDA is on its docking station, i.e. a support for allocating the PDA for synchronization and exchange of information.

The support center updates the video files, keeping in consideration user's suggestions, and transfers the new video files into the device. After the last video sequence, the software informs the user and returns to the stand-by mode.

2.2 Traditional assembly representation

The present solution is the use of slides or CAD drawings, with or without sections, which show the final resulting part exploded, with all the small pieces that will make up the part separated; in the following figure 3, for example, it is shown the drawing of a fuel pump of a car.

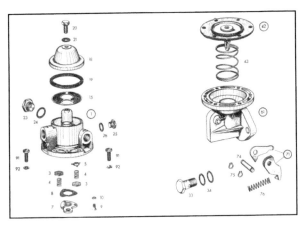

Figure 3. Fuel pump assembly drawing

As it can be easily noted, even with a relatively low number of parts, the assembly drawing is really complicated and there is nothing that explains which part has to be assembled sooner or later than the others.

Furthermore there is no sequence of how the parts have to be assembled and workers must deduce it. This means more time to produce; as mentioned above, time can be reduced with a

A003/007/2003 © IMechE 2003

specific help, as the animation of a three-dimensional model that shows the correct way to manage the parts.

2.3 The proposed approach

This kind of assembly representation concept intends to give a valid help to the workers in the assembling operations: instead of a static and bi-dimensional representation, a clear and solid sequence of the process is provided.

The operator disposes of a solid model instead of a simple drawing and all the solid parts are joined in a visualised spatial environment: nothing is left to the imagination and the worker can easily understand how the parts have to be assembled by watching as many times as he needs a video of the operation. In fact the video-sequence, that may be stopped at any time, shows to the worker each priority and each detail of the assembly operation.

Another key point of this approach is the possibility of a feedback that guarantees the continuous improvement of the process. After the display of each video, the worker can record a voice message, containing his suggestions about how to improve the process. This is a really important step because this part makes the process always dynamic: in fact the video designers do not have the experience of a worker and need to know better the operation steps in order to model the animations as much real as possible.

3 DEVELOPMENT OF A PROTOTYPE

Aim of this work was to analyse the feasibility of the proposed approach by means of a software prototype; eMbedded Visual Basic 3.0 has been successfully used.

At present a Pocket PC can play the most diffuse formats of video files but the MPEG format has been chosen; it should anyway be noted that for a device with a small memory quantity, like a PDA, it would be better to manage video files with the strongest compression, like the one of a DivX, in order to have more free memory. Unfortunately, when dealing with continuous movement video stream, for example those generated by a personal video camera, DivX format is not able to generate this continuity while MPEG format works really good; this happens because DivX videos require an adequate hardware to work really good but unfortunately Pocket PC processors have "only" 206 MHz. Furthermore, MPEG video player for Pocket PC, PocketTV, is better than the DivX player, Pocket DivX, which gives problems in the automatic display of the videos.

The first steps of the development of the prototype are the preparation of a database, which contains all the information about the operations, and the transferring of the video files into the device.

Since the database is the "engine" of the process, it has been preferred to have a good but at the same time easy solution: the Access Database has been chosen because it is very well integrated with eMbedded Visual Basic and, by considering that the application deals with approximately 60.000 records, faster than a simple text-file database.

In this prototype, input variables are entered manually because it was not possible the implementation of the bar code scanner because the one we tested was not compatible under Windows CE; anyway this is not a problem because the concept is absolutely the same.

Table 1. Database Records.

ID	NAME OF PARTS	BAR CODE1	BAR CODE2	NAME OF VIDEO	DESIGNER	PATH1	PATH2	SUPPORT CENTRE E-MAIL
1	Axis	12345	12346	Video1	Rossi	/video1_1.mpg	/video1_2.mpg	sc@x.com
2	Seat	12345	12347	Video2	Smith	/video2_1.mpg	/video2_2.mpg	sc@x.com
3	Light	12345	12348	Video3	Braun	/video3_1.mpg	/video3_2.mpg	sc@x.com

Once entered the two input variables, i.e. the two bar codes, the software looks for them into the database with ADOCE through the Connection and Recordset objects and through an SQL statement. Also the possibility of "no record" is provided through the BOF and EOF properties of the Recordset object.

About the voice suggestions recording, top Pocket PC devices, through an integrated microphone, can record a WAV file. In order to manage these WAV files and to associate them to the right video files, and in order to implement everything in an automate process, "Voice Recorder Control" has been used; this allows the use of the voice recorder inside an eMbedded Visual Basic or Visual C++ application.

At the moment; the only way to generate and send e-mail, containing attachments, for a Pocket PC, is to use MAPI in a simple eMbedded Visual C++ application (5). This application, recalled through the code, automatically generates an email to the address contained in the database, with the voice file attached and with a small description about the video, whom the voice file is referred to, and the person in charge for that video. Since all the single steps of the process are feasible, the last step is to put everything together, automate the process and realize a Graphical User Interface (GUI), shown in Figure 4.

Figure 4. 3DSP GUI

A003/007/2003 © IMechE 2003

3DSP is the name given to the application and it is the acronym of "Three Dimensional Simulation on Pocket PC". The GUI is user friendly but, at the same time, it contains all the necessary information and helps to the workers. There are no command buttons but only labels because everything is done by hardware buttons.

Buttons at the same actual position in the Pocket PC were reproduced, by means of rounded rectangles. At the top-right are visible the name of the selected video and a label which shows microphone status. The software has a "steps" conception, not sequential: once recognized the operation, the application goes to the first video and then is possible to do all the operations about that video such as displaying it as many times as it needs, recording eventual voice suggestions about that video and sending the relevant e-mails.

4 RESULTS AND DISCUSSIONS

The software application for the prototype works very well in all its parts, with a reasonable speed for a Pocket PC. The association with the operation, done with a Pocket Access database, is much faster than one done with a simple text database. The display of the video sequences is very good also considering that a very small screen is employed, even if sometimes the movement is slow, due maybe to not enough memory available on the device: just rebooting or deleting temporary files, the video sequence is fluid again. The voice suggestions recording produces a good quality WAV file, called with the name of the operation and the name of the video whom is referred to; then automatically the file is attached to the generated e-mail, and deleted from the device.

The part of the process that presents many possibilities to be improved is the data transfer module: the present solution for this prototype, by docking station, can be acceptable for a first implementation period but technology offers today many interesting opportunities, such as for example wireless data transfer.

Wireless connection means less memory occupied in the device because there is no e-mail with attachment stored and it could also mean no video stored in the device but transferred "just in time": the two main possibilities are represented by WiFi, a wireless LAN, up to 11 Mbps, and by Bluetooth, up to 1 Mbps maybe not really suitable for this aim because of the video files size. More free memory available on the device means consequently more speed and less problems.

5 CONCLUSIONS

A new approach, based on devices like PDAs, has been analysed and proposed in order to assist the workers on the assembling operations; in particular a process chain has been projected and a fully working prototype has been realised, at this time without a bar code scanner implementation, even if this is not a problem because for the software there is no difference between the input variables entered manually or the ones scanned.

Every step of the project can be done with the present technology; devices in some situations are still not yet as powerful as it is needed and memory quantity available is not really enough; this problem could any way be avoided by endowing each PDA with a Microdrive hard disk.

Probably a fast evolution of processors power, like what happened to desktop computer processors, will be realized and at the same time there will be more memory available on the devices.

Furthermore it is not difficult to think that Augmented Reality, a new form of man-machine interface where computer-generated animations are superimposed to the user vision of the real world, can be used also in an assembly process and maybe in some years, when this technology will be brought to perfection, it will happen.

REFERENCES

1 **Tacke C., Bassett T**. (2002) eMbedded Visual Basic: Windows CE and Pocket PC Mobile Applications. Published by Sams, ISBN: 0672322773.

2 **Tiffany R.** (2001) Pocket PC Database Development with Embedded Visual Basic, ISBN: 1893115658.

3 **Grattan, N.** (2001) Pocket PC and Handheld PC Developer's Guide. Prentice Hall PTR; ISBN: 0130650773.

4 **Forsberg C., Sjostrom A.** (2001) Pocket PC Development in the Enterprise, ed. Addison Wesley; ISBN: 020175079.

5 "Introduction to MAPI in Pocket PC 2002 C++ Applications". (2002) Microsoft tutorial

A003/007/2003 © IMechE 2003

Study of the optimization of a tendering process in warship refit

D A FLEMING, G A FORBES, L E HAYFRON, A H B DUFFY, and **P D BALL**
Department of Design Manufacture and Engineering Management, The University of Strathclyde, Glasgow, UK

ABSTRACT

The optimisation of a tendering process for warship refit contracts is presented. The tendering process, also known as the pre-contract award (PCA) process, involves all the activities needed to be successfully awarded a refit contract. Process activities and information flows have been modelled using Integrated Definition Language (IDEF0) and a Dependency Structure Matrix (DSM) with optimisation performed via a Genetic Algorithm (DSM-GA) search technique. By utilising this approach the process activities were re-sequenced in such an order that the number and size of rework cycles were reduced. Initially, a 57% reduction in the Scott partitioning criterion has been obtained.

1 INTRODUCTION

Babcock BES designs and refits both naval and commercial vessels, with more than 90% of the work involving one main customer (1). Facing the dilution of future turnover in this competitive sector, due to the reduction of available contracts, the company is responding by accelerating its improvement programme. It is accepted that Babcock BES contains the expertise and knowledge to continue supplying the customer with high quality, leading edge products. The challenge lies in reducing cycle times and costs.

To obtain the necessary performance improvements within the pre-contract award (PCA) process, a strategic business process, it was decided to apply a business process improvement (BPI) methodology that reflects the philosophy of Cook (2) and Ziari (3). The methodology will use techniques to identify key areas for improvement. These improvement areas will be benchmarked and changes in the process's performance monitored, via key performance indicators. Based on early findings, process rework was identified as a target area where performance improvements may be obtained. This paper describes the application of an optimisation technique to the PCA process. The findings of this study led to recommendations that aim to reduce unintentional rework cycles. A qualitative modelling tool was used to formalise the 'As-Is' process. This was then transferred into a quantitative modelling and optimisation tool. The results of which led to the following recommendations:

- Re-sequence activities, given the information constraints, so that the size and number of iterative cycles is reduced.
- Greater congruency with 'customer' and 'design' processes.
- Improve activities that have the most positive impact on streamlining overall process flow.

The above recommendations are discussed in more detail, as are the particular strengths and weaknesses of the dependency structure matrix-genetic algorithm (DSM-GA) method. The paper details some cycle time reduction challenges, in an information intensive process, and demonstrates how the DSM-GA method was applied. Advice is given on technical and social issues that have to be overcome when performing a similar project.

2 DESCRIPTION OF CASE STUDY PROCESS: PRE-CONTRACT AWARD (PCA) PROCESS

Make/Engineer-to-order companies spend a significant amount of time and effort in putting together tenders (4). In Babcock BES this is known as the PCA process and involves all the work activities required to obtain a refit contract. This includes converting the customer requirements into a product specification and executing the design work, through to estimating (material & labour requirements), tendering and contract negotiation. These functions were modelled to a level of abstraction that showed 86 work activities and 460 information links. Hence, the size and importance of the PCA process necessitates that it is structured and supported to allow the generation of both accurate and timely tender bids and in addition communicate the rationale, upon which the tender is based, to the rest of the business (4).

The selection of the PCA process for improvement was based on an internal Babcock BES study, which explained that a significant degree of company difficulties, in refitting a ship, lay in the early contract stages (i.e. precisely the activities that are covered by the PCA process). This is a common situation in many companies and is discussed further by O'Grady, et al (5) and Nevins and Whitney (6). This can often be attributed to two generic problems, new/late information being injected into the project and/or activity products failing to meet downstream requirements. Therefore, it is important to understand the PCA process's key characteristics, information transfer between the various departments and the customer. If the information transfers can be formalised and understood there is an increased probability that changes, aimed at improving the efficiency and effectiveness of the PCA process, will be focused where they would have most positive impact (7).

3 PROCESS MODELLING AND ANALYSIS TECHNIQUES

Practitioners and academics have recognised that the key to process improvement lies in how well the process is understood (8), decomposed and subsequently restructured (9). This realisation has led to many process modelling tools and techniques, each possessing certain strengths and weaknesses. These were evaluated against process characteristics and project objectives, to ensure that an appropriate tool was selected for the study. The PCA process is characterised by a considerable exchange of information, both internally and with the customer. It was important that the tool selected could capture these information requirements and allow the process to be streamlined. It was decided that the tool should have the capability to:

- Illustrate activities and the information flows between them.
- Demonstrate the size and number of iterative cycles.
- Demonstrate the degree of importance of information in relation to each activity.
- Handle large information intensive processes.
- Methodically decompose a process.

Integrated Definition Language (IDEF0), Structural Analysis Design Technique (SADT), Gantt Charts, DSM-GA and Program Evaluation and Review technique (PERT) were considered, as they are commonly used tools for modelling and managing processes. None of these techniques were uniquely appropriate for this study. Therefore, it was decided to use IDEF0 in tandem with the DSM, in order to fulfil the above modelling requirements. The IDEF0 technique, discussed further in (10), was used to gather the required information and populate the DSM, whilst the DSM-GA was used for process optimisation and analysis in this study.

The DSM is a generic process modelling tool. Developed by Stewart (11), the DSM was developed, primarily to formalise complex information flows and iterative cycles, a process characteristic often overlooked when modelling processes (12). Figure 1 shows a DSM model of the 'As-Is' PCA process.

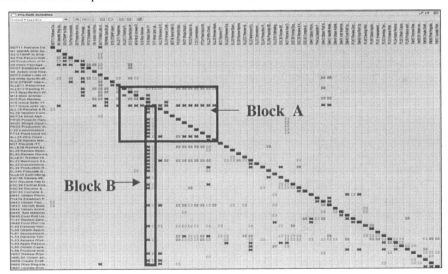

Figure 1 'As Is' matrix

The 'As-Is' model contains a series of activities in identical order, both along the horizontal and vertical axes. An 'X' represents information being passed on from an activity in a column to an activity in a row. An 'X' above the leading diagonal represents iterative blocks (see section four for a definition of iterative blocks). The darker 'X' signifies more critical information.

Whitfield et al (13), built on the strengths of the DSM modelling technique with the introduction of a Genetic Algorithm (GA) optimisation search technique. This technique allows the described process to be quantitatively modified, i.e. computationally re-sequenced. There are currently 6 performance criteria that may be selected when optimising the process. Scott partitioning criterion was selected as it is a measure of the reduction in iterative cycles (7).

The GA works on the basis of searching various combinations of process sequence of which, there are a potential $2.42*10^{130}$ within the PCA process. A sequence is evaluated with respect to the Scott partitioning criterion. At this stage the sequence with the highest performance characteristic is selected for the next generation. Crossover and mutation operators then act on the selected sequence. This increases the probability that the next generation has less iterative cycles. A full description of the GA and its structure can be found in (13).

The number of activities within the process dictates the maximum GA search space and can be calculated by n!, where n is the number of activities. Hence, the computational time increases with the number of activities, however this may be limited by the use of a population size and generation count. Instead of searching through all possible combinations, the search space is limited to a practically reasonable yet effective domain size, as proposed by Whitfield et al (13).

Despite the DSM-GA having had considerable use in product design, (14, 15), it has had limited application in projects similar in size and context to the PCA case study, (7). Hence, a number of novel issues are described in the following sections, such as cross-functional optimisation, information validation techniques and the suitability of the DSM-GA for applications outside design.

4 DSM APPLICATION OF PCA PROCESS

For the case study it was decided to adopt three different dependency weightings, represented by coloured crosses within the matrix. The darker crosses represent the most critical information; the interested reader is referred to (10, 15). The GA re-sequences the activities and their associated dependencies, to reduce the effects of untimely information across the process. The aim being to reduce the overall number and size of iterative blocks within the process.

Two workshops were arranged to formalise and verify the 'As-Is' process's activities, information links and their specific weightings. These workshops, each lasting three hours in duration, allowed the key staff within the process to verify that the DSM matrix reflected their current process. The process owner signed off the 'As-Is' process, to confirm its accuracy and completeness. This signified the end of the information gathering and process modelling stage. Important features of this type of model are the iterative blocks, represented by an 'X' above the leading diagonal. This is information that is required by an activity to allow its completion, yet its availability is constrained by the fact that the feeder activity has not occurred and/or the information it provided was incomplete. In such cases the activity owner will often make an intelligent estimate, and at a later time when the required information is available, confirm the accuracy of their estimate. Figure 1 shows an example of an iterative block, driven by product specification updates, marked on the matrix as 'block A'. These updates are caused by critical activities occurring too late in the PCA process. These activities provide knowledge and experience from previous contracts, customer requirements and engineering drawings. This is even more significant as the product specification is used as the main input to many of the downstream activities, such as material, labour and sub-contract estimating- see 'block B' in Figure 1. Each product specification update requires these activities to be reworked according to the new information. This can lead to significant problems for the estimating and planning functions e.g. time constraints and managing changes. Further examination of the matrix in Figure 1 highlights other areas in the process with a high concentration of 'X's above the leading diagonal i.e. areas where untimely information is causing rework. The performance criterion, Scott Partitioning, measured the existing rework at 9.29×10^7.

5 OPTIMISED AND VALIDATED PCA PROCESS

The 'As-Is' PCA process was optimised using the DSM-GA. A population size and generation count of 3000 was used in tandem with independent position crossover and shift mutation operators. This allowed $9*10^6$ possible solutions to be evaluated over a period of 4 days using a standard personal computer. A 'new' sequence was produced that exhibited a reduction in iteration i.e. there are less information dependencies above the leading diagonal. This reduced the Scott Partitioning performance criterion by 51% to 4.55×10^7. The reduction in Scott Partitioning, brought about by computationally optimising the sequence, represents a reduction in the overall process 'rework' cycles.

There are factors that may result in an impractical sequence, these include, no specific process heuristics, a limitation in the applicability of the GA, an insufficient search space and/or incorrect dependencies in the matrix. Hence, the optimised sequence was presented to the process operators for validation. This was an important step to gain 'buy-in' and prevent the project becoming a paper-based exercise.

A workshop was organised as it provided an appropriate environment to encourage the process operators to interact and participate fully in the validation (17). It allowed the process operators to critique the practicality of the DSM-GA solution and explore 'What If' scenarios. The following points provide insight to how the workshop was performed:

- The optimised process, showing a 51% reduction in the Scott partitioning criterion, was projected in front of the group. Each individual was invited to start at the top of the matrix and identify tasks they felt were impracticably ordered along with the rationale. As a result, 54 separate changes were made, for example the 'collate and compile tender' activity was repositioned before the 'submit tender' activity. It is important to note that changes made were only accepted if a further reduction in the Scott Partitioning performance criterion was achieved and/or there was a fundamental requirement for activities to be repositioned elsewhere. The optimisation results should however be treated with caution until fully validated.

- The DSM-GA is a tool that allows the dynamic manipulation of the matrix sequence. This enabled the user to manually 'drag' an activity to a new position, the Scott Partitioning criterion is re-calculated at the same time. A number of 'What If' scenarios were performed and the resulting changes to the global impact assessed and discussed. In doing so the process operators built on their existing expertise by learning from each other. The end result was a further 6% reduction in the Scott Partitioning value. (See Figure 2, for optimised and validated process).

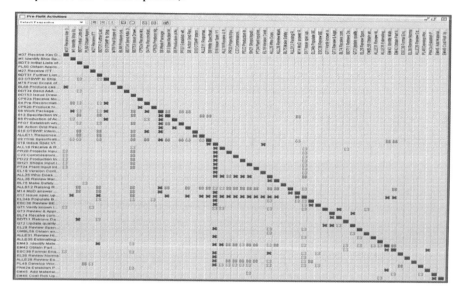

Figure 2 Matrix – Optimised and validated process

It has been argued that companies should move from silos of expertise and become more process focused (16). This is not an easy transition for any company, especially those of a traditional nature. The workshop brought together the various representatives, from each department within the process, and raised awareness of information needs and constraints. The new sequence meant that in some instances process operators would now receive information they previously estimated, whilst others would have to estimate information they previously had. It was agreed that some tasks would suffer due to some additional rework, but the overall process would benefit by the global reduction in rework. This was a step towards Babcock BES becoming a more process focused company.

The optimised and validated process reduced the Scott Partitioning criterion, which measures rework due to ineffective activity sequencing, by 57%. However, non-conforming work products and delivery mechanisms can also cause rework. These are issues that should not be ignored as they may have a significant effect on the new process.

A003/039/2003 © IMechE 2003

6 IMPROVING EFFECTIVENESS OF KEY ACTIVITIES

In addition to improving the sequence of the process this study has identified key areas that need to be improved. These key areas were identified, based on the PCA process' specific characteristics and some analysis of the various matrices. (See Figures 1 & 2)

The PCA process's key characteristic is information transfer, both with the customer and internal departments. Modules of information are used to create the product specification and the company generates additional information, which is in turn exchanged with the customer for approval. In addition to every module exchanged with the customer, many more transfers of information occur within the company. All this information is required to fully prepare the specification and the subsequent tender. These information transfers need to be enhanced from their present state. The following recommendations address some of these issues.

- The information needs to be co-ordinated and transferred more effectively internally within Babcock BES, and externally between the company and customer. This will help reduce the occurrence of late and unreliable information, such as estimators working from different versions of the specification, a possible contributor to rework. Communication with the external customer, who is dispersed geographically, is currently done via traditional modes, such us email, phone and post. Improvement areas will involve looking at adopting better transfer media e.g. a shared database could be considered.

- Better integration of the company's design cycle and the customers review cycle should be considered as many of the approvals and associated information are received late. When the company's design cycle has reached its peak the customer is not following in step. In fact the customers peak is closer to when physical refit begins, forcing overlap between the PCA process and refit processes. A recommendation for improved congruency could be achieved by having senior customer engineers onsite and involved in the creation of the product specification.

7 DISCUSSION

The PCA process captures customer requirements and translates them in to a tender bid. It is therefore critical that the process is efficient, in terms of resources it uses, and the products meet the processes high level goals. An improvement area that was identified as a constraint on these elements of performance, is rework. This project looked at the PCA process' information dependencies and the unintentional rework caused by the order in which they are performed. The DSM-GA performed a quantitative optimisation on the PCA process that reduced the Scott partitioning criterion by 51%. This criterion was reduced by a further 6% during the validation stage. The DSM process models provided a rich source of information on blocks of activities that are dependent on each other. These blocks represent the key areas that should be looked at for improvement. Examples of the future improvements include better communication mechanisms and greater congruency with the supplier processes.

Application of this modelling and optimisation technique, to the PCA process, has unearthed some interesting insights. The final IDEF0 models proved realistic and represented the activities and information requirements. However at times the process operators found it difficult to comprehend this formalism. This was due to varying levels of process decomposition and the concentration of information flows within each model. If a more structured approach had been utilised to capture the 'As-Is' process then it is believed that less time and effort would have been expended. Future projects that require IDEF0 will follow a more structured procedure. This will include, setting the context for IDEF0, identifying the sub-processes and placing them into a main path. The diagram will be detailed using the 80/20 rule, i.e. 80% of information is gathered initially and the remainder at subsequent interviews. Finally the IDEF0 model will be reviewed and validated by process operators.

Verification and validation workshops were designed as control stages, to ensure the 'As-Is' model was a true and accurate representation of the PCA process and the optimised process was practical. Although these control points were necessary to get the desired information integrity they were time intensive. To help streamline these stages a glossary of terms and definitions was produced. In future a simple flow diagram will be created to accompany the DSM process model.

The validation workshop allowed the cross-disciplinary staff to view the whole PCA process and observe the information needs and constraints of others. This was important in allowing the company to move toward being more process focussed and allowed a consensus to be gained on what was perceived to be an optimum solution. This solution will be used in the next phase of this project. However, the number of changes made, 54, took away from the main power of the DSM i.e. quantitative optimisation of the process sequence. Future work will involve a further investigation of the relationship between project size and search space.

During the validation workshop it was highlighted that some areas of the process still do not receive products or information required. For example procurement do not currently get the specification, tender schedule, quality or risk plan. However, the DSM-GA will not solve these types of issues. Future improvement initiatives will address these types of issues. This could cause a further need to optimise the process.

The next stage of this project will focus on the extraction of the optimised and validated sequence and lead to a new process that has less iteration and planned concurrency. Performance indicators will also be introduced to monitor the current and future performance of the process.

8 CONCLUSION

Information dependencies affect the order in which activities can be performed. Therefore, inappropriate sequences can cause iterative cycles and limit the process performance. This paper demonstrated a DSM-GA technique that, when fully validated, resulted in a 57% reduction in the Scott Partitioning criterion, which indicates re-work cycles. In practical terms it is recommended that any optimised sequence be fully validated by process operators before considering implementation. It is suggested that this be done through a series of group workshops. The DSM-GA has proved useful in visualising the information relationships and resultant constraints between activities. This study recommends that process improvement could be achieved through, re-sequencing, greater congruency with the customer and streamlining of key activities.

A003/039/2003 © IMechE 2003

9 ACKNOWLEDEGMENT

The work presented in this paper was conducted under Teaching Company Scheme (TCS) programme 3615. The authors particularly benefited from the guidance and support provided by Keith Kirkpatrick, Tom Dane and Dare Awobiyi of Babcock BES. The authors also thank members of the estimating and tendering team in Babcok BES, especially John Mitchell, Brian Mann, Alan Duch, Alan Deas, Paul Stephen, David Henderson, Alan Murray and David Gibb for their continual help in gathering information, conducting workshops and validating the case study. The optimisation of the process also benefited greatly from the expertise provided by Dr Ian Whitfield of The University of Strathclyde.

10 REFERENCES

1   **Colquhoun, K,** (2002). "Voyage to New Territory", Professional Engineering, 18 September Vol. 15, No.17, pp34-35.

2   **Cook, D.** (1996) Process Improvement: a Handbook for Mangers, Gower, Aldershot, Brookfield, VT

3   **Ziari, J.** (1997), "Business process management: a boundaryless approach to modern competitiveness", Business Process Management Journal, Vol. 3, No 1, pp 64-80

4   **Henderson, I.** (2001) "Why is making to order so different"? Control, December – January, pp 19-23

5   **O'Grady, P., Ramers, D. and Bolsen J.** (1988) "Artificial intelligence constraints applied to design for economic manufacture". Computer Integrated Manufacturing, Vol.1, No. 4, pp 204 –209

6   **Nevins, J. L. and Whitney, D. E.** (1989) Concurrent Design of Products and Processes (New York: McGraw-Hill)

7   **Scott, J. A.** (1998) A strategy for modelling the design-development phase of a product. Ph.D Thesis, Department of Marine Technology, University of Newcastle upon Tyne, UK.

8   **Browning, T. R.** (1998) "Use of dependency structure matrices for product development cycle time reduction". Proceedings of the Fifth ISPE International Conference on Concurrent Engineering: Research and Applications, Tokyo, Japan, July 15-17

9   **Hippel von, E.** (1990) "Task Partitioning: an innovation process variable", Research Policy, vol.19, pp 407-418.

10 **Forbes, G. A., Fleming D. A., Hayfron L. E., Duffy A. H. B., Ball P. D**. Modelling and Optimisation of an Engineering Design Process, Submitted to 1st International Conference on Manufacturing Research, University of Strathclyde, 2003

11 **Stewart, D. V.** (1981) "The design structure system: a method for managing the design of complex systems", IEEE Transactions on Engineering Management, Vol. EM-28, No.3, August

12 **Rogers, J. L.** (1997), Reducing design cycle time and cost through process resequencing, International Conference on Engineering Design ICED Tampere, August 19 – 21

13 **Whitfield, R. I. Duffy, A. H. B. Coates, G. Hills.** (2002), "Efficient Process Optimisation", accepted for the Journal of Concurrent Engineering Research and Applications

14 **Coates, G., Duffy, A. H. B., Hills, W. and Whitfield, R. I.** (2000). "A generic coordination approach applied to a manufacturing environment", Journal of Materials Processing Technology, 107: 404-411

15 **Eppinger, S.D., Whitney, D. E., Smith, R. P., Gebala, D. A.** (1994) "A model-based method for organising tasks in product development", Research in Engineering Design, 6: 1-13

16 **Armistead, C. and Machin S.** (1998). Business process management: implications for productivity in multi-stage service networks, International Journal of Service Industry Management, Vol. 9, No.4, pp. 323-336

17 **Gilgeous, V.** (1995) "Workshops for improving manufacturing effectiveness" Integrated Manufacturing Systems, Vol. 6, No 6 pp. 23-30

A003/039/2003 © IMechE 2003

Computer-aided process design system for axi-symmetric deep drawing

P V VIJAYAKUMAR and **N VENKATA REDDY**
Department of Mechanical Engineering, Indian Institute of Technology, Kanpur, India

ABSTRACT

In the present work, an attempt is made to develop a computer aided process design system for axisymmetric deep drawing of flat-bottomed cups. The model presented by Sonis et al. [1] to determine limiting drawing ratios is modified by Vijayakumar [2] to incorporate bending and unbending at the die profile and is used for analysis. Developed system is tested with process sequences available in the literature and they are in good agreement. Capabilities of the developed system are demonstrated with the help of a case study.

1. INTRODUCTION

Design and manufacture of dies/tools for producing deep drawing components is extremely costly and time consuming. It is necessary to cut down the time for developing, designing and producing the dies/tools for the production of sheet metal components so as to stay competitive. For designing automated process sequence, many researchers have developed hybrid Computer-Aided Engineering (CAE) systems, which consists of a knowledge based expert system module and an empirical process analysis module. Eshel et al. [3] developed automatic generation of forming process outlines (AGFPO) for axisymmetric deep drawing based on a rule-based model. Sing et al. [4] developed a knowledge based process layout system for axisymmetric deep drawing using decision tables with fuzzy interface. Park et al. [5] developed a knowledge based process design system for axisymmetric deep drawing products. Kang et al. [6] developed a CAPP system for the non-axisymmetric deep drawing components. Choi et al. [7] developed an intelligent design support system for multi-step deep drawing process along with FE simulation for thickness variation at each step. All these process design systems [4-7] were developed using a GTR strategy that was proposed by Eshel et al. [3]. This strategy *generates* geometrical process sequence as a first step, then *tests* the formability limits between successive geometric sequence stages with the limiting drawing parameters. Then *rectifies* the geometric sequence by introducing intermediate stages in the geometric sequence, wherever the formability violation was found. From the literature it is evident that there are some attempts [3-7] to develop hybrid computer aided engineering systems to generate the process sequence, but, they were developed using an empirical process analysis module/data tables to predict the limiting drawing ratio. Hence, the process sequence generated by these attempts may not be optimal.

Limiting Drawing Ratio (LDR) is defined as the ratio of the largest blank radius that can be successfully drawn (i.e., without failure) to the punch radius. In the deep drawing process, the limiting drawing ratio depends on the characteristics of the material, die and punch design and friction condition. Recently, Leu [8,9] studied the effect of process variables namely, normal anisotropy, strain-hardening exponent, coefficient of friction, yield strength and die arc radius on limiting drawing ratio using a model developed on the basis of force equilibrium. Many of the deep drawing components need multiple draws to achieve the required reduction with or without annealing. Very few attempts have been made to analyze the redrawing process either using analytical or numerical formulations. More recently Sonis et al. [1] have modified the first draw analysis presented by Leu [9] to suit the process planner's requirements and then they extended for the redraw analysis without considering the effect of sheet thickness (i.e., bending and unbending). None of the above attempts considered annealing requirements between stages.

Critical review of literature indicate that there is a need for the development of an aid to the deep drawing process planning to generate optimal process sequence. In the present work, a model presented by Vijayakumar [2] by modifying the analysis presented by Sonis et al. [1] to compensate bending and unbending effects along the die arc profile is used to predict LDR's. Annealing requirements are also taken into the consideration in process sequence design. Proposed system automatically select the die arc radii at each stage of drawing from the design rules and from the shape of the desired cup. Present work is an attempt towards developing an intelligent aid to the process planner that can generate a set of feasible forming process sequence for the purpose of die design by taking the input of required finished part.

2. COMPUTER AIDED PROCESS DESIGN SYSTEM

The block diagram of proposed product and process design system for axisymmetric deep drawn components is shown in fig. 1. Brief description regarding the various elements of the proposed system is presented in this section.

2.1 Object Modeling and Interface
A parametric feature based design module is developed to automate the input of geometric information to the process planning system. The final object geometry for which a process sequence needs to be generated is input to the system in the form of concatenation of line elements (features). The input geometry can be created by selecting the different elements (features) such as horizontal, vertical, incline, concave arc, convex arc etc from the feature library.

2.2 Blank Size Calculation
Pappu's second theorem is used to determine the required blank surface area based on the assumption that the blank thickness remains constant during the operation. The input for this

module comes from the product design module. In order to account for non-uniform flow of material and allowance for trimming, above calculated area is increased by 15%.

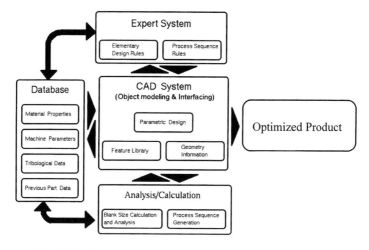

Fig. 1 Block diagram of process design system.

2.3 Database
The database of the present system contains the material data, tribological data and machine data. The database of the present system is flexible with an easily extendable structure.

2.4 Analysis

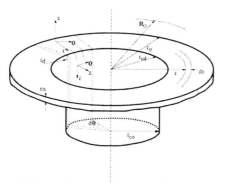

Fig. 2 Schematic representation of cup drawing

The analysis developed by Sonis et al. [1] is modified by Vijayakumar [2] with bending and unbending along die profile and same is used for the prediction of LDR. This model is based on the force equilibrium method. Limiting drawing ratio for first as well as redraws in the deep drawing of a cylindrical cup with a flat-nosed punch is obtained using an integral technique based on the load maximum principle for localization of the plastic flow. This analysis considered the effects of normal anisotropy, $\bar{R}$, coefficient of friction, μ, strain hardening, n, die arc radius, r_d and sheet thickness, t_0. In the first draw bending and unbending occurs at the die profile only, where as in redraws these effects are present at die profile as well as at blank holder profile. The final expressions used to obtain LDR for first draw as well as redraw are given below.

$$f(LDR_0) = \left(-\frac{C_1}{1 + r_d(LDR_0)/R_0} - \sigma_b \right) e^{-\mu\frac{\pi}{2}} + \left(\frac{C_3(LDR_0)}{1 + r_d(LDR_0)/R_0} \right) \left(\frac{r_{cd}}{r_0} \right) + C_2(1+n)\ln\left(\frac{r_0}{r_{cd}} \right)$$

$$- 2nC_2 \ln\left(\frac{(LDR_0)}{1 + r_d(LDR_0)/R_0} + \frac{r_0}{r_{cd}} \right) - \frac{2nC_2(r_0/r_{cd})}{\left(\frac{(LDR_0)}{1 + r_d(LDR_0)/R_0} \right)}$$

$$+ 2nC_2 \left(\ln\left(\frac{R_2}{r_{cd}} + 1 \right) + \frac{1}{(R_2/r_{cd}) + 1} \right) = 0 \qquad \ldots\ldots\ldots\ldots(1)$$

where,

$$C_1 = \left(\frac{1+\overline{R}}{\sqrt{1+2\overline{R}}} \right)^{1+n} (n^n e^{-n} K) \ ; \quad C_2 = K\left(\sqrt{\frac{2(1+\overline{R})}{1+2\overline{R}}} \right)^{1+n} \ ; \quad C_3 = 2\mu(1.1\sigma_y)$$

Bending stress [10], $\sigma_b = \dfrac{\overline{\sigma}t_0}{(2r_d + t_0)}$; where, $\overline{\sigma} = K(\varepsilon)^n$

$$LDR_0 = \frac{R_0}{r_{co}} \ ; \quad r_{cd} = r_{co} + r_d \ ; \quad \frac{r_0}{r_{cd}} = \sqrt{\left(\frac{R_0(LDR_0)^2}{R_0 + (LDR_0)r_d} \right) - \left(\frac{R_2}{r_{cd}} \right)^2 + 1}$$

where R_1 and R_2 are the intermediate radii of annular rings at the blank, which draws inward into the die to radii r_{co} and r_{cd} respectively. Here R_2/r_{cd} can be evaluated from the volume constancy condition in the die arc region.

The expression for the calculation of LDR in redraw is given by

$$f(LDR_i) = -\frac{C_1 r_{c_{(i-1)}}}{r_{c_{(i-1)}} + LDR_i r_d} + \frac{C_3 r_{c_{(i-1)}}}{r_d + r_{c_{(i-1)}}/LDR_i} + C_4 \ln\left(\frac{r_{c_{(i-1)}}}{r_d + r_{c_{(i-1)}}/LDR_i} \right)$$

$$+ C_5 \ln\left(\frac{r_{c_{(i-1)}}}{r_d + r_{c_{(i-1)}}/LDR_i} \right) - \frac{C_5}{(r_d + r_{c_{(i-1)}}/LDR_i)^2((R_{i+1}/r_{cd})^2 - 1)}$$

$$\left[r_{c_{(i-1)}} \sqrt{(r_d + r_{c_{(i-1)}}/LDR_i)^2((R_{i+1}/r_{cd})^2 - 1) + r_{c_{(i-1)}}^2} \right.$$

$$\left. - (r_d + r_{c_{(i-1)}}/LDR_i)^2(R_{i+1}/r_{cd}) - (r_{c_{(i-1)}}^2 - (r_d + r_{c_{(i-1)}}/LDR_i)^2) \right]$$

$$+ C_5 \ln\left(r_{c(i-1)} \sqrt{(r_d + r_{c_{(i-1)}}/LDR_i)^2((R_{i+1}/r_{cd})^2 - 1) + r_{c_{(i-1)}}^2} \right)$$

$$- C_5 \ln\left[(r_d + r_{c_{(i-1)}}/LDR_i)(1 + R_{i+1}/r_{cd}) \right] \qquad \ldots\ldots\ldots\ldots\ldots\ldots(2)$$

A003/051/2003 © IMechE 2003

where

$$C_4 = K\sqrt{\frac{2(1+\overline{R})}{1+2\overline{R}}}\left[\ln\left(LDR_{(i-1)}\right)\right]^n \quad ; \quad C_5 = 2Kn\frac{2(1+\overline{R})}{1+2\overline{R}}\left[\ln\left(LDR_{(i-1)}\right)\right]^{n-1}$$

$$LDR_i = \frac{r_{c_{(i-1)}}}{r_{c_i}} \quad ; \quad r_{cd} = r_{c_i} + r_d$$

constants C_1 and C_3 remain the same as in the first draw analysis. R_{i+1}/r_{cd} can be obtained by considering volume constancy in the die arc region. To obtain LDR values, Newton Raphson method is used for solving equations 1 and 2.

2.5 Expert system

This module consists of elementary design rules, standards and regulations for process design as well for process sequence. Rules suggested by Eshal et al. [2] are implemented in the present system.

2.6 Process sequencing

A process sequence module is developed for the automatic generation of the process sequence from the blank to the required final cup. This module consists of two sub-modules namely geometrical sequence module and rectification module. *Geometric sequencing* module automatically generates the process sequence in the backward direction starting from the component geometry based only on the geometrical features of the cup. Considering the shape of previous deformed zone is a straight and vertical non-flanged cup, this module generates current deformation zone (fig. 3). This procedure is repeated until the previous shape is the flat circular blank. The process sequence generated using the geometrical design module may not be feasible, as the formability of the material is not taken into consideration. Adding new intermediate stages to the geometric sequence where the process analysis module predicts formability violation modifies process sequence. Analysis module of the proposed system is used to generate the *rectified process sequence*, considering formability and the geometric sequence instead of testing the formability violation using the decision tables for a given material.

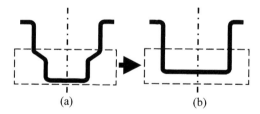

(a)   (b)

Fig. 3 Geometric sequencing, (a) current zone (b) previous zone

2.7 Feasible Sequences

This module generates alternative process sequences by performing annealing between stages. Annealing is required whenever the total reduction reaches a critical value. This value is different for different materials. The present system generates all possible sequences irrespective of the critical drawing ratio to minimize the number of stages. The selection of optimum process sequence depends on the minimum number of stages and strength requirements of the final cup.

3. CASE STUDY

The LDR values predicted by the present system are in good agreement with the experimental and theoretical values available in the literature [9,10]. The process sequences generated by the proposed system are also in close agreement with the case studies presented by Park et al. [5] and Choi et al. [7].

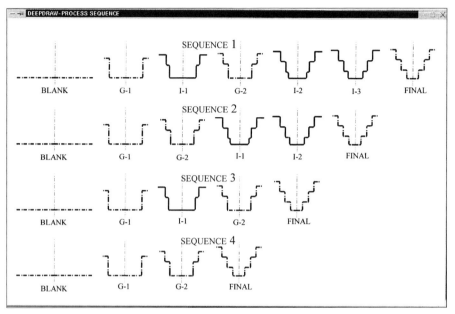

Fig. 4 Possible sequences generated by the system

To determine the capabilities of the present work a case study is presented in this section. The material chosen for the study is CA-DDQ steel as it has better draw ability. Thickness of the material is taken as 1.0 mm. Initial blank radius, R_0 required to produce chosen cup geometry is obtained as 207.77 mm. The output of the present system with possible process sequences is shown in fig. 4. Geometric details of the process sequences generated by the present system (fig. 4) are given in table. 1. The formability details of the processes generated by the present system (fig. 4) are given in table. 2. The term "total drawing ratio" can be used to determine the stages at which annealing is required. For a given material if total drawing ratio reaches a critical value, $DR_{critcal}$, then annealing is must. Present system has the capability of generating feasible process sequences by performing annealing at different stages (fig. 4). The number of draws required to produce a given cup cannot be reduced less than the number of geometric sequence stages. The total number of stages can be minimized by performing annealing operation to minimize the number of intermediate stages required between the successive geometric stages without violating formability conditions. However, these sequences may not be producing the cup in minimum time with required strength and cost. These considerations also have to be taken into the study to obtain optimal process sequence. Further study is in progress.

A003/051/2003 © IMechE 2003

Stage	h	r	v	c	h	r	v	c	h	r	v	c	h
						Sequence no. 1							
0	207.77												
1	95.00	5.0	105.00	5.0	40.00								
2	71.28	10.0	57.41	8.0	5.72	5.0	50.00	5.0	40.00				
3	65.00	5.0	72.86	5.0	20.00	5.0	50.00	5.0	40.00				
4	44.91	10.0	18.90	8.0	2.09	5.0	50.00	5.0	20.00	5.0	50.00	5.0	40.00
5	34.90	8.4	28.55	6.4	15.30	5.0	50.00	5.0	20.00	5.0	50.00	5.0	40.00
6	30.00	5.0	40.00	5.0	25.00	5.0	50.00	5.0	20.00	5.0	50.00	5.0	40.00
						Sequence no. 2							
0	207.77												
1	95.00	5.0	105.00	5.0	40.00								
2	65.00	5.0	72.86	5.0	20.00	5.0	50.00	5.0	40.00				
3	45.39	10.0	18.65	8.0	1.61	5.0	50.00	5.0	20.00	5.0	50.00	5.0	40.00
4	35.17	8.4	28.31	6.4	15.03	5.0	50.00	5.0	20.00	5.0	50.00	5.0	40.00
5	30.00	5.0	40.00	5.0	25.00	5.0	50.00	5.0	20.00	5.0	50.00	5.0	40.00
						Sequence no. 3							
0	207.77												
1	95.00	5.0	105.00	5.0	40.00								
2	71.28	10.0	57.41	8.0	5.72	5.0	50.00	5.0	40.00				
3	65.00	5.0	72.86	5.0	20.00	5.0	50.00	5.0	20.00				
4	30.00	5.0	40.00	5.0	25.00	5.0	50.00	5.0	20.00	5.0	50.00	5.0	40.00
						Sequence no. 4							
0	207.77												
1	95.00	5.0	105.00	5.0	40.00								
2	65.00	5.0	72.86	5.0	20.00	5.0	50.00	5.0	40.00				
3	30.00	5.0	40.00	5.0	25.00	5.0	50.00	5.0	20.00	5.0	50.00	5.0	40.00

Table 1. Geometric details of all sequences

Stage	Die Arc radius	Punch Arc radius	DR_i	DR_{total}	Cup Radius	&Red
		Sequence 1: Without annealing				
1	5.00	5.00	2.08	2.08	100.00	51.87
2	8.00	10.00	1.23	2.56	81.28	18.72
3	5.00	5.00	1.16	2.97	70.00	13.88
4	8.00	10.00	1.27	3.78	54.91	21.55
5	6.40	8.40	1.27	4.80	43.30	21.16
6	5.00	5.00	1.24	5.94	35.00	19.16
		Sequence 2: Annealing after stage no.1				
1	5.00	5.00	2.08	2.08	100.00	51.87
2	5.00	5.00	1.43	1.43	70.00	30.00
3	8.00	10.00	1.26	1.81	55.39	20.87
4	6.40	8.40	1.27	2.29	43.57	21.33
5	5.00	5.00	1.24	2.86	35.00	19.68
		Sequence 3: Annealing after stage no. 3				
1	5.00	5.00	2.08	2.08	100.00	51.87
2	8.00	10.00	1.23	2.56	81.28	18.72
3	5.00	5.00	1.16	2.97	70.00	13.88
4	5.00	5.00	2.00	2.00	35.00	50.00
		Sequence 4: All possible annealings				
1	5.00	5.00	2.08	2.08	100.00	51.87
2	5.00	5.00	1.43	2.97	70.00	30.00
3	5.00	5.00	2.00	5.94	35.00	50.00

Table 2: Formability information of all the sequences

4. CONCLUSIONS

Presented case study demonstrates how the process planner can utilize the developed system effectively along with other design rules. Further work is in progress to enhance the capabilities of the system presented in this work.

ACKNOWLEDGEMENTS

The authors gratefully acknowledge the financial assistance from Department of Science and Technology, India for carrying out this work.

REFERENCES

1. **Sonis, P., Reddy, N.V., Lal, G.K.** (2002) On Multistage Deep Drawing of Axisymmetric Components, ASME Journal of Manufacturing Science and Engineering, (In press).

2. **Vijayakumar, P.**V. (2002) Process Sequencing in Axisymmetric Deep Drawing, M.Tech., thesis (under preparation), IIT-Kanpur.

3. **Eshel, G., Barash, M., Johnson, W.** (1986) Rule Based Modeling for Planning Axisymmetrical Deep Drawing, Journal of Mechanical Working Technology, Vol. 14, pp. 1-115.

4. **Sing, W.M., Rao, K.P.** (1997), Knowledge-Based Process Layout System for Axisymmetrical Deep Drawing Using Decision Tables, Computers and industrial Engineering Vol. 32, no. 2, pp. 299-319.

5. **Park, S.B., Choi, Y., Kim, B.M., Choi, J.C.** (1998) A Study of Computer-aided Process Design System for Axisymmetric Deep-drawing Products, Journal of Material Processing Technology, Vol. 75, Issues 1-3, pp.17-26.

6. **Kang, S.S., Park, D.H., Choi, B.K.** (2002) Application of Computer-aided Process Planning System for Non-axisymmetric Deep Drawing Products, Journal of Material Processing Technology, Vol. 124, Issues 1-2, pp. 36-48.

7. **Choi, T.H., Choi, S., Na, K.H., Bae, H.S., Chung, W.J.** (2002) Application of intelligent design support system for multi-step deep drawing process, Journal of Material Processing Technology, (In press).

8. **Leu, D.K.** (1997) Prediction of the Limiting Drawing Ratio and the Maximum Drawing Load in Cup Drawing, International Journal of Machine Tools and Manufacture, Vol. 37, pp. 201-213.

9. **Leu, D.K.** (1999) The Limiting Drawing Ratio for Plastic Instability of the Cup Drawing Process, Journal of Material Processing Technology, Vol. 86 no. 2, pp. 168-172.

10. **Lange, K.** (1985) Hand Book of Metal Forming, McGraw-Hill, New York.

Metal working process and the associated skill modelling in UML – a case study of grinding troubleshooting

S ITO and **T KOJIMA**
Digital Manufacturing Research Center, National Institute of Advanced Industrial Science and Technology, Ibaraki, Japan
K AI
APTES Laboratory, Kanagawa, Japan

SYNOPSIS

The skill has an important role at the metal working shops. This paper discusses a skill model diagrammatically represented and implemented as database and programs. The model is based on the information collected from the metal working cases. UML is introduced as definition and communication tool. In the model, all the information concerned is represented by class diagrams. The skill is identified as a set of relations among the objects, represented by collaboration diagrams. The model has functionality of multiple representations reflecting to the differences in situation and community. As a case study, grinding trouble is selected and the practical applicability of the model is examined.

1 INTRODUCTION

Determining the processing conditions and solving the troubles at the machine shops, especially those of the small and medium enterprises, are substantially left to the judgments of skilled engineers, though the related technology has been advancing steadily in recent years.

The manufacturing skill treated in this paper is defined as information processing which is obtained and belongs to a person through his/her long experiences. This definition does not imply that skill as a whole is to be processed as a black box. For example, company standards of operation procedures as well as the reference data can be a model of skill and they are used effectively at the machine shop for high quality and cost reduction. The model is explicitly expressed and can be shared among engineers to be observed. Skill and technology can be applied in the same way in this paper, but skill differs from technology in the aspect that skill itself is derived basically from human experiences directly and applicable conditions are not defined explicitly beforehand.

There are several levels of the manufacturing skill. In this paper, the skill models obtained can be represented as data and programs. This model is basic and considered as a first step of the skill modelling.

This paper discusses and evaluates computer model of the manufacturing skill through the case studies how to solve the grinding troubles. Firstly, UML [1] is introduced to describe the cases. Next, the representation of the skill and the technology is proposed. Then, the proposed representation method is examined as to the effectiveness and handiness.

2 MODELS OF GRINDING TROUBLES

2.1 Basic concept
The skill modelling process of troubleshooting in grinding is applied as follows. Firstly, grinding cases are collected to form a data model. This model is a fundamental data model of grinding operations, and the specifications for the ground parts, grinding wheel, grinding machine, lubricant as well as the grinding conditions. It also includes the observations of the grinding troubles such as defects and quality of the work piece, as well. From a set of grinding troubles collected, experts of the area examine and derive the causes for each case. This process is represented to put a set of relations on the data model. The relation is verified by the actual practices if it is good, acceptable or wrong. In the model, other generic cause and effect relations are specified. The data model with technology based and skill based relations is defined as grinding model in the paper.

When a new grinding trouble is brought in, it is examined if there are similar cases in the model and if the grinding model can be applied to solve the trouble. The derived model will be corrected and improved constantly through the applications of the model to new troubles. This process can be summarized as follows:

Collect cases -> Analyze and systematize to make grinding model (1a)

Apply the model -> Evaluate the case -> Improve the model   (1b).

These two processes are mutually connected in practice. Through the process above, the skill is to be identified and modelled as a family of proper subset of the grinding model.

2.2 Representation methods of grinding troubles
There are several methods to model and systematize the grinding troubleshooting. Description of the trouble in the form of question and answering using natural language is popular and it will be implemented as Q&A database. So-called fish-bone diagram is also used widely [2]. These can easily be understood and is to be applied directly to solve the trouble. But, it is not suitable to use them directly and effectively in digital environment, as they are originally records in print. UML is a set of diagrammatic languages used for the specification definition of object-oriented approach and UML definition can be an executable program finally.

UML is a de facto standard of specification description language and the software tools to build the model are commercially available. UML is not supposed to be applied to the skill modelling. But, the skill discussed in this paper can be represented and modelled as data and programs similar to the application of the related technology, and it is expected to be useful in the modelling. A typical application of the technology is the simulation of the grinding process of (1a). It is to set a relation among a set of physical and chemical parameters in the

A003/056/2003 © IMechE 2003

model by computing the thermal change of the work piece to be ground. It will be used to estimate the causes of the trouble. Use of experimental results is another technological application [3].

As to the skill model, it is used in the same way and corresponds to the processes of (1b) and is discussed as follows. The process (1a) is also important in skill modelling, but it is a model with higher level and left to future studies.

3 CASE STUDIES

3.1 UML diagrams
Twelve practical troubles examples are collected by factory visits. Then, they are analyzed by a group of experts of grinding wheel, lubricant, grinding machine, etc. and the trouble causes are examined and identified. Then, the expected solutions are summarized through their discussions. Several articles and handbooks are referenced in the development [4], [5]. In this chapter, the results of the work are described. Firstly, three charts mainly used in the modelling are outlined and then detailed description follows.
Class diagram:
It is to represent all the information required for the purpose as a set of classes. This chart is like EXPRESS-G [6] developed by ISO to define product data model. Class diagram can also define the associated operations to specify the relations as message passing. For example, loading and other grinding trouble types are defined as classes with attributes representing the detailed conditions in the package. The information to be used for avoiding the troubles and solving the troubles are represented as class attributes and represented collectively.
Collaboration diagram:
It is used to represent the results of the analysis in the aspect of grinding troubles. Specifically, a line with an arrow connects between object of grinding trouble and object for control. The arrow has label number, showing the relation number of cause and effect. (But message does not pass sequentially in the order of this number on our diagram.) The information contents are described separately. This relation is similar to the fish-bone diagram, but UML can define n-ary relations and network type relations.
Statechart diagram:
Statechart diagram (or state diagram) is a supplementary diagram of collaboration diagram. Typical use is to describe the changes of attributes in the course of grinding processes. This is to reproduce the grinding troubles in the model.

3.2 Example definition of grinding troubles in UML

3.2.1 Class diagram
Figure 1 shows the class diagram expressing information related to the grinding. This diagram includes all classes to represent the process totally. The whole number of the classes defined is 50 and the number of the attributes is 330. To avoid complexity, classes with close relationship are collected as package. Class is drawn as a rectangle with 3 compartments. The top compartment holds the class name; the middle compartment holds a list of attributes; the bottom compartment holds a list of operations. Package is shown as a rectangle with a small rectangle on the left side of the top line. In the package, name of the class is indicated after plus sign. As the major elements of the grinding process, the grinding conditions and the

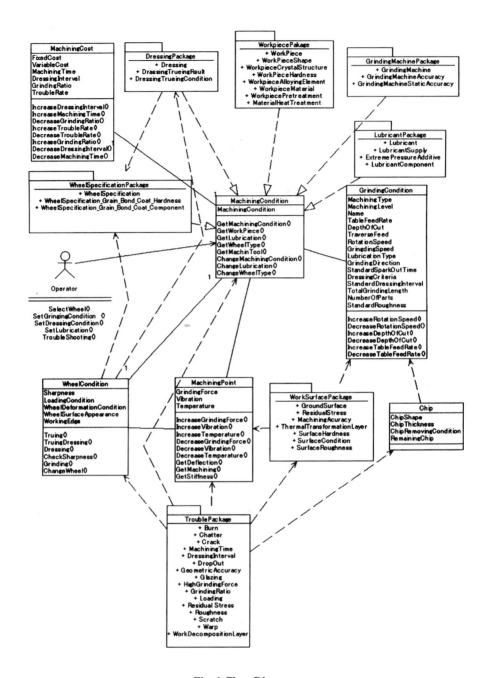

Fig. 1 Class Diagram

A003/056/2003 © IMechE 2003

specifications for work piece, grinding wheel, grinding machine and so on are described.

The grinding trouble is classified into 16 types such as "loading". The combination of troubles can be defined, as well. They are placed collectively in the package named *TroublePackage* on the lower part of the diagram.

The classes and objects including the attributes to be controlled by operator to prevent and solve the grinding trouble are shown on the upper part of the diagram. These classes correspond to the grinding condition and manufacturing environment, such as the specifications for machine, grinding wheel, lubricant. They are generic classes and common to other machining methods.

The attributes of the objects are set and modified by the accompanied operations in the course of troubleshooting. The change will call another operation. This corresponds to the execution of programs of simulations and other calculations.

3.2.2 Collaboration diagram

Collaboration diagram for the troubleshooting is shown in figure 2. This is obtained from the analysis using the class diagram of figure 1.

Name of a class is placed after colon ':' and a list of selected attributes of class is placed under the horizontal line. To avoid complexity, name of attribute related to sending the massage is indicated with an arrow on our diagram.

On the left side of the diagram, objects that can be adjusted are placed. Grinding trouble object are collected on the right side. On the middle part of the diagram, objects which can be observed or estimated but cannot be controlled directly are placed. When some values of these objects exceed some criteria, the related troubles are to be caused. This part of the class diagram is added for the purpose. One of the newly added classes is *MachiningPoint* class that is to record the changing values at the contacting area of grinding wheel and work piece, grinding point. To recognize the trouble in the very early stage to prevent serious problem is a typical skill, but it is not expressed here.

The lines of no. 4 and no. 36 in the diagram show that the cause of crack is maximum temperature at the contacting zone and cooling power of lubricant, respectively. This means that heating and rapid cooling is to a candidate to solve this trouble.

As to the crack, the analysis goes as shown the diagram. The lines of no. 7, no. 10, no. 18 and no. 13 show that to change the temperature, grinding condition (depth of cut, work speed, grinding speed and so on), work piece material, sharpness of wheel (wheel condition), lubricant supply and lubricant cooling power can be adjusted, respectively. However, to solve the trouble, we must consider compound effects of the changes above mentioned. So, we must select one element for the first step. If we select the decrease of the lubricant cooling power, there is a possibility to cause another trouble relating to the temperature. So, skill is required to avoid wrong selection order, although the skill to choose best control item such as depth of cut from other elements is not described in the figure.

In the early stage of the collaboration diagram development, it is described that burn is caused mainly by the temperature at the grinding point and that to improve the temperature condition,

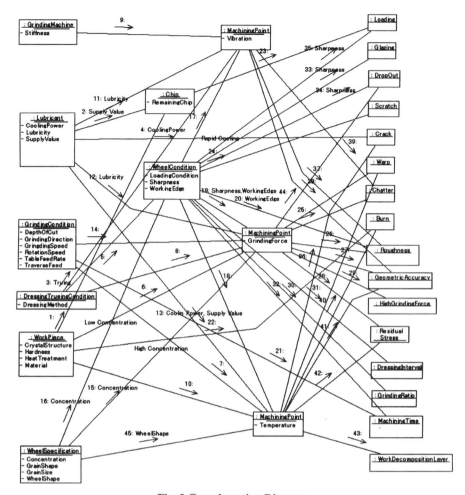

Fig. 2 Corroboration Diagram

change of the grinding condition (depth of cut, the work speed, wheel speed), the dressing condition, the lubricant supply and the lubricant type are considered.

However, from the examination of the actual cases, two side effects are found. One is that the change of the lubricant takes cost and the second is that the change of the grinding condition will cause less efficiency. But, the change of the grinding wheel shape, forming ditch on the wheel surface has no side effect at all. So, the third change was selected. At this point, the relation was added to the collaboration diagram as the line of no. 45. This is the expression of the skill.

When factors related to the trouble cases are described by several engineers, it become clear that the more skilled and experienced, the wider factors the engineer has. For example, as the solution of the scratch caused by the drop out grain attached on the grinding wheel surface, one engineer pointed out that heavy cavitation in the flow of lubricant on the grinding wheel surface would remove a drop out grain immediately. This is the effect of lubricant flow, but the skill is beyond the scope of this paper.

3.2.3 Statechart diagram
Figure 3 shows a statechart diagram describing how the *Sharpness* attribute of the *WheelCondition* class changes.

The rectangle with round corners shows each state of the same object. The *WheelCondition* class includes five attributes (see figure 1) and one of them is *Sharpness* attribute. The arrow shows the direction of state transition of *Sharpness* attribute. Above the line, operation name of *WheelCondition* class is indicated. And the restriction for the operation is indicated between the square brackets, '['and ']'. The rectangle with small triangle on upper right corner is the note. On our diagram, messages from *WheelCondition* object are described as the note.

It starts from the black circle that indicates start of wheel setting and transit to *InitialConditionWheel*. In general condition, super abrasive wheel get dressing after truing, but the conventional abrasive wheel get truing and dressing at the same time. Therefore, the grinding wheel changes to *DressedWheel* in the right of *InitialConditionWheel*. It changes to *SharpEdgeWheel* shown on the right side in the figure if the sharpness of cutting edge is right. When wheel is used for grinding the work piece, worn grain increases and becomes close to warning zone, and sharpness of cutting edge degrades, the state transfer to *DullEdgeWheel* occurs. If the wheel is kept grinding in this state, a message is sent to set increased values of *GrindingForce, Vibration, Temperature* attributes in the *MachningPoint* object and to change

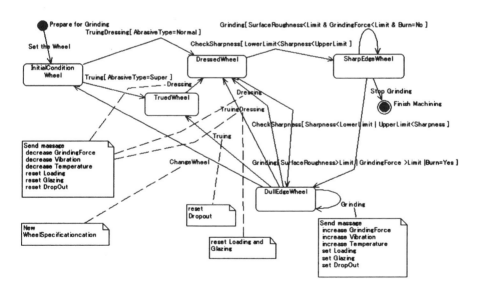

Fig. 3 Statechart Diagram

the value of *Loading, Glazing* and *DropOut* objects of the *TroublePackage*. When the sharpness recovers by dressing, the state transits back to *DressedWheel* again by receiving a message of *Dressing* and a message of *decrease GrindingForce, Vibration* and *Temperature* to *MachiningPoint* object is send.

The *Sharpness* attribute in the object of *WheelCondition* class will change between *SharpEdgeWheel* state and *DullEdgeWheel* state when wheel is dressed/trued and used for grinding.

4 DISCUSSIONS ON THE SKILL MODEL

In the preceding chapters, we specified a skill/technology model of grinding troubleshooting using three UML diagrams. The model concerned covers the skill of primitive level in the aspect that it may include knowledge, which is interpreted substantially and practically by experimental results and theories, though it is not done so at the machine shop. From the model developed, several features in the skill model are derived.

(1) More than two sets of relations exist, which are different from each other and is used for alternative actions.

(2) Cause and effect relations are sometimes bilateral, which are helpful to define context dependent relations uniformly.

(3) New objects can be defined and included for additional relationship specific to the application context.

Other type of skill model can be formulated. For example, selection priority form a set of candidates is not included in the paper and left for future studies. As to the troubleshooting, anticipation of the trouble and method how to avoid troubles are also models of higher levels. From the definition of the skill compared with the technology, several levels of manufacturing skill exist.

The metal working process is in principle, physical and/or chemical process. So, most of the troubles can practically be interpreted by technological achievements and they can be reproduced digitally. In this context, skill will be replaced by technology. But, this is not our approach. We are subject to practical situations. For example, if best effort to solve the encountered trouble is to adjust grinding conditions to some value theoretically, there may be a situation that the value is not acceptable from the tool life point of view. This is not covered by technology completely, as the situation will change from occasion to occasion and cannot be specified exactly. The evaluation measure is not to select best but satisfactory. All of this sort of knowledge can be defined skill based.

From the technological aspects, exact parameter values obtained are not so important, as the real process includes fluctuations.

The value of the skill is to solve the problem practically and effectively. Another type of skill is used in the process. They are to compare with feasible methods to select proper one and to find out alternative method from a set of known methods, for example.

UML expression has advantage to handle the skill. When similar troubles are identified, similar relations of tool, machine, work piece, etc. will be applied to. This comes from the

                                             A003/056/2003 © IMechE 2003

object-oriented property in UML. When useful tool is provided for the description, the correction and modification of diagram is easy. Specifically, classes defined previously and their attributes could be reused partly. This is also helpful to apply the results to another processing method like cutting and polishing in the future. For this purpose, the generic part of the modelling method can be packaged and will be used as templates.

5 CONCLUSIONS

Using grinding troubleshooting as example, a skill model is proposed, which can be used in the same way as the technological achievements can be applied to. Obtained results are summarized as follows:

(1) We proposed a description method using UML, in which grinding troubles and their controlling parameters are defined by class diagrams, the causality among them by collaboration diagrams, the detailed process changes by statechart diagrams, respectively.

(2) We specifically showed that portions of the collaboration diagram correspond to the expression of the skill. This is mainly determined if the functions are based on the human judgments and entrusted partly to human, as well.

(3) The method is useful for the communication among engineers at the machine shops. It can be implemented as software system systematically from the design specification phase to operational phase.

ACKNOWLEDGMENTS

This work is done in the Digital Meister Related Projects supported by Small and Medium Enterprise Agency, METI and New Energy and Industrial Technology Development (NEDO). The authors extend their sincere thanks to the members of the Grinding WG established for the projects.

REFERENCE

1   http://www.omg.org/technology/documents/formal/uml.htm
2   **Ishikawa K.** (1988) What is total quality control? Prentice-Hall, Inc.
3   **Kojima, T., Ohtani, S.** and **Ohashi, T.** (2002) Digitalization of manufacturing skill - A case study of welding operation planning-. Proc. 1st CIRP Seminar on Digital Enterprise Technology, pp. 49-52.
4   **Lindsay, R. P.** (1985) Principles of Grinding. Metals Handbook ninth edition Vol.16 Machining, pp.421-429, ASM International.
5   **JSPE** Precision Machining handbook (in Japanese), Nikkan-Kogyo-Shinbunsha, 2000, Tokyo.
6   Industrial automation systems and integration -Product data representation and exchange-Part11: Description methods: The EXPRESS language reference manual, ISO 10303-11, 1994.

Inventory sorting rules and dispatching strategies for scheduling jobs in FMSs using a real-time genetic algorithm

M A SHOUMAN
Faculty of Computers and Informatics, Zagazig University, Egypt
A A ABOUL-NOUR, H AL-AWADY, and **M ABDEL-FATAH**
Faculty of Engineering, Zagazig University, Egypt

ABSTRACT

The basic concepts of scheduling process are commonly based on the allocation of limited resources to tasks and jobs at certain specified times. Scheduling techniques aim commonly to minimize, makespan, number of tardy jobs, cost, and/or maximizing revenue. In the current paper a dynamic scheduling system is introduced. The system is capable of generating optimized schedules of jobs in FMS environments using genetic algorithms. The proposed approach seeks the best processing sequence at loading station and takes into account the buffer capacity and the time of transport operation constraints for minimizing the makespan objective. The performance of the solution approach is compared with the results of using traditional dispatching rules. A comparison with other works shows that the proposed genetic algorithm leads to promising results.

Keywords: Scheduling, FMSs, Genetic Algorithms, and dispatching strategies.

Notation

$bin(a, p)$	Binomial distributed random variate with probability p.
By turn	A dispatching rule that rotate among locations.
C_{max}	The make-span of part type set.
M	Total number of stations.
mg	Maximum number of generations.
n	The total number of jobs.
Newest part	A dispatching rule that selects the location with the part in the system of least time.
O_i	The i th operation.
Oldest part	A dispatching rule that selects the location with the part in the system of longest time.
P_c	Probability of crossover.

pop_size	Population size.
Random	A dispatching rule that randomly selects the location.
$U(a,b)$	Uniformly distributed random variate.

1 INTRODUCTION

Scheduling is a decision-making process that plays an important role in manufacturing and service industries. It uses mathematical techniques or heuristic methods to allocate limited resources to processing number of tasks within a certain specified time domain. Scheduling in Flexible Manufacturing Systems (FMSs) has many factors under consideration such as increasing the number of available resources as Automated Guided Vehicles (AGVs), routing flexibility, and the possibility of different types of disruptions that may occur during the system operation. In order to obtain an initial feasible schedule, one should solve many types of problems such as part type selection, process plan selection, detailed assignment of AGVs and intermediate storage buffers, and deadlock prediction problem. The presence of a rescheduling mechanism due to uncertainties presented in the operation environment transfer the problem from static to dynamic scheduling problem.

Two methods have been adopted for dynamic scheduling in FMSs. The first is rule-oriented algorithms that use dispatching rules to prioritizes all the jobs that waiting for processing on a machine. These dispatching rules may be classified according to time-dependency to static or dynamic and according to the information they are based on to local or global rules. This approach is useful when one attempt to find a reasonably good schedule with regard to a single objective. Composite dispatching rules consider a number of jobs and machine parameters to address more complicated objective functions. The schedule obtained by these rules used as initial schedule for the other artificial intelligence techniques such as neural network (1), discrete-event simulation (2), fuzzy logic (3), knowledge-based systems (4) and hybrid systems (5). The second approach includes the job-oriented algorithms that generate the schedule in the form of an arrangement of job sequences on each machine. This approach works to optimize complex scheduling objectives. Rossi and Dini (6) have been concluded that this second approach generates better results; nevertheless, they use time-consuming algorithms, which can be critical when the system has to give the output in a very short time.

In this paper, a genetic algorithm is presented using a new chromosome-coding scheme that introduce the job sequencing at the load/unload station and the set of dispatching rules at stations buffer in order to achieve a minimum make-span objective. The performance of the proposed genetic algorithm and traditional dispatching rules is compared using a hypothetical example and some research works.

2 FMSs SCHEDULING PROBLEM DESCRIPTION

The FMSs scheduling problem may be stated as "Given a set of part types, given set of available resources (station, AGVs and intermediate storage buffer capacities), and the process plan for each part, the objective is to find a feasible and near-to-the-optimal schedule for the given set of part types to optimize the make-span objective". The data for the scheduling problem are number of stations, number of job types, route for each part type, number of tasks/operations for each job, processing times for each task/operation, time of transport operation between two neighbor operations locations, the size of intermediate storage buffers, and material handling devices (AGVs). In our case, the objective is to minimize the total time required to produce all jobs or the completion time of the last finished job (make-span) C_{max}.

$$C_{max} = \max(C_i) \qquad \text{where, } i = 1,2,3,............n.$$

The following constraint are identified for FMSs environment studied in this paper.

1. All operations for each job must be performed and precedence relationships must be maintained for each job type.
2. Any station includes one machine and can perform one operation at a time.
3. Tool magazine capacity constraint and tooling are negligible.
4. Material handling availability and transportation time is considered.
5. A random dispatching rule is considered for material handling device service.
6. Intracellular movement is not considered in the proposed genetic algorithm.
7. A part can only be stored in a buffer when there is enough space available.
8. A common storage buffer capacity at each station is fixed.

3 OVERVIEW OF THE PROPOSED GENETIC ALGORITHM

A genetic algorithm is a stochastic searching algorithm that randomly hops from a solution point to another systematically. It has been used to solve many problems. Chen and Chang (7) investigated the dispatching problem. Liepins et al. (8) investigated the simplest scheduling problem with minimal lateness as a criterion. Falkenaur and Bouffouix (9) applied genetic algorithms to small, medium, and large job shop problems. Gupta et al. (10) addressed an n–job, single machine problem to minimize the flow time variance. Bagchi et al. (11) addressed the problem of alternative process plans and developed a new two-chromosome representation. ElMaraghy (12) applied genetic algorithms in FMS scheduling of six different performance measures and twelve dispatching rules. Yokota et al. (13) investigated the interval nonlinear integer programming problems. Lin et al. (14) investigated the inspection policy problem. The structure of a genetic algorithm is depicted in fig 1 blew. In this paper, a new chromosome-coding scheme that combine between the advantages of the job-oriented and rule-oriented algorithms is presented (15,16,17).

The chromosome coding consists of three main parts. The first includes the sequence of jobs at the loading station where it consists of n sites, each one corresponding to a place for a job that define its order in the sequence. The second part corresponding to the inventory sorting rules that is adopted in the intermediate storage buffers in the front of each station. The third part of the chromosome corresponding to the manner by which each station draw a certain job from the intermediate storage buffers. Figure 2 is a typical example of chromosome-coding scheme.

Feasible chromosomes are generated randomly in their parents to represents the initial population. The reproduction operator selects specific solution strings to construct the next generation according to their fitness function value. This operator can be implemented in many ways. In this paper, The roulette wheel with slots sized according to fitness is used. The genetic operators that have been used in the genetic algorithm are Partially Mapped Crossover operator (PMC) that perform the crossover between the jobs of the first part of the chromosome and the simple crossover operator that used to perform the crossover in the second and the third part of the chromosome.

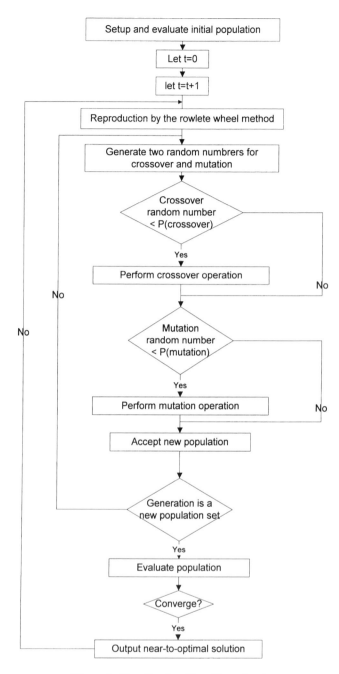

Figure 1. Genetic algorithm flow chart.

   A003/074/2003 © With Authors 2003

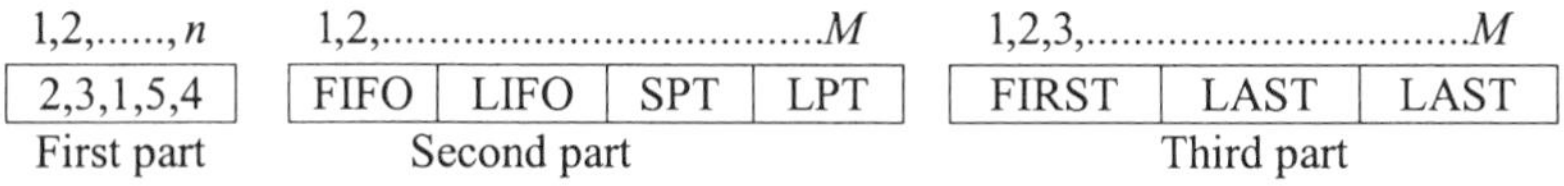

1,2,......,n	1,2,...........................M			1,2,3,.............................M		
2,3,1,5,4	FIFO	LIFO	SPT	LPT	FIRST	LAST	LAST
First part	Second part				Third part		

Figure 2. Chromosome-coding scheme.

4 ILLUSTRATIVE EXAMPLES

The following illustrative examples tend to use the rule-oriented algorithm and the proposed genetic algorithm to demonstrate the validity of the proposed GA.

4.1 A hypothetical illustration example

The hypothetical FMSs system model consists of five stations each with an intermediate buffer, an automated guided vehicle, and a load/unload station as shown in fig 3. The system used to produce three different parts with process plans exhibited in Table 1. Using SIMFACTORY II.5 (18), 6.25 time units as the best make-span has been achieved through a combination of inventory sorting rules and machine dispatching rules that are available in the software. These results are summarized in table 2. Applying the proposed genetic algorithm, a make-span of 5.65 time units has been achieved as shown in fig 4.

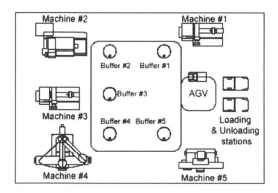

Figure 3. The hypothetical FMS model.

Table 1. Process plan for part types.

Part Type	First operation		Second operation		Third operation		Forth operation		Fifth operation	
	Time	Location	Time	Location	Time	Location	Time	Location	Time	Location
1	0.50	MC1	0.60	MC1	0.85	MC2	0.50	MC5	--	--
2	1.10	MC1	0.80	MC1	0.75	MC3	--	--	--	--
3	1.20	MC1	0.25	MC5	0.70	MC1	0.90	MC4	1.00	MC3

Table 2. SIMFACTORY II.5 solution.

Ws/buffer DR	Random	By turn	Oldest part	Newest part
FIFO	6.25	6.80	6.25	6.25
LIFO	7.05	6.25	6.25	7.05
Oldest part	7.05	6.25	7.05	6.25
Newest part	7.05	6.25	6.25	7.05
SPT	7.05	7.05	7.05	7.05
LPT	6.25	6.25	6.25	6.25

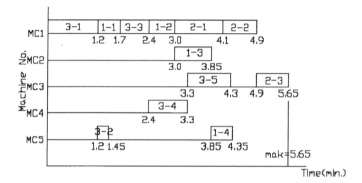

Figure 4. Genetic Algorithm solution.

4.2 Rossi and Dini illustration example

In this model, the proposed FMS was used for batch oriented scheduling of flexible manufacturing systems that include eight CNC machining centers and eight CNC lathes. Table 3 lists the production plans of 14 different parts used for the scheduling of 120 operations in a batch of 70 parts. For this example, Rossi and Dini concludes that the rule-oriented algorithm that use traditional dispatching rules like FIFO, LIFO, SPT, and LPT give a make-span of 56 minutes and the Genetic Algorithm without the Gradient-Based Parameter Optimization give a make-span of 57 minutes, and REGAL (Real-time genetic algorithm) performs a real-time scheduling in 0.08 minutes (= 4.8 seconds) and gives a make-span of 49 minutes. Solving the example by SIMFACTORY II.5, a make-span of 37 minutes is obtained as shown in table 4. Applying the proposed genetic algorithm, a make-span of 36 minutes is achieved as shown in figure 5.

    A003/074/2003 © With Authors 2003

Table 3. Production cycles and part-routings of the 70 parts.

Parts	Lathe Operation	Machining center Operation	Part routings	
1-5	$(O_2,4)$	$(O_1,7)$	(O_1,MC)	(O_2,L)
6-10	$(O_3,3)$	--	(O_3,L)	--
11-15	$(O_4,3)$	$(O_5,6)$	(O_4,L)	(O_5,MC)
16-20	$(O_6,3)$	--	(O_6,L)	--
21-25	$(O_7,9)$	$(O_8,9)$	(O_7,L)	(O_8,MC)
26-30	$(O_9,3)$	--	(O_9,L)	--
31-35	$(O_{11},5)$	$(O_{10},9)$	(O_{10},MC)	(O_{11},L)
36-40	--	$(O_{12},3)$	(O_{12},MC)	--
41-45	$(O_{14},2)$	$(O_{13},5)$	(O_{13},MC)	(O_{14},L)
46-50	$(O_{15},3)$	--	(O_{15},L)	--
51-55	--	$(O_{16},3)$	(O_{16},MC)	--
56-60	$(O_{17},3)$	$(O_{18},7)$	(O_{17},L)	(O_{18},MC)
61-65	$(O_{19},9)$	$(O_{20},4)$	(O_{19},L)	(O_{20},MC)
66-70	$(O_{21},6)$	$(O_{22},2)$	(O_{21},L)	(O_{22},MC)

Table 4. SIMFACTORY II.5 solution of Rossi and Dini example.

WS/buffer DR	Random	By turn	Oldest part	Newest part
FIFO	37	37	41	37
LIFO	40	40	38	40
Oldest part	40	38	40	38
Newest part	38	40	38	40
SPT	41	39	39	41
LPT	41	40	40	41

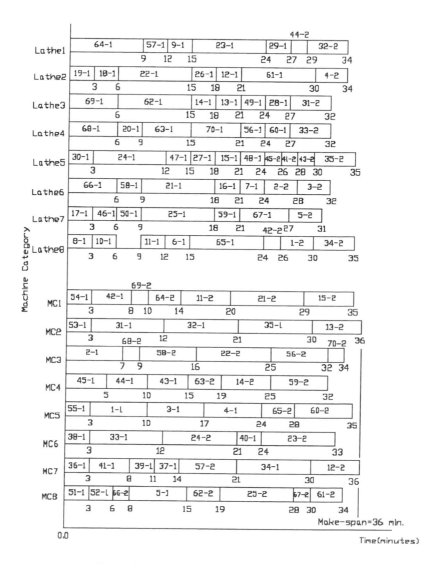

Figure 5. GA solution of Rossi and Dini example.

4.3 Liu AND MacCarthy ILLUSTRATION EXAMPLE (19)

In this model, the problem has four jobs in a FMC with two stations. Each job has two operations that can be processed on either station. Operations processing times are given in table 5. Local buffers are used for the machines and a central storage area with infinite capacity is considered for storing preprocessed parts and finished products. There is only one material-handling device in the system. The transport time between any two stations, or

   A003/074/2003 © With Authors 2003

between any station and the central storage area is 4.0 time unites. The objective is to minimize the mean completion time. The make span and mean flow time of SEDEC (Sequential Decomposition Algorithm) and CODEC (Coordinated Decomposition Algorithm) are 203.9,144.075, and 191.2, 133.3 respectively.

Table 5. Data for Liu and MacCarthy example.

Job	Time for operation 1	Time for operation 2
1	41.4	40.4
2	43.1	55.3
3	48.2	45.2
4	35.5	37.4

Simulating the problem using SIMFACCTORY II.5 of job assignment by the two heuristics, the feasible schedule is shown in figure 5 and figure 6.

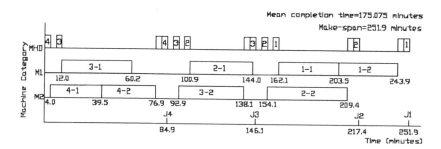

Figure 5. SIMFACTORY II.5 solution for example problem solved by SEDEC.

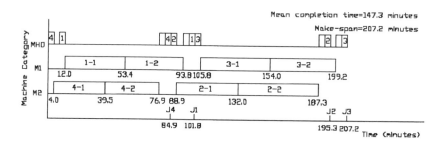

Figure 6. SIMFACTORY II.5 solution for example problem solved by CODEC.

The make-span of SEDEC is 203.9 minutes and is 251.9 minutes by SIMFACTORY II.5 with a difference of 48 minutes representing 23.54% of the solution given by SEDEC. The make-span of CODEC is 191.2 minutes and is 207.2 minutes by SIMFACTORY II.5 with a difference of 16 minutes representing 8.34% of the solution given by CODEC. This difference is due to the two heuristics ignore the transportation time of parts after finishing an operation on a certain machine to the storage buffer. Also, the schedule ignores MHD unaviliability. Figure 7 and 8 show the solution of the two heuristics assignment using the rule-oriented algorithm. Figure 9 and figure 10 show the solution by the proposed genetic algorithm. However, table 6 exhibites the summary comparative results.

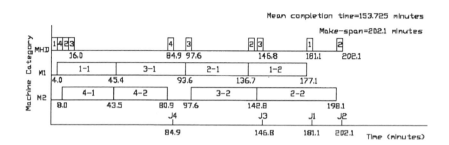

Figure 7. Solution of SEDEC assignment by rule-oriented algorithms.

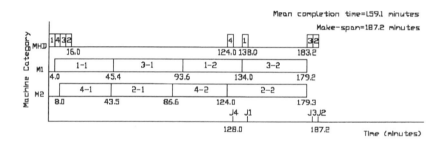

Figure 8. Solution of CODEC assignment by rule-oriented algorithms.

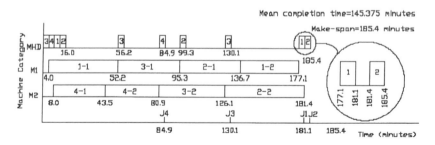

Figure 9. Solution of SEDEC assignment by the proposed genetic algorithm.

                          A003/074/2003 © With Authors 2003

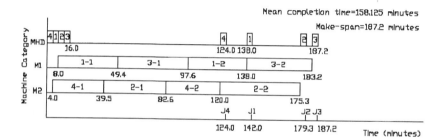

Figure 10. Solution of CODEC assignment by the proposed genetic algorithm.

Table 6. Summary comparative results.

FMS Model	Minimum make-span (time unites)			
		Rule-oriented algorithm		Genetic-based algorithm
	Base value	SIMFACTORY II.5	Developed ROA code	
The hypothetical example	---	6.25	6.25	5.65
Rossi and Dini example	49.00	37.00	37.00	36.00
Liu and MacCarthy example				
SEDEC assignment	203.90	251.90	202.10	185.40
CODEC assignment	191.20	207.20	187.20	187.20

5 EXPERIMENTS ON DIFFERENT FMSs PROBLEMS

Shouman et al. (20) have been concluded that different factors affecting the FMSs scheduling problem are those describe the attributes of the manufacturing system, resource tightness, flexibility and complexity. These factors are number of stations, number of machines in each station, average number of part per job type, tool magazine availability and capacity, availability and type of material handling system, types an availability of fixtures, types and availability of pallets, types and availability of tools, size of storage buffers, average number of operations per job, average number of tools per operation, average number of permissible machines per operation, range of processing times, and number of simultaneously produced jobs

In this paper, only seven factors each with three levels A, B, and C as shown in table 7 are considered for generating a set of representative FMSs problem instances. Given a level for each factor, the data for a test problem can be generated randomly. the test problems are generated using the principle of orthogonal experimental design of SPSS. Within the range set by the factors, a parameter in a problem is generated randomly by sampling from an appropriate uniform distribution. Computational experiments have been carried out on the selected problems to compare the performance of the rule-oriented algorithm and the proposed genetic algorithm, each represent a different strategy for solving FMS scheduling problems.

Table 7. Factors and levels for experimental designed problems.

Factor definition	Level A	Level B	Level C
Number of stations	2	6	12
Number of jobs	10	50	100
Route of job operations	$U(1,M)$	$bin(M,0.2)$	$bin(M,0.5)$
Average number of operations per job	5	20	60
Range of processing times	$U(20,80)$	$U(35,65)$	50
Time of transport operation	0	2.0	4.0
Number of material handling devices	1	2	4

The solution results of the selected problems showed the following:

- The superiority of the proposed genetic algorithm on the rule-oriented algorithms where the average make-span is 25148, 25464, 25293, 25494, and 23768 time units for FIFO, LIFO, SPT, LPT, and the proposed genetic algorithm respectively.
- The computation time is high for the proposed genetic algorithm. It is in average equal to 8.92 seconds in GA while equal to 0.11 seconds for solving the same scheduling model by all rule-oriented algorithms.
- It is shown that increasing the average number of operations per jobs increases the difference between the average performance of genetic-based algorithm rather than rule-oriented algorithms.
- The percentage of scheduling models that FIFO dispatching rule provides a solution equivalent to the proposed GA is 37%, the percentage of scheduling m odel t hat i t provide a worst solution than the GA is 50%, and the percentage of scheduling models that it provide a better solution than the GA is 13%. These percentages for LIFO are 0%, 94%, and 6% respectively and for both SPT and LPT are 11%, 83%, and 6% respectively.

6 CONCLUSION

A genetic algorithm that uses chromosome-coding scheme consisting of three main parts has been developed. The first part of the chromosome represents the sequence of jobs at load/unload station of the FMSs, the second part represents the inventory sorting rules used at each station, and the third part represents the manner by which jobs drawn from the intermediate storage buffers. The analysis of the performance of the proposed genetic algorithm verified its superiority to the other rule-oriented algorithms. Also, t he d eveloped GA is more realistic in its treatment of more sophisticated FMSs scheduling problems.

REFERENCES

1 **Wang, L., C.** (1995) Intelligent scheduling of FMSs with inductive learning capability using neural networks. In *International Journal of Flexible Manufacturing Systems*, Vol. 7, pp. 147-175.
2 **Ishii, N.,** and **Talavage, J., J.** (1991) A transient-based real-time scheduling algorithm in FMS. In *International Journal of Production Research*, Vol. 29, pp. 2501-2520.

A003/074/2003 © With Authors 2003

3   **Perrone, G., La Commare, U., Lo Nigro, G.,** and **Nuccio, C.** (1995) Dynamic scheduling in a multiple objective production environment using a fuzzy adaptive controller. In *Proceedings of the 11th International Conference on Computer Aided Production Engineering*, London, pp. 143-148.

4   **O'Kane, J. F., Harrison, D. K.** and **Gentili, E.** (1994) The analysis of reactive scheduling issues in a FMS using a dynamic knowledge-based system approach. In *Proceedings of the 10th International Conference on Computer Aided Production Engineering*, Palermo, pp. 545-554.

5   **Rabelo, L. C., Jones, A.** and **Yih, Y.** (1994) Development of a real-time learning scheduler using reinforcement learning concepts. In *Proceedings of the IEEE International Symposium on Intelligent Control*, pp. 291-296.

6   **Rossi, A.** and **Dini, G.** (2000) Dynamic scheduling of FMS using a real-time genetic algorithm. In *International Journal of Production Research*, Vol. 34, pp. 1-20.

7   **Chen, P.,** and **Chang, H.** (1995) Large-scale economic dispatch by genetic algorithm. In *IEEE Transactions on Power Systems,* Vol. 10, pp. 1919-1926.

8   **Liepines, G. E., Hilliard, M. R., Palmer, M.,** and **Morrow, M.,** (1987) Greedy genetics: genetic algorithms and their applications. In *Proceedings of the Second International Conference on Genetic Algorithms* (Cambridge, MA: Erlbaum).

9   **Falkenauer, E.,** and **Bouffouix, S.** (1991) A genetic algorithm for job shop. In *Proceedings of the 1991 IEEE International Conference on Robotics and Automation*, Sacramento, CA.

10 **Gupta, M. C., Gupta, Y. P.,** and **Kumar, A.** (1993) Minimizing flow time variance in a single machine system using genetic algorithms. In *European Journal of Operational Research*, Vol. 70, pp. 289-303.

11 **Bagchi, S., Uckun, U., Miyabe, Y.,** and **Kawamura, K.** (1991) Exploring problem-specific r ecombination o perators for job shop scheduling. In *Proceedings of the Fourth International Conference on Genetic Algorithms*, San Diego, CA, pp. 10-17.

12 **Jain, A. K.,** and **ElMaraghy, H. A.** (1994) Manufacturing scheduling using genetic algorithms. In *Proceedings of the CSME Forum*, McGill University, Montreal, pp.712-727.

13 **Yokota, T., Gen, M., Taguchi, T., Li, Y.,** and **Kim, C. E.** (1996) A genetic algorithm for interval nonlinear integer programming problem. In *Computers in Industrial Engineering,* Vol. 31, pp. 913-917.

14 **Lin, C., Yeh, J. M.,** and **Ding, J. R.** (1998) Design of inspection policy in an assembly oriented flexible manufacturing system: a heuristic genetic algorithm approach. In *Microelectronics Reliability,* Vol. 38, pp. 545-551.

15 **Syswerda, G.** (1991) Schedule optimization using genetic algorithms. *In Davis, L. (ed.) Handbook of Genetic Algorithms* (New York: Van Nostrand Reinhold), pp. 332-349.

16 **Forrest, S.** (1993) Genetic algorithms: principles of natural selection applied to computation. *Science*, 261, pp. 872-878.

17 **Park, L.** and **Park, C. H.** (1995) Application of genetic algorithm to job-shop scheduling problems with active schedule constructive crossover. In *Proceedings of the IEEE International Conference on Systems, Man and Cybernetics*, Vol. 1, pp. 530-535.

18 **CACI Products Company.** (1990) SIMFACTORY II.5 user's and reference manual. Version2.0, La Jolla, Calif.

19 **Liu, J.,** and **MacCarthy, B., L.** (1999) General heuristic procedures and solution strategies for FMS scheduling. In *International Journal of Production Research*, Vol. 37, pp. 3305-3333.

20 **Shouman, M., A., Aboul-Nour, A., A., Elawady, H.,** and **Abdel-Fatah, M.** (2002) A comprehensive classification scheme for FMSs scheduling problems. In *AJE*. pp. 57-69.

Quality in Manufacturing
and Design

A neural network approach for predicating roundness error in turning

M A YOUNES
Production Engineering Department, University of Alexandria, Egypt
M A SHOUMAN
Faculty of Computers and Informatics, University of Zagazig, Egypt

ABSTRACT

Form and roundness errors greatly affect the performance and lifetime of cylindrical parts. Evaluation of roundness is time consuming and usually done in laboratory. In modern manufacturing there is a growing need for a rapid technique to evaluate roundness in the workplace. In this work, a back-propagation artificial neural network (ANN) is used to predict the roundness error of turned parts. The inputs to the ANN are the workpiece material and the machining parameters namely, the cutting speed, depth of cut, and rate of feed. The output is the out-of-roundness of the turned parts. The ANN was trained using a set of experimental results obtained when turning copper, brass, and aluminum under specified machining conditions. The ANN structure is able to correctly predict the out-of-roundness error when turning the materials used. The neural network was later trained to predict the appropriate turning parameters that would yield a predetermined error in roundness. Using the same set of experimental results with the roundness error as input and the machining parameters as outputs, the ANN was able to propose values for the machining parameters that are very close to the experimentally used ones.

1. INTRODUCTION

Accurate evaluation of circularity error is very critical because it greatly affects the performance of cylindrical parts. The ASME Y14.5 M-1994 defines circularity as a condition of surface where all points of the surface intersected by any plane perpendicular to an axis are equidistant from the axis. Circularity tolerance specifies a tolerance zone bounded by two concentric circles within which each circular element of the surface must lie [1]. Circularity error measurement is usually performed in laboratory using either intrinsic or extrinsic datum methods. Some attempts have been made for on-line evaluation of circularity error [2].

However, the presence of machining chips, the use of cutting coolants and the relatively high costs limit the application of these techniques.

This paper presents a new technique for the automatic evaluation of roundness error of turned parts. The technique is based on a back propagation (BP) artificial neural network (ANN).

ANN approximate continuous nonlinear functions well, based on the mathematical principles and models of biological neurons and the nervous system [3]. ANNs are dynamic systems that consist of many processing units with weighted connections to each other. Each unit receives input from other units (neurons) and after processing, sends its output to other units through weighted connections. The output of a unit determines its level of activity, while the weight value of the connection between two units determines the strength of the connection [4]. The ANN has the advantage of quicker pattern recognition, real-time learning application, the ability of unsupervised learning and competitive learning rule for modifying the weighing factor in the network [5].

ANN-based techniques have been successfully used to monitor tool wear in turning. David A. Dornfeld [6] combined the outputs of several sensors (Acoustic Emission (AE), force, and spindle motor current) for monitoring progressive tool wear in a single point turning operation. Multichannel autoregressive series model parameters and power spectrum amplitudes were used as inputs to the neural network. S. Das et al [7] tried an ANN for on-line cutting tool condition monitoring to decide when to change a worn tool. Shiuh-Jer Huang et al [5] proposed an approach based on an ANN to the problem of constant turning or cutting in order to increase metal removal rate and tool life. Qiang Liu et al [3] used a three-layered feed-forward neural network to predict tool wear based on the model of cutting mechanics. A ratio of the cutting force, which is most sensitive to tool wear, the cutting speed, and feed rate were the inputs to the ANN, while the output was the flank wear.

Attempts to apply ANN for prediction of surface quality have been reported. In this regard P.G. Mathews et al [8] used parameters extracted from force, vibration, and acoustic emission sensors `as inputs to a multi-layer neural network, the outputs were the average roughness value Ra, roundness error and residual stresses generated in reaming process. Using spindle speed, feed rate, depth of cut, and vibration average per revolution as inputs. An ANN model based on BP was developed to predict Ra roughness value in end milling [9]. E. Mainsah [4] proposed an approach for on-line 3D surface characterization/ classification based on ANN. The system has been able to identify worn and unworn cylinder bore surfaces. S.K. Ozdemir et al applied an ANN for the classification of surfaces made from different materials by different manufacturing conditions [10]. A compact sensor using the self-mixing interference inside a semiconductor laser was designed to collect data which were fed to the ANN as inputs, while the output was the type of surface examined.

In the present work a back-propagation, feed-forward artificial neural network (ANN) has been used to predict the roundness error of turned parts. The inputs to the ANN are the workpiece material, cutting speed, depth of cut, and feed rate. The output is the roundness error.

     A003/075/2003 © With Authors 2003

2. MATHEMATICAL MODEL

In a turning process, the error in roundness of machined parts depends on several factors such as: part material, cutting parameters, and machine condition. An empirical model was developed [11] which relates out-of-roundness resulting in a turned part of a chosen material to the turning parameters namely, the cutting speed (V m/min), feed (F, mm/rev) and depth of cut (d, mm).

$$E_r = C * V^\alpha * F^\beta * d^\gamma \qquad (1)$$

Where:

E_r: roundness error, μm
C: constant
V: cutting speed, m/min
F: feed rate, mm/rev
d: depth of cut, mm
α, β, and γ: power coefficients

A design of experiment approach was applied to study the effect of different turning parameters on the resulting error in roundness for each type of the considered materials.

A set of experiments was performed for each type of material. The three variables considered in each experiment were, the speed, the rate of feed, and the depth of cut. Three levels were chosen for both the speed and the depth of cut (high, medium and low). For the rate of feed, five levels were taken. Parameter levels were chosen to cover the recommended values in case of turning as given in the Machining Data Handbook [12], taking into consideration the speeds and feeds available on the machine used. The lathe machine used in the study was tested, both statically and dynamically, and all test results were well within the specified tolerances. In the experimental work, the length to diameter ratio was kept constant. The workpiece fixation was also monitored.

The stock material used was 50 mm in diameter. The machined diameters were so chosen as to provide the chosen levels of the considered parameters. Forty-five experiments were performed on each type of material to cover the effect of the three parameters at the considered levels. Table (1) shows the design of experiment used.

Table (1) The factor levels for the design of experiment used.

Parameters (Factors)	Factor Levels				
	1	2	3	4	5
Speed, V, m/min	75	63	51		
Depth of cut, d, mm	0.50	0.30	0.10		
Feed, F, mm/rev	0.25	0.20	0.16	0.10	0.05

After machining, the out-of-roundness was measured on a Talyrond-200 instrument. Three different sections were inspected for circularity on each part representing a specific experiment (i.e. a specific value for V, d, F). The average readings were taken as the estimated roundness error for that experiment.

A least squares method was used to determine the mean circle, then the out-of-roundness error was evaluated for each inspected section. Tables 2, 3 and 4 give the average values for the out-of-roundness as measured for the 45 turned parts of each type of used materials, Aluminum, Brass and Copper, respectively.

Table (2) Roundness error for Aluminum parts

Roundness error, (µm)	V= 51 m/min			V= 63 m/min			V= 75 m/min		
F, mm/rev d, mm	0.10	0.30	0.50	0.10	0.30	0.50	0.10	0.30	0.50
0.05	2.60	4.60	6.00	2.10	3.20	5.10	1.70	3.00	4.50
0.10	3.20	6.80	8.50	2.60	4.20	6.50	2.20	3.30	6.00
0.16	5.00	7.70	9.50	4.00	6.00	8.70	3.00	4.50	7.20
0.20	5.50	8.90	12.50	4.10	6.70	11.50	3.30	5.30	10.00
0.25	6.00	10.00	13.00	5.00	8.20	12.00	3.50	6.00	11.00

Table (3) Average values of roundness error – case of Brass parts

Roundness error, (µm)	V= 51 m/min			V= 63 m/min			V= 75 m/min		
F, mm/rev d, mm	0.10	0.30	0.50	0.10	0.30	0.50	0.10	0.30	0.50
0.05	3.50	6.20	8.00	3.10	4.50	6.30	2.90	8.00	5.60
0.10	3.80	8.80	11.00	3.70	7.40	9.60	3.60	6.00	8.00
0.16	4.70	11.00	13.50	4.60	10.00	12.20	4.40	9.00	11.50
0.20	5.30	12.50	15.00	5.10	11.50	13.60	5.00	10.00	13.00
0.25	6.10	13.60	16.00	6.00	13.00	15.00	5.50	11.00	14.00

Table (4) Average values of roundness error – case of Copper parts

Roundness error, (µm)	V= 51 m/min			V= 63 m/min			V= 75 m/min		
F, mm/rev d, mm	0.10	0.30	0.50	0.10	0.30	0.50	0.10	0.30	0.50
0.05	4.40	8.10	11.00	3.80	6.00	10.00	3.00	5.50	8.50
0.10	6.50	13.00	17.60	5.30	10.00	15.00	4.50	10.00	13.80
0.16	8.50	18.30	23.00	7.00	16.00	21.00	6.30	13.50	19.20
0.20	9.00	20.00	26.00	7.50	18.50	24.00	6.70	15.90	22.00
0.25	9.50	22.00	30.00	8.00	20.00	27.50	7.50	16.90	26.00

The experimental results were used to estimate the coefficients α, β, and γ, for the emprirical model (1). A response surface methodology [13] was applied to estimate the coefficients α, β, and γ. Table (5) gives the estimated values of the three coefficients for the materials under consideration.

Table (5) Estimeted values of the three model coefficients α, β, and γ for the materials used.

Coefficients Material	Aluminum	Brass	Copper
α	-0.7567	-0.5068	-0.6616
β	0.5400	0.4358	0.5914
γ	0.5401	0.5255	0.6758

   A003/075/2003 © With Authors 2003

3. APPLICATION OF ANN

A feed-forward back-propagation artificial neural network (ANN), Figure (1) was used to predict the out-of-roundness of turned parts. Two automated tools namely, EasyNN and Neuro-solutions were used for the training process which give the same trends and tendencies. This part of the work was completed in two main phases as follows:

The set of data collected experimentally was used to train the ANN to predict the error in roundness. The speed V, feed F, and depth of cut d, were used as inputs to the ANN, and the output was the roundness error E_r. The same set of data was used to train an ANN to predict the turning parameters (feed and depth of cut) that will produce a given value of the roundness error when using a recommended cutting speed. The roundness error and the cutting speed were used as inputs to the ANN, the output are the speed V, feed F, and depth of cut d. The target error, i.e. the difference between the estimated and the desired values of an input parameter, was set at different levels, namely, .05, .02, and zero μm . The training curve and the training cycles required to give a specific value of the target error were recorded.

The verified empirical model (1) was used in the present work to train the ANN. Values for V, F, and d were chosen from the recommended ranges given by the Machining Data Handbook [12]. The model was then used to estimate the out-of-roundness expected when turning each of the considered materials at specific values of the turning parameters (V, F, and d). The chosen values for V, F, and d were used as inputs to the ANN and the output was the estimated error in roundness E_r. After training, the ANN was examined by feeding the values of V, F, and d used in the experimental work as inputs to the ANN. The obtained values of the out-of-roundness were then compared with the corresponding values obtained by measurement.

4. RESULTS AND DISCUSSION

Figure (2-a-1) shows the ANN learning curve for the case of turned aluminum parts, using the experimental results. The target error (the difference between the desired and the estimated values of the output parameter) was set at 0.05 μm. The ANN reached an average error 0.0498 μm after 88 trail cycles, with a minimum error of 0.0005 μm and a maximum error of 0.1568 μm.

As the target error is decreased to 0.02μm, the learning cycles are increased from 88 to 1261 cycles. The average recorded error is 0.02μm, the minimum-recorded error in this case is 0.000, while the maximum-recorded error is 0.0616μm, Figure (2-a-2). If the target error is set at 0.000μm, the learning cycles are increased to 1030000 cycles, Figure (2-a-3).

Figure (2-b) shows also the learning curves for the case of aluminum turned parts, but when using the data generated by the empirical model (1). The performance of the learning curves is fairly similar to that obtained using the experimental data but with shorter training cycles. Similar results were also obtained for copper and brass turned parts however, the number of the training cycles required to reach the average target error differs from one case to another. Table (6) gives the number of training cycles required to reach the specified target error for the cases under consideration.

Table (6) Number of training cycles required to reach the specified target error.

Material	Aluminum		Copper		Brass	
Target error, μm	Experimental	Theoretical	Experimental	Theoretical	Experimental	Theoretical
0.05	88	67	146	139	33984	1650
0.02	12061	132	1200000	272	550000	10818
0.00	1030000	570000	1500000	134000	1237351	846000

It is quite obvious from Table (6) that the training cycles required to reach the target error is much smaller in the case of the data generated by the empirical model, especially for smaller values of the target error.

This fact suggests a new approach for handling the problem of roundness error evaluation in turning. A design of experiments approach can be applied to evaluate the coefficients α, β, and γ of the empirical model (1). After verification, the model can be used to generate the data necessary to train the ANN. The ANN will then be tested using the experimental data. When the validity of the ANN is proven, it can then be used for predicating roundness error when turning a part made of a specific material at given turning parameters V, F, and d.

To investigate the efficiency of the ANN used a set of experimental runs were performed at specified values for V, F and d. The roundness error for the machined parts was measured experimentally (E_{des}). The same values for V, F and d were fed to the ANN as inputs, and the estimated error in roundness (output) was recorded (E_{est}). Figure (3) shows the desired values of (E_{des}) and the corresponding estimated values of (E_{est}) for 25 runs. It is obvious that the values of the roundness error evaluated experimentally and those estimated using the trained ANN are fairly equal in magnitude.

In practice it is more useful to have the possibility of determining the most suitable machining parameters that will yield a desired value of the error in roundness when turning a specified material. The ANN was trained to estimate the recommended values for the rate of feed (F) and the depth of cut (d) that will yield a desired out of roundness at a given cutting speed V. The inputs to the ANN were the out of roundness and the cutting speed V. The outputs were the rate of feed (F) and the depth of cut (d). Figure (4) shows the desired and the corresponding estimated values for the two parameters F and d. The same figure shows clearly the correlation between the desired and the corresponding estimated values of the two parameters.

5. CONCLUSIONS

1- A neural network approach was adopted to predict the out of roundness error resulting when turning parts from different materials.
2- The results obtained from a set of experimental runs, based on a design of experiments approach, were used to train the ANN to estimate the out of roundness error expected at specified values of the cutting parameters V, F and d.

A003/075/2003 © With Authors 2003

3- An empirical model –based on the results of the experimental results – was also used to train the ANN to estimate the out of roundness error expected at specified values of the cutting parameters V, F and d.

4- The ANN was capable of predicting the error in roundness, when turning aluminum, copper and brass, to values that are very close in magnitude to the actual values obtained by measurement using a stylus instrument.

5- It was also proved that the ANN can be trained to propose the most suitable machining parameters (rate of feed and depth of cut) that will yield a desired value of the roundness error, when turning aluminum, copper and brass, at a specified cutting speed.

6- Comparison of the output values obtained by the trained ANN with the corresponding experimental values shows a high degree of correlation.

REFERENCES

1. **ASME Y14.5 M**-1994, "Standard on Dimensioning and Tolerancing"
2. **A.M. Shawky and M.A. Elbestawy**, (1996). In-process evaluation of workpiece geometrical tolerances in bar turning. Int. J. Mach. Tools manufact., vol 36, No. 1, pp. 33-46.
3. **Qiang Liu, and Yusuf Altinas**, (1999). On-line monitoring of flank wear in turning with multi-layered feed-forward neural network. Int. J. Mach. Tools Manufact., vol. 39, pp. 1945-1959.
4. **E. Mainsah, and D.T. Ndumu**, (1998). Network applications in surface topography. Int. J. Mach. Tools Manufact., vol. 38, pp. 591-598.
5. **Shiuh-Jer Huang and Kao-chang Chiou**, (1996). The application of neural networks in self-tuning constant force control. Int. J. Mach. Tools Manufact., vol. 36, pp. 17-31.
6. **David A. Dornfeld**, (1990). Neural Network sensor fusion for tool condition control. CIRP Annals, vol 39/1, pp. 101-105.
7. **S. Das, R. Roy and A.B. Chattopadhyay**, (1996). Evaluation of wear of turning carbide inserts using neural networks. Int. J. Mach. Tools Manufact., vol. 36, pp. 789-797.
8. **P.G. Mathews and M.S. Shunmugam**, (1999). Neural Network approach for predicting hole quality in reaming. Int. J. Mach. Tools Manufact., vol. 39, pp. 723-730.
9. **Yu-Hsuan Tsai, Joseph C. Chen and Shi-Jer Lou**, (1999). An in-process surface recognition system based on neural network in end milling cutting operations. Int. J. Mach. Tools Manufact., vol. 39, pp 583-605.
10. **S.K. Ozdemir, S. Shinohara, S. Ito, S. Takamiya and H. Yoshida**, (2001). Compact optical instrument for surface classification using self-mixing interference in laser diode. Opt. Eng. 40(1), pp. 38-43.
11. **M.A. Shouman and M. Bou-Zeid**, (2000). Design of an experimental approach for optimization of some form errors. Alexandria Engineering Journal, Vol. 39, No. 1, pp57-63.
12. **Machining Data Handbook**. 3rd Ed., Merchantability Data Center – Metcut Research Associates Inc, Cincinnati, Ohio , 1986.
13. **D.C. Montgomery**. Design and Analysis of Experiments. 3rd Ed., John Wiley and Sons. Inc, 1991.

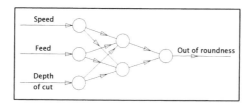

Fig. 1 Structure of the neural network applied.

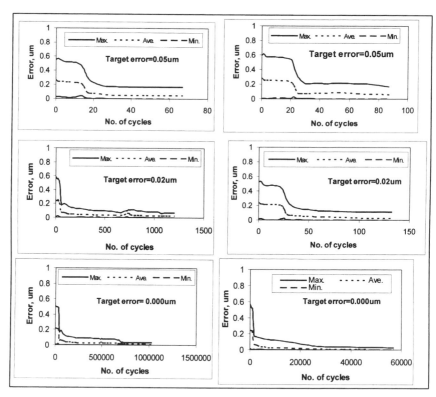

Fig. 2 Training curves for both experimental (A) and empirical data (B) at three different target errors (0.05, 0.02 and 0.00 μm) – case of aluminium turned parts.

A003/075/2003 © With Authors 2003

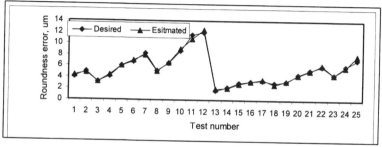

**Fig. 3 Desired and estimated values of the roundness error for
Aluminum turned parts.**

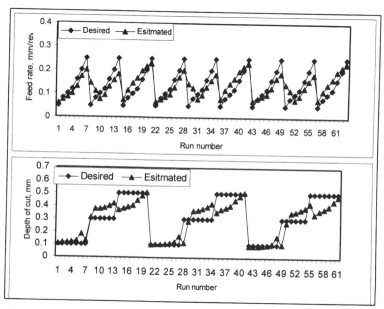

**Fig. 4 Relationship between desired and estimated feed rates and depth
of cut, when using 2-inputs, 2-outputs neural network.**

A procedure of controlling the quality of cast iron parts using ultrasound technique

A ATTANASIO, E CERETTI, C GIARDINI, and **E GENTILI**
Department of Mechanical Engineering, University of Brescia, Italy

ABSTRACT

For many companies non destructive techniques [1] are very important to guarantee the integrity of the produced parts during the quality control. The aim of the research developed by our team, was the definition of an easy and as general as possible procedure to control the quality of cast iron parts by means of ultrasound technique. In such a way, both costs and time for verifying the quality of the cast iron parts decrease. The procedure is based on several actions described in the paper. The last part of paper reports the results obtained using the designed procedure to control several automotive pieces.

1 INTRODUCTION

Within quality control, non destructive techniques are very important to guarantee the integrity of the produced parts, especially if the controlled component is important for the people safety. If we consider the cast iron control, the most used ND technique is X-rays. This exam is very expansive and time consuming for a single company so that it is impossible to accomplish a 100% control of pieces.

The aim of the research developed by our team, was the definition of an easy and as general as possible procedure to control the quality of the cast iron parts by means of ultrasound instead of X-rays. In such a way both costs and time for verifying the quality of the cast iron parts decrease.

The procedure is based on several actions. The most important are: the calibration of the instruments, which consists of the determination of all parameters which link the probe geometry and the material coefficients with the UT signal; the scanning of the part, consisting of two types of scanning, the quick scanning to detect the presence of internal cracks and the detailed scanning to define the crack geometries; the final report, to give a detailed description of all the executed operations.

To obtain good results the cooperation between the quality departments and the design department is fundamental. In particular this latter must furnish the part project with the critical zones highlighted and a table where all the control limits are reported. On the other hand it is very important to define a suitable procedure explaining to the operator how to proceed (especially for the first two actions reported above) and how to consider the results obtained from the control tests.

In order to validate the developed procedure, it was applied in controlling several automotive parts. The results obtained are very interesting: in fact all internal defects have been found and identified in their dimensions.

2 THE ULTRASONIC TECHNIQUE

Non-destructive testing with ultrasonic waves is widely used for internal defect detection. This technique is based on theory that a reflection signal is generated by the interaction between the UT-waves and the defects. Ultrasonic waves are sound waves with high frequency and this allows obtaining a defined geometry of the ultrasonic beam which detects the presence of small discontinuity in the metal matrix thanks to reflection [2].

To generate an ultrasonic wave into a workpiece, it is necessary to impose a periodic oscillating motion to the piece material. The so obtained wave front made of particles that shift in phase, spreads inside the material.

The ultrasonic wave, used to detect defects, can be classified by two parameters: the direction of the wave front propagation and the direction of particle vibration. So, different wave types exist: longitudinal, shear, superficial or Lamb.

The physical quantities that define the acoustic wave are:
- frequency (f), is the opposite of the time that a wave takes to go from a maximum (or minimum) in amplitude to the next one. The frequency range in ultrasonic testing varies from 0,5 up to 25 MHz;
- wave length (λ), is the distance between two consecutive maximum (or minimum);
- wave velocity (v) in the material, is the ratio between distance and time;
- acoustic pressure (P), is the applied pressure in the unit surface;
- wave intensity, is the energy that goes through the unit area in 1 second.

In general the UT technique [3] uses a probe which can generate the wave beam and receive the waves reflected by all discontinuities present into the material. So the probe is an element which plays a fundamental role in obtaining precise results. Also the material type greatly influences the obtained results. In fact the internal structure of piece material could attenuate the signal until it became not receivable by the probe. For instance, UT technique is not able to find defects into lamellar cast iron parts, owing to its graphite structure [4].

The technique described is usually used to control stainless steel pieces and to do this a control procedure already exists. As far as the spheroidal cast iron components are concerned, nowadays a procedure control does not exist. In the following the development of the procedure control is reported.

3. THE DESIGNED PROCEDURE

The aim of the conducted research was the definition of a UT-procedure usable in industrial environment to detect internal defects into spheroidal cast iron parts.

The procedure designed is similar to the classic UT-procedure [5, 6, 7], but it is customized following the piece material requirements. The contents list is the following:
1. objective of testing;
2. scope;
3. responsibilities;
4. test procedure;
5. test report.

3.1 Objective of testing
This procedure concerns the ultrasonic testing of spheroidal iron casting for the detection of internal defect using the UT-technique.

Comments: the objective of the test is simply outlined.

3.2 Scope
The guideline of this procedure is valid to write manuals of UT testing for spheroidal iron casting with different geometries and wall thickness between 10 and 300 mm. For wall thickness outside this range, the procedure must be validated by means of experimental tests.

Comments: the scope and the validity limits of the procedure are defined.

3.3 Responsibilities
The responsibilities are ascribed to:
- engineering department, in order to define the critical areas and the defects dimensional tolerances for every single part or set of parts;
- quality department, in order to write the manuals for every single part or set of parts;
- quality laboratory, in order to correctly execute the procedure defined by the quality department.

Comments: the different responsibilities of procedure definition and execution are set.

3.4 Test procedure

3.4.1 Instrumentation
The instrument must visualize the defect signals on a CRT screen in form of an echo signal versus time (A-scan). It must allow to operate with a frequency ranging from 0.5 to 6 MHz and it must allow the power setting. The signal amplification must have a dynamic range of 80 dB.

The instrument must be verified every year.

3.4.2 Probe

A T/R probe with the following specifications must be used:

- frequency: equal to 4 MHz;
- diameter: max 10 mm;
- transducer material: barium titanate or lithium sulphate.

The use of longitudinal waves for the contact technique is normally considered; all the other probe type can be used to have more information in defect valuation. The probe must have a right resolution capacity and its choice is influenced by the instrument set and the piece material, in order to obtain the best ratio between signal and attenuation.

3.4.3 Operator

The operator executing the UT testing must make an extensive training to carry out the tests correctly according to this procedure. It is necessary that the operator had obtained the first level certificate of UT technique according to standard UNI EN 473/93 or equivalent.

3.4.4 External piece condition

The piece surfaces must be clear and free of contamination which can generate interferences with the UT signal.

3.4.5 Coupling agent

The coupling agent must cover the testing surface in order to ensure acceptable sound transmission. Possible coupling agent can be: oil, grease, gels and other commercial agents. It must not be corrosive.

3.4.6 Reference piece for calibration

In order to correctly calibrate the probe, a spheroidal cast iron reference piece with spheroidizing equal to 90% must be used. Two sets of flat holes, whose diameter is equal to 1 and 5 mm and with different depths, have to be performed on the reference piece. Figure 1 shows the piece geometry.

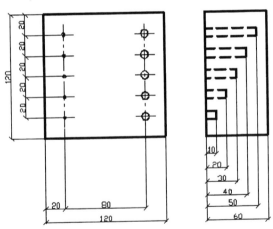

Figure 1. Reference piece geometry

The reference piece must be without defects.

 A003/022/2003 © IMechE 2003

3.4.7 Scanning

To scan the piece the operator must move the probe upon the piece surface finding defects in the volume under the probe. The scanning speed must be sufficiently low to ensure a clear evaluation of the defect signals. The distance between two consecutive scans must be less than the probe diameter. When curved instead of flat zones are scanned, it is preferable to perform the scan along both internal and external surfaces.

3.4.8 Calibration

The calibration must be carried out on the reference piece described in paragraph 3.4.6 and checked on the casting itself. The operation sequence to calibrate the probe is the following:

- the operator places the probe on the surface without the holes, in the middle of the piece in order to don't have any hole under the probe. Using the instruments control the operator must set the echo signal on 60 mm (reference piece thickness), as shown in Figure 2;
- the operator places the probe upon the most deep hole (50 mm), with diameter equal to 3 mm, and using the amplification control set the signal height on 80% of screen (Figure 3);
- using the previous setting the operator find all the other holes; he makes the necessary tuning and writes the DAC diagrams relative to the probe applied.

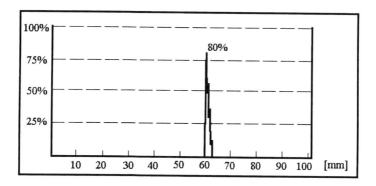

Figure 2. Calibration of back echo

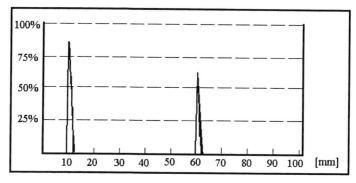

Figure 3. Gain calibration

3.4.9 Ultrasonic echo signal types

Two different types of ultrasonic echo signal must be taken into consideration and evaluated in the testing of the piece:

- first signal type: intermediate echoes (single or multiple) with attenuation of back echo (Figure 4a); it is necessary to consider all the signals which have a peak higher than 20% - 30% of the screen ;
- second signal type: several echoes one near to each other with attenuation of back echo; in this case the operator must individuate the maximum signal and set the gain to amplify the signal up to 80% of the screen; after this he must move the probe around the maximum point while the echo signal is upper the 50% of the screen. This operation furnishes the defect extension (Figure 4b).

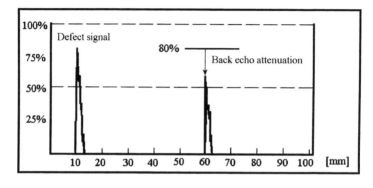

a)

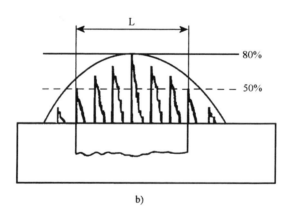

b)

Figure 4. Signal types

3.4.10 Defect limits

The engineering department must furnish a table where the tolerances of the defect detected are reported. In particular the highest extension and the maximum acceptable number as a function of the area (critical or not) and the quality level.

 A003/022/2003 © IMechE 2003

If the operator is not able to define the importance of the detected signal, it is possible to cut the piece to evaluate directly the defect.

Comments: in this procedure the instrument, the probe, the operator, the piece condition and the coupling agent are firstly identified; then the calibration phase is described both in terms of reference piece to be used for the calibration and in term of how to proceed. All the calibration procedure is described highlighting how to operate and what the operator must control and correct. Once the calibration is completed the two different echo signals are described together with the defect tolerances.

3.5 Test report
The results obtained must be recorded. The test report must contain: casting identification, test data, name and qualification of the operator responsible of the test, extent of testing, equipment used and calibration status, probes used, areas tested, signals detected, probe position that furnished a defect signal, comments.

Comments: finally how to prepare the report of the conducted test is described. In order to have a complete test record, the list of the things to be recorded is furnished.

4. Examples of procedure application

Different typologies of defects can occur in cast iron pieces. The complete classification is reported in UNI EN specifications [8]. The identified procedure is able to find all the internal defect typologies [9, 10, 11, 12].

In the following a comparison between the defect and the echo signal (Figures 5, 6 and 7) for the analysed parts is reported.

4.1 Shrinkage cavity and blowhole
These defects consist into a single cavity into the material, caused by the gas presence in the casting or by the material contraction during the cast solidification.

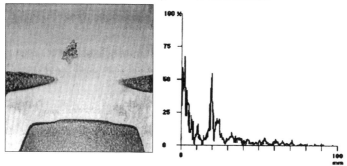

Figure 5. Shrinkage cavity and blowhole signal

4.2 Spongy contraction and pinhole

These defects consist into a cloud of microcavities; also in this case the causes can be the gas presence in the casting or the material contraction during the cast solidification.

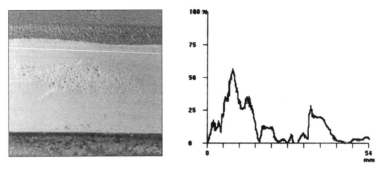

Figure 6. Spongy contraction and pinhole

4.3 Exogenous and endogenous inclusions

These defects consist into presence of slags in the pieces. The inclusions cause discontinuities in the material detectable by ultrasounds.

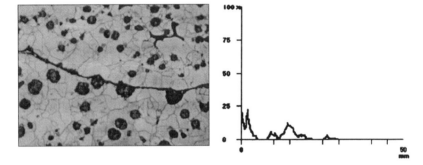

Figure 7. Inclusions

5. UT technique versus X-rays techniques

To validate the procedure defined a comparison with the results obtained from the X-rays technique has been made. The results are reported in Figure 8. It is possible to observe that the defect extension detected is the same. The UT technique furnishes also the depth position of the defect. This is an advantage versus the X-rays technique, in fact, repeating the control procedure on opposite and orthogonal surfaces, it is possible to know the three dimensional extension of the defect and its position into the part.

A003/022/2003 © IMechE 2003

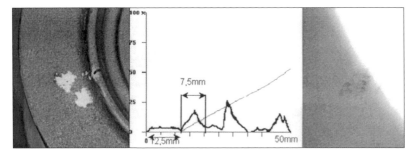

Figure 8. UT Vs X-rays

6. Conclusions

In the conducted research a working procedure to control the internal integrity of spheroidal cast iron products has been developed. This has allowed the quality control operators to carry out the testing based on a detailed procedure to be referred to. The procedure validity has been tested introducing it in a company that produces automotive suspensions with cast iron parts and the results reported above show that all defect types can be found.

Moreover, the defined procedure has been tested with X-rays results showing that this procedure is able to identify the geometry and the extension of internal defects detected into cast iron parts. This technique can substitute destructive techniques and X-ray, controls in order to decrease the cost for the quality control. Moreover, it is possible to execute it on a 100% lot control within the company.

REFERENCES

1 **AA. VV.** (1991) Nondestructive tesing, ENEA italian commission for nuclear and alternative sources.

2 **AA. VV.** (1991) Controllo non distruttivo con ultrasuoni, ENEA italian commission for nuclear and alternative sources.

3 **Calcagno, G., La Torre, C., Ansaldi, E., Re Fiorentin, S.** and **Magistrali, G.** Corso teorico sui controlli non distruttivi con ultrasuoni. Fisica degli ultrasuoni – Apparecchiature per controlli con ultrasuoni, C. R. F., Torino, Italy.

4 **Magistrali, G.** (1996) Controllo delle ghise sferoidali con tecniche ultrasoniche. 27th ISATA paper number 94EN014.

5 prEN 190/411-2: 1993.

6 prEN 12680-2: 1996.

7   Iveco Standard 18-5602: 1993.

8   UNI-EN 6047: 1967.

9   **Remondino, M. S.** Corso per Specialisti in Tecnologia di Fonderia. Identificazione dei Difetti. Diagnosi e Terapia, Associazione Nazionale delle Fonderie.

10 **Barton, R.** (1981) Nodular (SG) Iron: Possible Structural Defects and Their Prevention. 4th International Conference of Licensees for the GF Converter Process, Schaffhausen, Switzerland, pp. 340 – 353.

11 **Goodrich, G.M.** (1997) Cast Iron Microstructure Anomalies and Their Causes. In: *AFS Transaction.* Vol.105, pp. 669 – 683.

12 **Ceretti, E., Giardini, C., Attanasio, A.** and **Maccarini, G.** (2002) Use of Ultrasound to Detect Internal Defects of Spheroidal Graphite Cast Iron Products. 6th International Conference on Advanced Manufacturing System and Technology, Udine, Italy, pp. 833 – 842.

A003/022/2003 © IMechE 2003

A systems approach to manufacturing process optimization, including design of experiments and Kaizen techniques

G HOWELL
Manufacturing Division, Salton Europe Limited, UK

ABSTRACT

Field failures were being caused by a complex engineering problem that was known to be influenced by large numbers of variables ranging from product design features through to aspects of the manufacturing processes, including plastic injection moulding and assembly.

The Taguchi approach was employed to optimise the performance of the system in a minimum number of tests without having to understand the mechanisms by which each parameter affected the system. Kaizen was then used to look in further detail at the variability within the assembly process. The combined result was to improve the performance of the system by 80%, eliminate all field failures without any increase in product cost.

1 INTRODUCTION

The best selling kettle in the UK was a Russell Hobbs kettle made by the Manufacturing Division of Salton Europe Ltd and it suffered from unacceptable levels of field returns. The predominant reason for return was leaking but the root cause of the leaking was unknown because extensive life testing of the product had not replicated the failure mode. Even though the loss of sales volumes would have been very large, all the major customers of Salton Europe Ltd were threatening to de-list the kettle and so there was huge pressure on the Manufacturing Division to rapidly eliminate the problem and restore customer confidence.

2 BACKGROUND

The Russell Hobbs kettle was highly innovative when it was launched because it was the first 3kW rapid boil kettle and it incorporated a revolutionary thick film heating element printed onto a flat stainless steel plate; in addition, it offered a new and aesthetically pleasing industrial design.

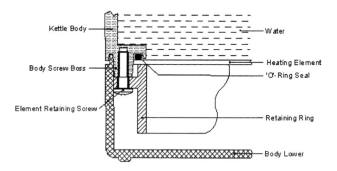

Figure 1.   Kettle heating element sealing arrangement

Figure 1 shows the element sealing arrangement of the new heating element where the retaining ring clamps the heating element against an 'O'-ring seal located in the kettle body. The retaining ring is held in place by 5 equi-spaced screws which are pneumatically driven into 5 bosses in the polypropylene (PP) body moulding to a specified torque value.

Poor detail mechanical design of the sealing arrangement had led to the element retaining screw being under-specified – it was too small in diameter and length for the loading; in addition, the plastic screw bosses were undersized. However, extensive life testing of the product before and after launch over a simulated 5 years' service had not resulted in any leaking kettles.

On the other hand, after a relatively short period in the field, some of the kettles were being returned for leaking. The returns were quickly analysed and it was identified that loosening of the element retaining screws was the cause of the leaking.

The diameter of the screw was increased and its specification changed to include a "rough" surface finish. Unfortunately, the industrial design defined the external surfaces of the product and, as can be seen in Figure 1, it left no room for increasing the dimensions of the screw bosses.

Although the field returns rate decreased, returns with loosened screws still persisted. Further analysis of the returns revealed a complex situation where random production periods experienced very high returns, against a general background of lower, but still unacceptable, returns. At the same time, whilst the greatest returns rates were seen from the original screw,

                                        A003/057/2003 © IMechE 2003

some high usage kettles incorporating this screw were operating perfectly and showing no loosening of the element retaining screws.

3   LIFE TEST SEQUENCE

Life testing was carried out on automatic test rigs but life testing of large numbers of kettles had failed to replicate the loosening of the element retaining screws.   The life test sequence comprised filling the kettle with cold water, switching the kettle on, leaving the kettle to come to the boil and switch itself off, pouring away the hot water, refilling with cold water and then repeating the test cycle. After a careful study, it was identified that the plastic screw bosses in the kettle body were never cooling down to room temperature and so the plastic/screw interface was not being exposed to the full thermal cycling that was experienced in normal domestic operation.

A dwell period was introduced after refilling the kettle with cold water so that the water could act as a heat sink and cool down the plastic body moulding, including the screw bosses. In addition, the quench effect of the cold fill and cool down was more extreme than the gradual cool down typical of normal usage, resulting in the life test being more aggressive than normal domestic use.

As soon as this dwell period was added to the life test sequence, kettles began to leak during life test due to loosening of the element retaining screw and so life testing was able to replicate the field failures.

4   DESIGN OF EXPERIMENTS

The fundamental causes of the screw loosening were not understood but it was decided that, rather than conducting a detailed investigation into why certain batches performed very badly, a pragmatic engineering approach would be adopted. This approach was to make the design and manufacturing processes robust so that screw retention could be guaranteed on 100% of products using a Design of Experiments (DoE) approach.

4.1 Brainstorming
A multi-disciplinary team comprising engineers and representatives from the shopfloor was formed and a brainstorming session held to identify potential causes of screw loosening. The causes generated were recorded and then ranked in perceived order of importance so that they could be used as variables in the Design of Experiments.

4.2 Control variables
From the causes generated in the brainstorming session, all the parameters that could be controlled were extracted as Control Variables for the Design of Experiments. As there were no interactions predicted between the 7 most highly ranked parameters, it was decided to investigate the top 7 Control Variables and employ an L_8 array for the experiment :

		Configuration	
Ranking	**Parameter**	**Standard (Std)**	**Alternative (Alt)**
C1	Body - plastic material	10% talc filled PP	High crystallinity PP
C2	Screw - underhead design	Plain	Serrated
C3	Retaining ring - screw hole	Transition	Clearance
C4	Screw - surface finish	Rough	Smooth
C5	Seal – hardness	Hard	Soft
C6	Air gun – speed	1100 rpm	500 rpm
C7	Air gun – torque setting	0.70Nm	0.55Nm
C8	*Heating element – notched*	*Un-notched*	*Notched*
C9	*Mould flow index of body plastic*	*-*	*-*
C10	*Screw – thread form*	*-*	*-*
C11	*Screw – plating process*	*Mechanical*	*Electro-plating*

4.3 Noise variables

When the Design of Experiments approach is used, the most robust optimisation comes from incorporating all major uncontrollable variables into the experiments. From the brainstorming, the Noise Variables were identified and ranked as follows :

Ranking	**Parameter**	**Noise Levels**	
N1	Body – age of moulding	16 hours	72 hours
N2	Colour of body	White	Dark
N3	Screw – angular position around body	Position 1 to 5	
N4	*Heating element - shape*	*Flat*	*Domed*
N5	*Body – mould tool*	*#1, #2 or #3*	

Noise factors N1 and N2 both affect the hardness of the plastic moulding and, hence, affect the retention of the element retaining screw during thermal cycling :

"Soft"	**"Hard"**
White	Dark colour
16 hour	72 hour

These noise factors were combined as Compounded Noise Factors in order to reduce the total number of experiments required.

A003/057/2003 © IMechE 2003

The Millennium kettle heating element is clamped by 5 equi-spaced element retaining screws and, by measuring the performance of all 5 screws, the DoE was set up to also cover the noise variable N3 - screw position.

4.4 Quality characteristic

A test procedure was developed that was proven to reliably differentiate between a good and a bad screw retention configuration. The Backloose Torque test involved boiling a kettle for a defined number of cycles, incorporating the dwell period between each cycle to allow the plastic screw bosses to cool down. Upon completion of the specified number of cycles, the peak torque required to unscrew each element retaining screw was measured. This test provided a quantitative result and so was ideal for use as the Quality Characteristic in the Design of Experiments.

4.5 L_8 test matrix

The L_8 array was constructed to include the Compounded Noise Variables and resulted in a total of 16 experiments :

Expt	CONTROL VARIABLES							NOISE VARIABLES	
	Body material	Screw head	Screw finish	Retention ring hole	Seal	Gun speed	Gun torque	Age of body	Colour
1A	Std	Std	Std	Std	Std	Std	Std	16h	White
1B	Std	Std	Std	Std	Std	Std	Std	72h	Dark
2A	Std	Std	Std	Alt	Alt	Alt	Alt	16h	White
2B	Std	Std	Std	Alt	Alt	Alt	Alt	72h	Dark
3A	Std	Alt	Alt	Std	Std	Alt	Alt	16h	White
3B	Std	Alt	Alt	Std	Std	Alt	Alt	72h	Dark
4A	Std	Alt	Alt	Alt	Alt	Std	Std	16h	White
4B	Std	Alt	Alt	Alt	Alt	Std	Std	72h	Dark
5A	Alt	Std	Alt	Std	Alt	Std	Alt	16h	White
5B	Alt	Std	Alt	Std	Alt	Std	Alt	72h	Dark
6A	Alt	Std	Alt	Alt	Std	Alt	Std	16h	White
6B	Alt	Std	Alt	Alt	Std	Alt	Std	72h	Dark
7A	Alt	Alt	Std	Std	Alt	Alt	Std	16h	White
7B	Alt	Alt	Std	Std	Alt	Alt	Std	72h	Dark
8A	Alt	Alt	Std	Alt	Std	Std	Alt	16h	White
8B	Alt	Alt	Std	Alt	Std	Std	Alt	72h	Dark

In each experiment 3 kettles were assembled and, after 300 cycles, tested for backloose torque and 5 torque measurements taken per kettle.

4.6 "Larger-the-better" analysis

The DoE Quality Characteristic can be classified as "Larger-the-Better" with the objective of the optimisation being to maximise the backloose torque.

There were a total of 240 backloose torque measurements taken and the backloose torques measured in each experiment were used to calculate the signal-to-noise ratio (S/N) for "Larger-the-Better" data (1) :

$$\text{Signal/Noise (dB)} = -10 \log_{10} [(\textstyle\sum \{1 / \text{torque}^2 \}) / n] \qquad \text{where n = no. of torque results}$$

Experiment	Signal-to-noise ratio (dB)
1	-14.6
2	-17.5
3	-15.5
4	-11.9
5	-19.7
6	-19.0
7	-15.2
8	-14.2

A signal-to-noise ratio Response Table was created to enable the Control Variables to be ranked :

Configuration	CONTROL VARIABLES						
	Body material	Screw head	Screw finish	Retention ring hole	Seal	Gun speed	Gun torque
Standard	-14.9	-17.7	-15.4	-16.3	-15.8	-15.1	-15.2
Alternative	-17.0	-14.2	-16.5	-15.6	-16.1	-16.8	-16.7
Difference (dB)	2.2	-3.5	1.1	-0.6	0.3	1.7	1.6

Ranking of the Control Variables was performed and the size of the effect of each variable categorised :

Rank	Difference analysis (dB)	Control variable	Optimum configuration	Size of effect
1	3.5	Screw head	Serrations	Large effect
2	2.2	Body material	10% talc filled PP	)
3	1.7	Gun speed	1100 rpm	) Medium effect
4	1.6	Gun torque	0.7Nm	)
5	1.1	Screw finish	Rough	)
6	0.6	Retaining ring	Clearance	) Small effect
7	0.3	Seal	Hard	)

A003/057/2003 © IMechE 2003

Using the S/N values for the large and medium effects only, the prediction of optimum S/N performance was calculated :

Predicted optimised S/N performance   = - 11.5 dB

The backloose torque of the optimised configuration can be calculated from the optimum signal-to-noise ratio (1) :

Predicted optimum mean backloose torque $= \sqrt{[\,1\,/\,10^{\,-\,\{\,(\text{optimised S/N})\,/\,10\,\}}\,]}$

$= \mathbf{\underline{0.27\ Nm}}$

4.7 Standard build performance
By comparison, the performance of the standard build configuration of the kettle was represented by Experiment #1 of the DoE :

Standard build (Expt #1) S/N performance   = - 14.6 dB

Mean backloose torque of the standard kettle = **0.19 Nm**

Hence, the DoE was predicting an increase in mean backloose torque of greater than 40% by solely implementing the change to underhead serrations on the element retaining screw.

4.8 Confirmation run
A Confirmation Run was performed to check that the DoE had included all the influential factors that affected the system being tested. If a discrepancy between the DoE prediction and the measured performance of the Confirmation Run was found, then a major variable would not have been included in the DoE that was influencing backloose torque values.

The optimum configuration recommended by the DoE comprised the serrated screw and a clearance hole in the retaining ring; however, the magnitude of the effect of the clearance hole was only small and so did not justify the cost of modifying the retaining ring mould tool. The Confirmation Run was performed with the serrated screws assembled into an otherwise standard kettle and again the Compounded Noise Variables were used. 5 off kettles were built using white, 16 hour old mouldings and 5 off kettles were built using dark, 72 hour old mouldings.

The results were used to calculate the signal-to-noise ratio for "Larger-the-Better" data (1) :

Confirmation Run S/N performance   = -11.2 dB

Confirmation Run mean backloose torque = **0.28 Nm**

Hence, the difference between the DoE prediction (0.27Nm) and the measured optimum configuration torque was 3%. This indicated that the DoE had successfully tested all the major system variables and had achieved a measured increase of nearly 50% in mean backloose torque.

5 PROCESS REPEATABILITY – CONTINUOUS IMPROVEMENT

In the case of loosening screws, the mean backloose torque optimised by the DoE is a poor indicator of screw retention because the failures that occur result from instances of low backloose torques. Figure 2 shows two statistical populations with differing mean values and standard deviations. Although the average performance of Population A is better, the reduced variance of Population B results in a higher minimum value. When applied to backloose torques, Population B would offer the better screw retention performance.

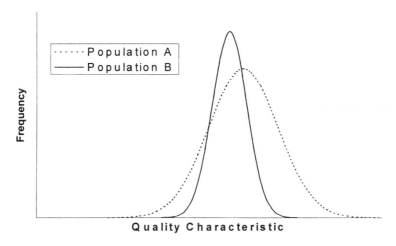

Figure 2.   Comparison of 2 normal distributions

5.1 Analysis of variability
Detailed analysis of the assembly process highlighted areas that were introducing variability into backloose torques :

5.1.1   Gapping between retaining ring and screw boss
After assembly of the Confirmation Run kettles, each kettle was visually inspected and one of the element retaining screws was found to have not clamped the retaining ring down onto the top of the screw boss. When measured for the Confirmation Run analysis, this screw produced a low backloose torque. It was found that as the element retaining screw was cutting its thread into the plastic screw boss, it was raising a burr in the top surface of the boss. In extreme circumstances, this burr was sufficient to prevent the retaining ring from tightening down onto the top of the screw boss.

A design change was implemented that added a chamfer to the bottom of the screw hole through the retaining ring. The chamfer was sized to accommodate the largest burr and ensured that the retaining always screwed down onto the top surface of the boss.

                                                      A003/057/2003 © IMechE 2003

Whilst the pin in the mould tool was being replaced to create the chamfer, the opportunity was taken to increase its diameter at zero extra cost in order to provide a clearance hole for the element retaining screw. The original DoE had identified this as a Control Variable with only a "small effect" and so the cost of the tool modification had not been justifiable at that time.

5.1.2 "Cam-out" of the screwdriver
It was found that the screwdriver bits on the air guns were prone to wear and the bit could "cam-out" from the Pozi recess in the screw head. The assembly operators changed the bits when they observed a "cam-out" with a typical frequency of 2-3 days. There was no guarantee that the operator would notice all instances of "cam-out" and, whenever a cam-out occurred, the torque applied to the screw was indeterminate.

The screw was re-specified with a Torx head that eliminated the potential for "cam-outs" and the Torx bits were replaced every month, as a purely preventative measure.

5.1.3 Heating element scraping the screw bosses
There was a small chance that the element could scrape down the plastic screw boss during assembly and prevent the retaining ring from being tightened down onto the screw boss. This variable had been identified in the brainstorming session but it had not been ranked as one of the 7 most important Control Variables and so had not been part of the DoE tests.

A design change was implemented that notched the heating elements adjacent to each of the 5 screw bosses to prevent this occurrence.

5.1.4 Screw surface finish
The supplier of the element retaining screw was unable to supply screws with a consistent "rough" finish and occasionally a batch of screws would fail Goods Inwards inspection when tested for backloose torque. It was found that the finishing process for the screw was time-dependent and difficult to control to provide a consistent surface finish.

In conjunction with the supplier, a new finishing process was proven that was highly consistent and has since 100% successfully passed the Goods Inwards checks.

5.2 Backloose torque performance through the design improvements

	Backloose torque				
	Mean (Nm)	Range		Minimum (Nm)	Incremental change in minimum
		(Nm)	(%)		
Standard configuration	0.19	0.05	25%	0.15	-
Optimised DoE build	0.27	0.23	55%	0.19	+ 27%
Final configuration	0.32	0.09	25%	0.27	+ 42%

The DoE provided a step increase in mean backloose torque (42%); at the same time, the variability may have increased but the sample size of standard configuration kettles tested was very small. The net effect on the minimum backloose torque was to increase the value by 27%.

When compared to the DoE optimisation, the Continuous Improvement exercise yielded a small increase in mean backloose torque but its major effect was to increase the minimum measured backloose torque by a further 42%.

The total increase in minimum backloose torque of the final configuration when compared to the standard build was 80% and this directly relates to improved element retaining screw retention.

6   EFFECT OF IMPROVEMENTS

- The kettle retails at between £25 and £30. Product on-cost was less than 1p.

- 450,000 kettles have been built to the final design and not one kettle has been returned for leaking due to a loosened element retaining screw.

- Customer confidence in the Manufacturing Division of Salton Europe was re-established

- New and innovative derivatives of the kettle were able to be developed and launched with no risk of the element retaining screws loosening, further extending the product life span of the kettle.   One of the derivatives even had to be moulded in a softer specification of polypropylene for the body moulding and would never have been feasible without the robustness of the improved sealing arrangement that had been implemented.

7 CONCLUSIONS

- The Design of Experiments approach can be successfully used to simultaneously optimise a complex system that includes variables ranging from product design features to aspects of the manufacturing and assembly processes.

- Where appropriate, the Design of Experiments approach offers robust solutions that can be optimised in a minimum number of experiments.

- Design of Experiments is a valuable tool in the toolkit of an engineer. However, as with any toolkit, the best results are obtained by possessing a broad range of techniques and then applying the most appropriate problem solving techniques to each individual problem.

8 REFERENCES

1 **ASI Quality Systems**   "Taguchi Methods – Practitioner Course Notes"

Orbital electrochemical finishing of holes using stationary tools

H EL-HOFY and **N AL-SALEM**
Mechanical Engineering Department, Qatar University, Qatar
M A YOUNES
Production Engineering Department, Alexandria University, Egypt

ABSTRACT

Electrochemical machining, ECM, is used to produce complex shapes in super alloys and other hard metals. The process has many advantages such as the independence of machining rate on hardness, the absence of tool wear and thermal stresses in the workpiece and, the ability to produce high surface quality. Typical applications of ECM include complex shapes in turbine blades, gun barrel rifles, forging dies, and moulds. Modern process applications include sawing, grinding and finishing of holes. Orbital electrochemical machining (OECM) is currently adopted in many areas using smaller power supplies, tools, and electrolyte circulation systems

In this paper OECM is used for finishing holes using stationary (non-feeding) tool. Effects of some process parameters such as orbiting speed, and tool diameter have been investigated. Material removal rates, current efficiency. Machined allowance, surface roughness and roundness error have been also evaluated.

1. INTRODUCTION

In ECM, the machining current is allowed to pass through an electrolyte solution that fill the gap (typically below 1 mm) between a cathodic tool and an anodic workpiece. With the application of a dc-voltage (10-30 Volt) high current density in the range of 0.5 to 5 Ampere/mm^2 is produced. Under these conditions the work surface is dissolved at a rate that can be described by Faraday's laws. The only reaction that takes place, at the cathodic–tool surface, is hydrogen gas evolution and, thus, the tool retains its shape throughout the process. The power transmitted across the machining gap is sufficient to cause electrolyte heating and evaporation. To keep the electrolyte at the required temperature, it is pumped through the gap at high velocity up to 30 m/s. The large electrolyte flow velocity flushes away the machining products, from the gap, thus ensuring reasonably constant electrolyte conductivity during machining.

The achievement of an enhanced material removal rate with high machining accuracy demands the use of high current density. In this regard, the application of a pulsating voltage leads to high instantaneous current density, without the need for elaborate electrolyte pumping system and a rigid machine frame. The main objective for using pulsed current is to improve the machining accuracy and surface quality [1-3]. For high accuracy requirements, hole drilling is often followed by reaming. However, this technique is limited by the attainable increased surface roughness. Subsequent conventional polishing process induces non uniform residual stresses, surface cracks and depressions which reduce the product surface life. On the other hand, when such a hole is machined by EDM, the brittle heat affected surface layer, due to carburization and quenching, creates additional difficulty for the following conventional polishing. In order to tackle these problems, Masuzawa and Sakai [4] adopted the use of ECM mate electrode for removing the recast layer produced by wire-EDM in die and mold industry. However, the use of such stationary electrode needed large power supply since machining occurs allover the entire electrode area. Sadollah and El-Hofy [5] used an orbiting ECM electrode for finishing surfaces machined by EDM. They noticed a reduction in the surface protrusion heights at the flow ports and improvement in surface roughness with the increase of both orbiting eccentricity, time and, frequency. The orbital motion during ECM served the following purposes;

1. Reduced the area being machined and hence, the size of the power supply required.
2. Agitated the static electrolyte which eliminated the need for forced electrolyte flow.
3. Maintained high current density and ensured a smooth surface finish.

Yew and Goh [6] used ultrasonic vibration in enhancing the debris removal which improves the surface quality for micro holes. Ghabrial et al [7] adopted the use of air electrolyte mixture in stationary finish ECM to improve both the surface quality and geometry of holes. Kozak et al [8] used a rotating tool electrode that ensured an adequate electrolyte flow in the inter-electrode gap and eliminated the need for high electrolyte pressures. Using the rotating electrode ECM, pressure changes within the gap were quite small.

Hocheng and Pa [9] introduced the process of electro polishing and electro brightening of holes using different axial feeding electrodes. They concluded that electro brightening, after reaming, required quite a short working time, whilst electro polishing avoided the need for reaming, which made the total cycle time less than electro brightening. They recommended the use of electrodes of boring cutter type.The experimental and theoretical analysis of Rajurkar [10] indicated that orbital ECM distributes the electrolyte more uniformly and hence leads to a significant reduction in the flow field disrupting phenomena that adversely affect the machining accuracy. The proposed method has been found to offer substantial gains in the accuracy and the uniformity of the machined components.

The effetive utilization of ECM in achieving the optimal combination of enhanced material removal rate and the generation of good surface quality when finishing holes, necetates the use of orbiting tool electrode. OECM can be useful in deburring, embossing and in internal profiling and finishing of cavities. Al-Salem and El-Hofy, [11] introduced a simulstion model for orbital finishing of holes using electrochemical machining. The presented work focuses on the development of a precision OECM using a concentric orbital tool movement for sizing and finishing of holes.

2. EXPERMENTATION

An experimental system has been developed to study the process behavior. This system, consists mainly of a machining cell, power supply and electrolyte feeding unit. Fig. 1 shows

A003/076/2003 © With Authors 2003

a schematic of the process. The anodic workpiece is 0.15 % carbon steel having an initial reamed hole of 16 mm and a height of 12 mm. Brass cathodic tools of different diameters, 6, 8, 10, 12 and 14 mm were used, with a central electrolyte flow port of 3 mm. NaCl electrolyte 150 g/l is pumped at 6 l/min in the inter-electrode gap. An initial gap was set constant at 0.75 mm in the radial direction as shown in Fig. 1. A full wave rectified dc-voltage of an average 12.5 volts is applied across the electrodes for a machining time of 90 seconds. For each tool diameter orbiting speeds of 45, 95, 179 and 351 rpm were tested. During machining, the electrolyzing current was recorded. The diameter of holes, before and after machining, was measured using Sigmascope 500. At the end of machining, the workpiece was cleaned, dried and reweighed using a precision balance. Roughness of the machined holes was also measured using Form Talysurf instrument. Performance indices such as the diametral machined allowance, material removal rate and current efficiency were also determined.

3. RESULTS AND DISCUSSIONS

Fig. 2 shows the increase of the experimental removal rate with tool diameter. The larger the tool diameter, the greater will be the cathodic tool area and, moreover, the smaller the inter-electrode gap resistance. Under such circumstances the machining current rises as shown in Fig. 3. Fig. 4 shows a small effect of the orbiting speed on the machining current when using small tool diameter of 6 mm while no effect can be detected with larger tool diameter of 10 mm. Accordingly, no clear effect of the orbiting speed on both the average current and hence the experimental removal rate can be expected as shown in Figs 2, 3

The linear relationship between the average machining current and the removal rate is shown in Fig. 5. The lower experimental values, compared to the corresponding Faraday's ones, are related to the current efficiency. Fig. 6 shows a decrease of the current efficiency with tool diameter up to 10 mm beyond which the current efficiency remains unchanged. The same figure also indicates no clear effect for the orbiting speed in the current efficiency. The low current efficiency associated with larger tool diameters may be related to the increased hydrogen gas evolution that fill the narrow inter-electrode gap and limits the dissolution process.

Fig. 7 illustrates the effect of orbiting speed and tool diameter on the increase of hole diameter. Accordingly, as the tool diameter increases, larger machined allowance is removed from the inner workpiece diameter. The results also indicate very slight decrease of the machined allowance at higher orbiting speeds. The use of small diameter tools minimizes the material removal rate and hence the removed allowance. This has the potential of reducing the sludge generation besides improving the accuracy levels that enhances the industrial applications [13]. As explained earlier, the larger the tool diameter, the greater will be the machining current, material removal rate and consequently the machined allowance. Fig. 8 shows the linear relationship between the average machining current and the diametral removed allowance which has similar trend to that appeared in Fig. 5.

Regarding the effect of tool diameter on the produced surface roughness, It can be seen from Figs. 9 and 10 that the increase in tool diameter produces rougher surfaces in terms of Ra and Rt. However the effect of orbiting speed is not clear. This trend can be related to the evolution of more gas at larger machining currents. Under such circumstances the current distribution will be impaired which results on uneven material dissolution and hence a rougher surface. Fig. 11 shows the increase in the roughness average, Ra with the rise of the machining current. According to the roundness profiles shown in Figs. 12, 13 the increase of orbiting speed for smaller tool diameters reduces the roundness error.

It is believed that using cathodic tools of small diameters enhances the dissolution process in a small arc length facing the workpiece. The rest of tool area is expected to have greater resistance to current flow. Under such circumstances, the degree of localized dissolution is enhanced leading to higher current efficiency, better surface quality, and smaller roundness error.

4. CONCLUSIONS

From the experimental work conducted at the above mentioned conditions, the following can be concluded.

1. The use of orbital ECM for hole finishing is an attractive machining method for producing accurate holes having better surface quality, and smaller roundness error.
2. Small diameter tools that ensures larger orbiting radius, smaller orbiting speeds and, narrow machining gaps are recommended for an efficient machining process.

REFERENCES

1. El-Hofy, H (1998), Characteristics of pulse electrochemical machining, Alexandria Engineering Journal, Vol 35 No 1, 51-59
2. Kozak, J; Rajurkar, K. P. and Ross, R. P (1991), Computer simulation of pulse electrochemical machining, (PECM) J. of Material Processing Technology 28, 149-157.
3. Kozak, J; Rajurkar, K. P and Wei, B (1994), Modeling and analysis of pulse electrochemical machining (PECM), ASME Vol 116, 116-121.
4. Masuzawa, T and Sakai, S. (1978), Quick finishing of WEDM products by ECM using mate electrodes, CIRP Vol. 36, no 1, 123-126
5. Sadollah, Z and El-Hofy, H. (2002), Orbital electrochemical machining of electro-discharge machined surfaces, AMST '02 conference, Udine Italy, 457-464
6. Yeo, S . H and Goh, K. M. (2001), The effect of ultrasound in micro electrodischarge machining on surface roughness, Journal of Engineering Manufacture, Vol. 215, No B2, 571 - 576
7. Ghabrial S. R and Ebied, S. J. (1981), Beneficial effects of air electrolyte mixtures in stationary electrochemical machining, Precision Engineering, 221-223
8. Kozak, J; Dabrowiski, L.; Osman, H. and Rajurkar, K. P. (1991) ,Computer modeling with rotating electrode, Journal of Materials Processing Technology, 28, 159-167
9. Hocheng H. and Pa, P. S. (1999), Electroplating and electro brightening of holes using different feeding electrodes, J. of Materials Processing Technology V 89-90, 440-446.
10. Rajurkar, K. P. (1999), Improvement of electrochemical machining accuracy by using orbital electrode movement, CIRP Annals Vol. 48 no 1 139-142.
11. Al-Salem, N and El-Hofy, H (2002), Computer simulation of orbital electrochemical finishing of holes, to be published.
12. Rajurkar, K. P; Zhu, D and Wei, B(1998), Minimization of machining allowance in electrochemical machining, Annals of CIRP Vol. 47 no 1 165-168.

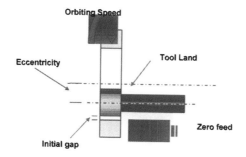

Fig. 1 Orbital ECM schematic

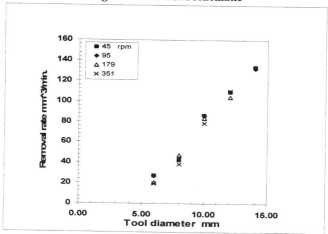

Fig. 2   Effect of tool diameter and orbiting speed on removal rate

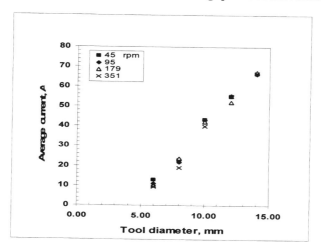

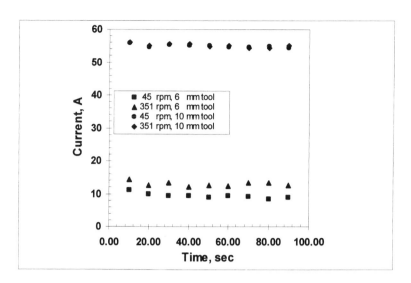

Fig. 4   Effect of tool diameter and orbiting speed on the machining current

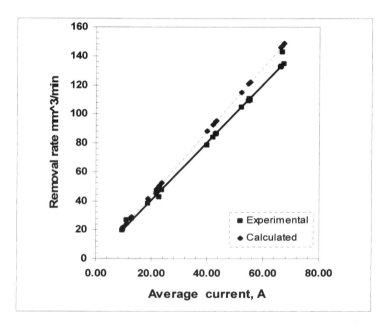

Fig. 5 Relationship between the average current and removal rate

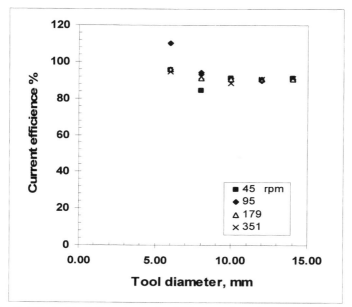

Fig. 6 Effect of tool diameter and orbiting speed on the current efficiency

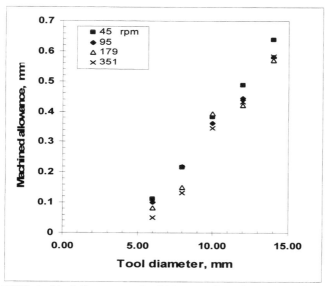

Fig. 7 Effect of tool diameter and orbiting speed on the current efficiency

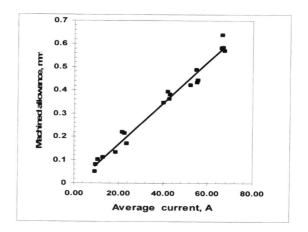

Fig. 8 Relationship between the average current and removal rate

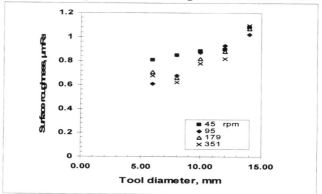

Fig. 9 Effect of tool diameter and orbiting speed on surface roughness, Ra

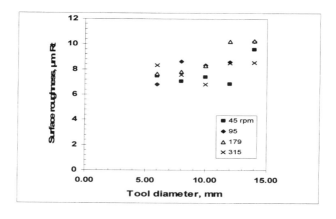

Fig. 10 Effect of tool diameter and orbiting speed on surface roughness, Rt

  A003/076/2003 © With Authors 2003

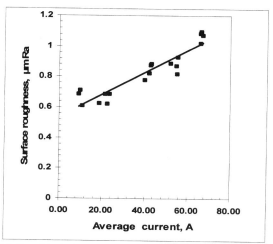

Fig. 11   Relationship between average current and surface roughness Ra

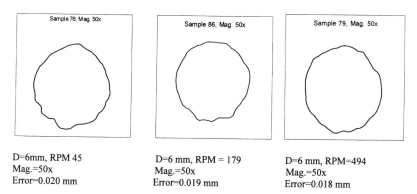

D=6mm, RPM 45
Mag.=50x
Error=0.020 mm

D=6 mm, RPM = 179
Mag.=50x
Error=0.019 mm

D=6 mm, RPM=494
Mag.=50x
Error=0.018 mm

Fig. 12   Effect of orbiting speed on roundness error

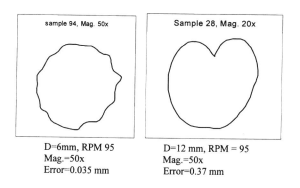

D=6mm, RPM 95
Mag.=50x
Error=0.035 mm

D=12 mm, RPM = 95
Mag.=50x
Error=0.37 mm

Fig. 13   Effect of tool Diameter on roundness error

A comparison of discriminator performance in fault diagnosis

C H WOOD and **D K HARRISON**
ESD, Glasgow Caledonian University, UK
T S SIHRA
UWIC, Cardiff, UK

SYNOPSIS

Acoustic emissions and on-line inspector quality grades were gathered to conduct both a grading of electric hand drill quality and then diagnose the root cause of any fault. Three types of discriminator were implemented; a Naïve Bayesian Classifier (NBC), Conjugate Gradient Descent (CGD) Artificial Neural Network (ANN) and Bayesian ANN. The first experiment graded the drill quality into four classes (Pass, Border Pass, Border Fail and Fail). The second involved a simpler Pass/Fail classification. The ANN methods were found to give better accuracy figures than the NBC in both experimental forms. The Bayesian performed better in the complex four-class problem, with the CGD performing better in the dual-class problem.

1 INTRODUCTION

The PREMIS project was part of a Department of Trade and Industry (DTI) funded initiative to encourage usage of the most modern condition monitoring and analysis techniques with regard to product quality. The project was intended to provide a platform for demonstrating working practices of potential condition monitoring (CM) techniques united with computer-based Artificial Neural Networks (ANNs). Glasgow Caledonian University (GCU), Black and Decker (B&D), Rolls Royce and Diagnostic Instruments were all involved in the project.

This paper describes an extension to work undertaken by McEntee (1) which involved the application of a multi-layer perceptron (MLP) to recognise the final product quality of several types of hand drill. The most appropriate drill for on-line quality monitoring was deemed to be the 'Phase 7 power drill' due to its better engineering characteristics. McEntee achieved 89% classification accuracy of final quality on the Phase 7 drill. That body of work did not extend itself to diagnosing the cause of failed drills; it was limited to pass/fail criteria.

McEntee laid down the foundations and showed how to employ the Acoustic data to discriminate between pass and fail drills, however, this paper extends that work. Both determination of failure and some estimation of the confidence in the predictions are needed. Diagnostics are achievable by better pre-processing of the data and application of an intuitive feature extraction routine. More accurate predictions are available by introducing both more powerful Gradient Descent (GD) algorithms, and the introduction of the novel Bayesian weight update training method. Application of the Bootstrap pseudo-sampling method enables the formation of Confidence Intervals (CIs) in the GD training routine predictions. However, despite the inherent formation of CIs based on the application of Bayesian statistical theory as applied to MLP training, this is only achievable in regression problems and could not be replicated in this classification problem.

This paper aims to illustrate the case for employing either GD or Bayesian theory for weight adaptation, or indeed if the application of ANNs is warranted by comparing both against the performance of a linear discriminator, Naïve Bayesian Classifier (NBC). The techniques outlined here may be applied to other CM applications.

2 CASE STUDY AND DATA MEASUREMENT

Drill quality assessment was conducted by the subjective human inspection of drill noise during a series of run up cycles. The drills were then transported to GCU for independent gathering of drill acoustic emission data with a view to correlating the acoustic emissions with the inspectors' quality assessment. Particular care was shown to correctly position the microphone in the most appropriate orientation and distance, ensuring that the cooling air expelled from the motor did not interfere with the probe. The inspectors' normal assessment is a simple pass/fail criterion, however, to aid this research the inspectors were encouraged to grade the drill quality between 1 and 5 (1 being very good quality and 5 being very poor quality). This process is not part of the standard production line quality procedure.

Acoustic emission data from 1225 drills was sampled using a portable signal analyser employing the Fast Fourier Transform routine. Diagnostically looking at the data revealed the types of failure mode and more significantly, where the peaks occurred in the spectra to aid in identification of the root cause of problems. McEntee simply classified the drills into one of seven separate groups, each representing a quality grade, and did not tackle the cause of failure. The new quality monitoring system will not only assign a quality mark to each drill, it will also attempt to classify each into one of seven possible failure modes:

- Motor Eccentricity (E)
- Motor Misalignment (M)
- Loose parts (motor) (L)
- Gear eccentricity and/or backlash for drive gear (G1A)
- Gear eccentricity and/or backlash for driven gear (G2A)
- Gear Misalignment for drive Gear (G1B)
- Gear Misalignment for driven Gear (G2B)

Fourteen input variables were identified as being highly significant indicators of the differing failure modes. These were identified using a manual visual examination and recording method, this elaborate process was felt necessary for optimum data reduction. The data set was split into one of four possible classes through a combination of inspectors' grade and spectra characteristics, where the latter is the determining basis. The four classes were representative of Pass (369 drills), Border Pass (329 drills), Border Fail (278 drills) and Fail (249) drills. For the dual-class problem, this resulted in 698 Pass and 527 Fail drills.

3 DISCRIMINATION METHODS

There are numerous ANN algorithms to select from and two of them, a common GD MLP employing Conjugate gradient descent and the Bayesian MLP, are compared and contrasted in this paper. However, it is often found in the literature that when proposing an examination of the most suitable pattern detection methods for a given problem, performance of advanced methods, such as ANNs, are usually contrasted against that of simple linear discrimination methods. This type of addition adds credence to the applicability of advanced techniques as the preferred choice of pattern recogniser for the given problem, provided they outperform the linear discriminator. Therefore, the application of Naïve Bayesian Classifiers (NBCs) (House et al (2) were employed for pattern discrimination.

3.1 Naïve Bayesian Classifier
Bayesian classification is founded on Bayes' theorem of conditional probability. The theorem is used to estimate the probability of an example belonging to each of the possible classes of a given classification problem. A Bayesian classifier will assign to a new example, x, the class value c_k that maximises $P(c_k|x)$. A very simple approach to performing this classification is the so-called "naïve" Bayesian classifier (NBC) (Duda & Hart (3)). This is named as such because it is a "naïve" assumption that all the input features, are conditionally independent given the class (i.e. the attributes are independent of each other). The joint, conditional probability is then just the product of the individual probabilities $P(X_i = x_i|y = 1)$ and thus, the probability that X, the input feature, represents class 1 is given by:

$$P(X = x|y = 1) = \prod_i P(X_i = x_i|y = 1)$$

For each separate class this probability is equated, and the class that has the greatest probability is the 'winner'. Dietterich (4) gives an introduction to the NBC, as well as an excellent introduction to the whole topic of machine learning. The specific software package employed here is a programme written by Wray Buntine entitled BAYDA[1]. This application employs a leave-one-out cross validation method, that constructs as many individual networks as there are input examples, taking the average of the set as the complete predictive model.

3.2 GD Conjugate Gradient ANN
Without a doubt the most frequently employed ANN learning method is the Multi-layer Perceptron (MLP) employing the Backpropagation learning routine. However, recent literature would recommend the use of a variant on this type not employing the Backpropagation weight update routine, instead employing the conjugate gradient method to update weights for learning. This change rids the algorithm of steepest descent weight adaptation in favour of the Conjugate Direction adaptation of weights. This is said to improve training time, ability to converge and smaller final errors when compared to the

---

[1] Software is available on-line for research purposes at www.cs.Helsinki.fi/research/cosco

Backpropagation algorithm (Reed & Marks (5)). As such it was employed as the benchmark ANN from which all other methods were compared.

3.3 Bayesian Learning ANN

An original method for training ANNs has been proposed recently, principally advocated by Neal (6) and MacKay (7). MacKay's implementation is a specific implementation known as the Evidence method that employs Gaussian functions to approximate the unknown posterior distribution; Neal's implementation is a generalisation of MacKay's Evidence implementation, which does not limit the distributions the posterior may take on. It is the form that Neal advocates that has been employed in this investigation.

Common MLP ANNs have their weights updated by some form of gradient descent weight adaptation, the aim is to achieve a single best set of weights that best models the relationship between inputs and outputs. The Bayesian form differs by not achieving a single best set; instead a probability distribution over the weights is achieved where sample values are drawn from the parameter distributions to construct a posterior probability distribution for the output vector. Numerous draws are required to construct the whole predictive distribution, thus requiring numerous individual networks. Therefore, the Bayesian method constructs hundreds of potential networks that satisfy the input-output relationship, these networks are combined (weighted by their respective posterior probability) to form a single probability distribution from which a single valued guess (the average) can be sought. This process is called integrating the posterior distribution over the weight space and is a very demanding computation. Neal (6) approximates this integration through the use of Markov Chain Monte Carlo (MCMC) methods. For a fuller account of all these matters refer to the work under taken by Neal (6) and MacKay (7) mentioned earlier. This posterior distribution is used in conjunction with a loss function (usually squared-error loss) to make a single valued guess at future input vectors, in this context these future events are held in the test set. In classification problems, it is not the actual values each output unit holds that matters, but their relative magnitude. Therefore, employing the *softmax* function results in a 'winning' class for each input vector (Marwala & Hunt (8)).

4 EXPERIMENTAL RESULTS

The data set was split into training (50%), validation (25%) and testing (25%) sets, however, it should be noted that the Bayesian method does not require a validation set, therefore, this data was used as training data. Moreover, the NBC implementation employs the leave-one-out cross validation method, therefore, there is no distinction between training/validation/testing; there is merely one complete data set. In addition to performance accuracy figures, the Bootstrap (Dybowski & Roberts (9)) pairs pseudo-sampling method was employed to construct CI bounds for the GD MLP network, which was not replicable in the Bayesian method. The software employed to construct both types of ANN used advanced routines to determine the optimum network architecture for the given problem, thus the architectures are unique. However, a great deal of time was spent determining the optimum priors and tuning parameters for the Bayesian ANN. Results for all three methods are given in Table 1.

A003/072/2003 © IMechE 2003

Table 1 General Classification Performance Comparison (4 class problem)

Failure mode	NBC	GD ANN	Bayesian ANN
E	73.8%	79.4% (95% CI: 75 to 84.1%)	87.7%
M	80.2%	90.9% (95% CI: 89.6 to 92.9%)	93.8%
L	80.2%	93.4% (95% CI: 92.5 to 94.8%)	94.5%
G1A	87.3%	90.7% (95% CI: 89.9 to 91.6%)	91.9%
G1B	86.2%	92.8% (95% CI: 89.9 to 94.5%)	93.5%
G2A	89.6%	92.8% (95% CI: 89.9 to 94.6%)	94.8%
G2B	88.6%	93.8% (95% CI: 92.2 to 94.8%)	95.1%

Detailed examination of the results revealed several interesting points, the most notable being the consistently better performance by the NBC algorithm in classifying the 'Border' classes. However, it was also consistently worse in classifying the extreme classes, Pass and Fail. Further examination highlighted that on small data sets (G1A, G1B, G2A & G2B) the extreme classification figures, for the NBC algorithm, were better, however, on larger data sets, (E, M & L), these figures were relatively poorer. One can assume then that, as the literature states, the NBC algorithm is best suited for small data sets and ANNs for larger data sets.

Disappointingly, the GD algorithm was bettered by both competing methods, and only managed on one occasion to have the greatest class predictive value. This is a poor situation and may be attributed to the introduction of averaging the performance of an ensemble of networks, which is required for the CI construction. Fortunately, the performance of the Bayesian implementation shone through, it is, by far, the best consistent classifier. It performed particularly well in the extreme classes, with the only blight being the G1B network. However, the performance appears to have come at a cost, the discrimination ability for the border classes is at best on a par with the NBC algorithm, but at worse it achieved several zero classifications. Worryingly, this ability was also reflected in the GD ANN algorithm. The problem is most acute for the smaller data sets.

Taking a broader viewpoint, the test set general classification results for the networks, illustrated in Table 1, range from 74% to 95%, where as McEntee (1), in his initial study, managed to achieve a rating of only 89%. It can be clearly seen that the most optimum classifier is the Bayesian ANN, followed by the GD ANN, however, as examination of the individual results illustrates, the general classification rate does not contain the whole picture. Despite this, clearly the Bayesian implementation is the best and given that both ANN methods performed either equally as well or even better than the Linear classifier vindicates the decision to employ ANNs in this context.

In general, however, it appears that all classifiers are having particular difficulty in distinguishing between borderline pass and borderline fail, which is not too great a surprise given that this is the area of maximum ambiguity as to what is a failed drill and one that should pass. In a bid to improve accuracy, a simplified classification problem is envisaged where the extreme classes, revealing a simpler dual class problem, will absorb the difficult border classes. Results are detailed in Table 2.

Table 2 General Classification Performance Comparison (Dual class problem)

Failure mode	NBC	GD ANN	Bayesian ANN
E	88.5%	91.1% (95% CI: 85.4 to 94.8%)	94.5%
M	87.6%	88.9% (95% CI: 69.8 to 96.4%)	97.4%
L	90.1%	89.3% (95% CI: 75.3 to 98.1%)	98.4%
G1A	90.9%	87.2% (95% CI: 74 to 94.5%)	96.1%
G1B	89.3%	81.2% (95% CI: 60.1 to 94.2%)	94.8%
G2A	93.9%	92.8% (95% CI: 86.4 to 96.8%)	98.1%
G2B	93.1%	93.7% (95% CI: 85.1 to 97.7%)	98.1%
Good	78.1%	84.6% (95% CI: 79.9 to 87.7%)	89.6%

If the main objective is to determine the nature of failed drills, then the question, which is the better classifier, produces an interesting answer. The best classifier appears to be the Bayesian method; it has higher general classification figures (Table 2), however the breakdown of these figures paints a different picture. Yes, the classification of the 'Pass', or other, drills is better but it comes at a cost because the GD method in the majority of cases has better Fail figures. In essence, the Bayesian classifier has become one that is able to determine good drills, through the filtering of the seven networks combined, and not, as one had hoped for, to identify specifically the type of failure. Thus, one would have to conclude that the GD ANN is the better classifier. The performance of the NBC classifier is noteworthy, through most failure mode networks it achieves comparable classification to the next best competing method and only on several occasions is it found wanting. Considering the ease with which this classifier was produced, one can only be impressed with its ability. Taking a broader viewpoint, comparison of the classification performance of both the four class and dual classification problems reveals general classification has improved (ranging from 3 to 15%). The grey area between a borderline pass example and a borderline fail example was clouding the overall ability of the classifier in its discrimination process. The improvement in predictive performance may be as a result of the simplification of the problem, however, one must not overlook the impact that this simplification process has had on the quantity of examples within each class. Considering especially the fail class, the redefinition has increased the number of examples and this factor probably lends a lot to these improved figures.

5 CONCLUSIONS

Three types of discriminator were trained using Acoustic emission data acquired from electric hand drills to determine final product quality, and, if applicable, determine the problem source. The initial investigation resolved to separate the drills into seven possible failure modes and four grades of quality within each mode. This was of limited success, but did serve to illustrate the power of the ANNs as a classifier for this problem. It was also apparent that the more powerful algorithms employed here improved on those figures achieved by McEntee. The redefinition of the problem to a simpler dual classification did result in improved figures, which would indicate that a larger data set is required if the four class problem is to be pursued. However, the dual class results have indicated that employing the GD ANN produced the best overall accuracy, and combined with the ease that CI bounds are produced clearly marks it as the preferred choice.

A003/072/2003 © IMechE 2003

REFERENCES

1   **McEntee, S.** (1996): *The Application of Intelligent Software for On-line Product Quality Monitoring in Manufacturing Processes*, PhD. Thesis, Glasgow Caledonian University, pp. 123-148.

2   **House, J. M.; Lee, W. Y.** and **Shin, D. R.** (1999): "Classification techniques for fault detection and diagnosis of an air-handling unit", *ASHRAE Transactions*, vol. 105, pt. 1, pp. 1087-1100.

3   **Duda, R. O.** and **Hart, P. E.** (1973): *Pattern classification and scene analysis*, New York: Wiley.

4   **Dietterich, T. G.** (1997): "Machine learning research: Four current directions", *AI Magazine*, vol. 18, pt. 4, pp. 97-136.

5   **Reed, R. D.** and **Marks, R. J.** (1999): *Neural Smithing: Supervised learning in Feedforward Artificial Neural Networks*, MIT press, Massachusetts, pp 166-169, ISBN 0 262-18190-8.

6   **Neal, R. M.** (1996): *Bayesian Learning for Neural Networks*, New York: Springer-Verlag, ISBN 0-387-94724-8.

7   **Mackay, D. J. C.** (1992): "A practical Bayesian framework for Backpropagation networks", *Neural Computation*, vol. 4, pp. 448-472.

8   **Marwala, T.** and **Hunt, H. E. M**. (2000): "Probabilistic fault identification usingvibration data and neural networks", *Proc. of the International Modal Analysis conference – IMAC*, vol. 1, pp. 674-680.

9   **Dybowski, R.** and **Roberts, S**. (2001): "Confidence intervals and prediction intervals for feed-forward neural networks", in Dybowski, R & Gant, V. (Eds.) *Clinical Applications of Artificial Neural Networks*, Cambridge: Cambridge University Press, pp. 298-326.

Concurrent Engineering and
Design for Manufacture

Controlling part variety through design reuse – retrieving two-dimensional drawings from a sketch

D M LOVE and **J A BARTON**
Integrated Design and Manufacture Research Unit, Aston University, Birmingham, UK

ABSTRACT

Reuse of existing designs offers large savings as the cost of creating a new part design is high and the potential for reuse significant. Conventional retrieval methods have very high operational costs due to the manual nature of the coding process. Automatic coding dramatically reduces the costs of implementation and also provides a means of simplifying the search process. This paper will describe a system in which the designer's CAD sketch of the desired component is automatically coded and matched against all the drawings in the database. Also described is its performance using a database of several thousand parts and the practical issues that arise in the automatic coding of real engineering drawings.

1   INTRODUCTION

Variety reduction is a popular objective and many companies use standards to control the proliferation of common parts such as fasteners etc. The reason for this can be appreciated from significance of the costs that have been estimated as being associated with the creation of a new part number. Creation of a new part number will generate downstream costs arising from administration, stockholding, obsolescence, service and reduction in the economies of scale. Actual figures are difficult to derive reliably but are often quoted as thousands of pounds per part per annum. For example, based on a study at Pitney Bowes in 1969-71, Sharma (1) reported a cost figure of $1882 for each of the company's 2400 new parts introduced each year. Dowlatshahi and Nagaraj (2) assert that part costs range between $1300 and $12000 although no evidence is presented to support these figures. A second issue concerns the amount of scope that exists to make these savings by reusing existing parts. A study by Hyer and Wemmerlov (3) found that 20% of parts could be reused unmodified and 18% needed some modification and a further 12% required substantial change before they could be used.

Another study at Scania reported by Johnson & Broms (4) suggested that design and development costs were directly proportional to the size of the parts range. The same study suggested that distribution and production costs also fall by 30% and 10% respectively with a 50% reduction in the parts range.

Most companies employ a procedure that ensures standardisation for common components, typically items that have a clear function and that are usually sourced externally. The procedures used, normally preferred-item standards books and catalogues, are much more difficult to apply to the control of non-standard parts where function cannot be defined so unambiguously. Re-use of parts that have been designed in house requires that the designer must be given a convenient means of checking whether an existing part can be used. It is important that the search process is significantly easier than the alternative of creating a new part or the process is likely to fall into disuse.

2   DESIGN RETRIEVAL TECHNIQUES

Part numbering systems exist in most companies and they can be used to retrieve information about parts. But since these systems exist primarily to provide unique identity, they offer very limited search facilities. Product data management (PDM) systems do allow searches on other fields, properties or special keywords and most CAD systems allow text searching on the drawing description. The effectiveness of all these techniques depends on rigorous application of naming conventions and none allow the search to be based on the part geometry. Geometry based searching has the considerable advantage that it avoids any consideration of function or textual description - when a designer is looking for a part he will already have a good idea of what it should look like.

Coding and classification systems are the longest established means of retrieving manufactured parts on the basis of their geometrical or other properties. Many systems have been developed for use in manufacturing industries, either for variety control or cell family formation (5). Once a part's shape information has been encoded into a code 'number', similar parts can be found since they will have identical or similar code numbers. However, whilst these systems may find parts very successfully, they have a number of drawbacks. They suffer from significant set-up costs that include the initial purchase, system tailoring, training of personnel and generation of the initial coded parts database. A typical small company database may extend to thousands or ten's of thousands of parts, and since, in our experience coding rates rarely exceed 10 drawings/hour, the latter cost may well exceed all the other set up costs combined. Resource costs have been cited as a major reason for companies abandoning coding and classification systems (3). To retrieve similar parts requires the designer to manually code the desired part. A designer, as a casual user, is unlikely to be able to code at even the rate quoted above and, more significantly, there is a further risk that the inexperience of the designer will lead to miscoding of the search request and a consequent deterioration in retrieval performance. Thus while coding & classification systems can retrieve parts effectively they are expensive to set up and maintain.

Automatic coding of drawings or models would eliminate the resource drawbacks discussed above, since it would drastically reduce the time needed to create the initial database and minimise the time needed to add new parts to the system. People can relatively easily extract

   A003/041/2003 © D M Love & J A Barton 2003

the geometric features from either 2D or 3D representations of a part, but automating this process is not a trivial. The 2D case is much less frequently addressed although it is likely that far more drawings exist in this format than have been produced in (or converted to) more modern 3D systems. There is an interesting example of the use of a neural network to identify features from a bit-mapped 2D drawings so as to populated a small part of the Optiz (6) shape code (7). It is not clear from the paper why Opitz was chosen although the segments of the code that were generated by the system were probably the most straightforward to automate. We could not find any report of subsequent work extending the process to encompass the remainder of the code and thus provide proof of the viability of the approach. There are more examples of systems that produce codes from 3D data sources, for example see (8) & (9). Perhaps this is because 3D systems, especially those based on features, contain much more information about the elements that make up the 3D model.

Compared to 3D, the 2D problem appears to be much more intractable since feature properties have to be inferred or derived rather than simply extracted from a 3D model. However the benefits of a viable 2D system would be significant because many companies have very large databases of legacy drawings in 2D formats. An effective re-use procedure would have to make use of such parts and not simply rely on a much smaller database of more recent 3D designs. In any case, we felt that an approach that worked for 2D could be extended to 3D in due course.

3   DEVELOPMENT OF A PROTOTYPE AUTOMATIC CODING SYSTEM

In order to assess the viability of the approach a prototype system has been built that utilises a geometrical feature extraction programme and a new version of the CAMAC (10) coding system. For the prototype system the range of data to be coded has been limited to geometric entities and properties that could be extracted from 2D drawings. The extraction programme processes the drawing files in AutoCAD's DXF format. This format is readily available in most popular CAD systems although any other open drawing format, e.g. IGES, could have been used. The extracted properties are then passed to the coding module that generates a CAMAC code that is then stored in an electronic catalogue along with the part details and drawing.

The prototype, version 6 of the CAMAC system, is designed around a simple catalogue metaphor. The user is presented with a browser that can be used to scan through the parts in a catalogue. Multiple catalogues can be created to organise collections of parts on any convenient basis that the user wishes to apply. The browser user interface is shown in figure 1. The designer can open a CAD window to load an existing drawing or create a sketch that can be used as the basis of a search. The search results are displayed in the browser window in similarity order with a thumbnail view of each part. A detail view of a particular part can be gained by double clicking the thumbnail. New parts can be added to the catalogue either singly or through a batch process.

There are two aspects to establishing that the approach used in the development of CAMAC v6 is viable and practicable. We have to be able to show that the system is able to automatically process typical engineering drawings, or at least deal with them much faster than conventional manual coding methods. In addition we have to show that the system can retrieve parts quickly and effectively. Ideally the retrieval should be based on a sketch of the desired part rather than a detailed drawing. The next sections will discuss these aspects of the system's performance.

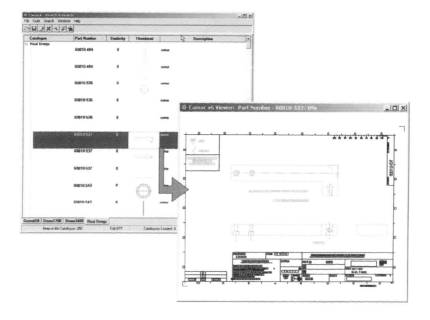

Figure 1 Camac v6 Parts Catalogue Browser

4 RETRIEVAL PERFORMANCE

4.1 Measurement of Retrieval Performance

Measuring the effectiveness of retrieval systems is not straightforward. For example, the definition of relevance (a key metric) is user and context dependent making repeatability and comparison difficult. While many measure of performance have been suggested in the literature, there is no one universally accepted measure (11).

A widely accepted measure of effectiveness is recall, which is the proportion of relevant items retrieved. To calculate recall requires that for each search item, the total number of relevant (i.e. similar) items in the database is known. This requires that a judgement is made of which parts in the database are 'similar' to the test component. Such a judgement is necessarily subjective therefore a group of observers was used to identify all the parts in the database that were similar. Since the database contained over 3400 parts this would be an arduous task. To reduce the work load, the parts to be reviewed were reduced by including only those items that had dimensions within 10% of the target item. This reduced each set to be reviewed to about

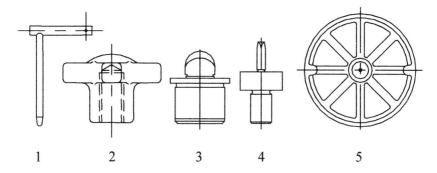

1   2   3   4   5

Figure 2. Search items used for testing

400 items. The rational for this simplification is based on the observation that in mechanical engineering the size of a component will generally be an important factor in determining its suitability. Eight people (a design lecturer, a manufacturing lecturer and six undergraduate and postgraduate students from mechanical engineering and the business school) were used to identify whether an item was geometrically similar to the search item. If more than 50% of the observers thought the part was 'similar' then it was included as part of the target set for the recall calculation. Five parts were randomly chosen for the comparison and are shown in figure 2.

The original CAMAC system had a useful search mode that calculated a similarity index between any two codes, usually the required part and a candidate component found by the system. This capability was used to list parts in decreasing order of similarity to the desired component and this facility is employed in the current version to list the retrieved parts in 'similarity' order.

Recall is dependent upon the size of the retrieved set e.g. a retrieved set of 1000 items is likely to contain a greater proportion of relevant items than a retrieved set of only 10 items. It was assumed that a designer would be prepared to look at up to 4 screens of information which is equivalent to 200 items, assuming each screen could contain 50 thumbnails. Thus if the part appeared in the first 200 returned by CAMAC it was considered a 'hit'.

4.2 CAMAC Retrieval Test Results

These tests measured the ability of CAMAC system to find the parts that were considered by the humans to be similar to the target component.

The recall performance of the five items is shown in figure 3. This gives the percentage recall (calculated after every block of ten items) against the cumulative number of items retrieved. In the legend box the values in parentheses are the total number of items that were considered similar by the humans to the target component. The results show that for all the items, CAMAC retrieved all the similar parts well within the 4 screens of information defined as the target benchmark and that in several cases 100% recall was achieved in a single 'screen' of 50 parts. The worst performance, item 3, is due partially to the small number of similar items to be retrieved (a total of 7), the final increase in recall is due to one item. These results are

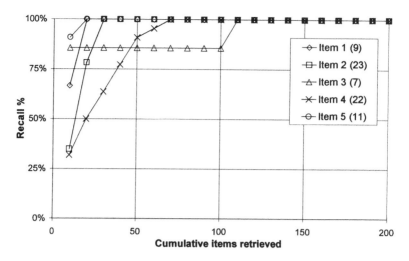

Figure 3 Recall performance of items shown in figure 1

limited by the small sample of test components but it should be noted that the database contained a large number of parts (3400) and that these varied considerably in nature from simple rotational parts to complex assemblies.

4.3 Retrieval using a Sketch

The originals used for these tests were fully detailed views of the parts concerned. Clearly if the system performed poorly in these tests then it could not be expected to work with less detailed sketches. A sketch of the component would save the designer's time but is likely to risk a reduction in the quality of the retrieval performance.
A sketch is likely to:

- omit detail (e.g. radii, fillets)
- have errors of size (e.g. overall length is too short)
- have errors of position (e.g. concentric holes are not concentric)
- have errors of form (e.g. parallel faces are not parallel)
- have errors of good drawing practice (e.g. at a corner lines do not meet)

Therefore, the retrieval system must to be tolerant of deviations if sketches are to be used for searching. We have carried out some preliminary testing using sketches of components known to exist in the database and the initial results are encouraging. Figure 4 shows the detail drawing of item 3 and its sketch form. For this example, 4 similar items were in the first 30 items retrieved. As would be expected this performance is not as good as when searching with the detailed view which retrieved 6 similar items in the first 10 items retrieved. However, we would envisage that, as long as the retrieval time of the system was relatively quick, one of the retrieved items (which would not contain the omission and errors listed above) would be used as the basis of a second, refining, search. Hence by using "two passes" the retrieval performance of a sketched item would be equivalent to that of a detailed representation.

 A003/041/2003 © D M Love & J A Barton 2003

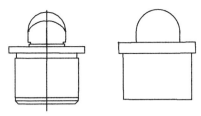

Figure 4 Detail drawing of item 3 and its sketch form

5 AUTOMATIC CODING OF REAL DRAWINGS

Even with the above issues dealt with, there are more mundane aspects that have to be addressed if a practical system is to be developed. The above discussion has generally assumed that the presentation of the item to be coded or searched for is a single view without any extraneous detail. In general, engineering drawings contain additional items beyond the geometric entities that represent the component. Such items include dimensions and tolerances, marks for machining, welding etc and textural notes (with or without leaders). All this is generally contained within a drawing frame or border. Thus, before the component can be coded it has to be identified within the overall drawing. The problem is further complicated in that a drawing will generally contain multiple views, some of which may be sectioned. Hence a method of removing extraneous material and detecting views is required.

CAD systems have a number of features such as layers, line types, blocks that enable attributes to be associated with specific drawing entities. Hence, if standard practices and conventions are adhered to, it offers a possible solution to the processing of complete engineering drawing e.g., the coded component geometry might always be drawn on a predefined named layer. The reality is, however, that many companies do not have such strict practices and hence drawing conventions vary within the same company. Where subcontract draughtsmen and designers are employed the variety is likely to be even greater. A company could obviously adopt standard drawing practices to enable processing of all new drawings, but a means of dealing with poor practice is still required if the legacy of existing drawings is to be processed quickly and cheaply.

Even if all the extraneous material has been removed from the drawing the problem of identifying the separate views in a multi-view drawing still remains. Our initial work suggests that this can be achieved with a simple algorithm providing the view regions do not overlap, which is generally the case in engineering drawing. Where this is not the case a more complex process will be required.

The system employs a 'filtering' procedure that removes extraneous material from the drawing. The user can set the filter mechanism to remove entities of a particular colour or line style, text, named layers and the drawing border. It can also be set to remove or ignore views containing a few entities and thus deal with machining marks and other symbols. The filters can be saved to a file and reloaded as required. During the coding process the system loads a drawing, applies the filter, detects the individual views, codes each view and saves the code and the full drawing into the catalogue.

The prototype system has not been optimised for speed and the rate of processing obviously depends on the complexity of the original drawings. Where drawings that consist of single views (thus requiring no filtering or view detection processes) coding rates can approach 1000 drawings/minute. With real, multi-view, engineering drawing the processing rate is much lower but even with coding rates one hundredth of this (about 6 seconds per drawing), the time to code a drawing is considerably less than by manual means. It should also be noted that processing is a batch process that can be run unattended. The difference in resources required can be gained from a simple example. Assuming a database of 30,000 drawings that have to be processed, not unusual for small to medium size companies, an automated system could do this in two calendar days (assuming only one machine is used). This compares to one and a half years using a single person manually coding 8 hours per day, 5 days per week.

6 FUTURE DEVELOPMENTS

The prototype version uses a CAD window in which the user can load an existing drawing or create a sketch in the prototype system CAD window. This arrangement requires that the user is familiar with the drawing protocols of the CAMAC CAD facility. However if the user's own package could be used to provide the CAD interface to CAMAC it would reduce the designer's learning curve and improve the system's ease of use. A version of the system is planned that will offer tight integration with a number of CAD systems.

More extensive testing of CAMAC's retrieval performance is underway. In particular the impact that sketching variation has on the efficacy of the system is being explored and also the viability of using originals traced from scanned paper drawings to create catalogues of old drawings or as seed parts for searches of conventional databases.

7 REFERENCES

(1) **Sharma, S. C.,** A critical study of the classification and coding systems in manufacturing companies, *MSc Thesis,* 1978. Lehigh University, U.S.

(2) **Dowlatshahi, S. &. N. M.,** 1998. Application of Group Technology for design data management, *Computers & Industrial Engineering*, vol. 34, pp. 235 255.

(3) **Hyer, N.L. & Wemmerlov, U.,** 1989. ., Group Technology in the U.S. manufacturing industry: a survey of current practices, *International Journal of Production Research*, vol. 27, pp. 1287 1304.

(4) **Johnson & Broms.** 2000, *Profit Beyond Measure*, Nicholas Brealey .

(5) **Gallagher, C.C.a.K.W.A.** 1973, *Group Technology*, Butterworth.

(6) **Opitz, H.** 1970, *A Classification System to Describe Workpieces*, Oxford: Pergamon Press.

(7) **Kaperthi, S. &. S. N. C.**, 1991, "A neural network system for shape-based classification and coding of rotational parts ", *International Journal of Production Research*, vol. 29, no. 9, pp. 1771-84.

(8) **Nadir, Y. C. M. M. C.**, 1993, "PROCODE - Automated coding system in group technology for rotational parts", *Computers in Industry*, vol. no. 23, pp. 39-47.

(9) **Ames, A. L.**, 1991, "Production ready feature recognition based automatic group technology part coding", *Proceedings. Symposium on Solid Modeling Foundations and CAD/CAM Applications,* New York, NY, USA, pp. 161-9.

(10) **Holmes, N. E. L. D. M.**, 1992, "The Role of Coding and Classification in a User-Oriented View of CIM", *Proc. 8th Int. Conference on Computer-Aided Production Engineering,* Edinburgh, UK, pp. pp. 137-141.

(11) **van Rijsbergen, C.J.**, 1997, *Information Retrieval*, 2nd edition London: Butterworths,

Application of integrated CAD/CAE systems in the process of hip joint reconstruction

M PAWLIKOWSKI, A M DĄBROWSKA-TKACZYK, K SKALSKI, and G WRÓBLEWSKI
Institute of Mechanics and Design, Warsaw University of Technology, Warszawa, Poland

SYNOPSIS

The hip joint is the most loaded joint in the human body. Therefore, it is most vulnerable to fractures (e.g. femur neck fractures), pathological changes (e.g. excessive bone tissue apposition on femur head) and diseases, such as arthritis, osteoporosis etc (1). Since the hip joint is also very important in human locomotion it is crucial that the joint works properly. In case of a severe trauma of hip joint, that cannot be treated pharmacologically, it is most often treated by implanting metal prosthesis in medullary canal or reconstructing the destroyed acetabulum (2, 3) – anatomical bearing of pelvis by a reinforcement cage. In spite of the fact that there are many types of hip joint prostheses in some cases only prostheses customised to a particular patient (custom-made prostheses) can be utilised.

1. INTRODUCTION

In this paper the process of hip joint reconstruction is presented. The reconstruction process consists of the two following steps: 1) modelling of the hip joint geometry (i.e. femur and pelvis) in CAD system and 2) simulation of the joint performed in the finite element method system. In the discrete model of the hip joint all the components of the joint are taken into account, i.e. the femur and the part of the pelvis with the acetabulum. Bone tissue was defined as a visco-elastic material (4, 5). Moreover, the remodeling properties of bone were also simulated (6). This approach is a novel one in the numerical calculations of bone-implant systems. The results of the numerical simulation of the stress and strain fields in the reconstructed pelvis and femur can give information about the conditions in which the hip joint works. Such analysis will surely be very helpful in the hip joint custom-made prosthesis design. It must be strongly emphasised that hip joint reconstruction process can be fully completed only by integrating the two systems: CAD and CAE.

2. METHOD OF ANALYSIS

2.1 Clinical case

A 34 years old female has been complaining about an acute pain of the left hip, which made it very hard to walk. The first pain occurred when she was 11 years old after a few-kilometer walk. At the age of 22, after giving a birth to a child, the pain increased. In recent years the suffering has been increasing and nowadays it makes her life significantly tougher.

The lower left limb shortened (relative shortening of 6 cm), positioned in a 10° adduction, 12° flexion and 22° external rotation. Movements in the left hip joint were very painful and significantly limited. Whilst walking the patient limped and tilted the upper body to the left. The CT images (Fig. 1) showed dysplasia of the left hip joint with the femur head frontal dislocation. The situation was complicated by necrosis of the femur head. The femur neck was displaced upwards and strongly rotated (anteroversion angle 60°). In addition to this significant rotation of the proximal part of the femur was also noted. The greater trochanter was close to the posterior edge of the hip acetabulum. The acetabulum was shallow and steep. There was also a big overload cyst near the upper edge of the acetabulum. The interstice of the hip joint significantly decreased. The joint surfaces were rough. There was also sclerosis of the under-cartilage layer.

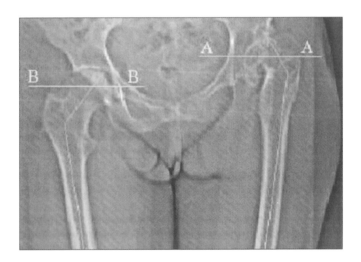

Fig. 1. The clinical case.

Having in mind the severe deformation of the proximal part of the femur and acetabulum and young age of the patient it has been decided to design and make an individual cementless femoral component of the hip joint prosthesis.

First the solid models of the two major bones that constitute the hip joint were generated in a CAD system on the basis of the CT data. Having created the models the prosthesis was then

 A003/066/2003 © With Authors 2003

designed. The designed stem was verified virtually whether or not it fits the medullary canal of the femur and if it is possible to fix it in the bone without causing any microdamages of the femur. If the stem design was not correct it was redesigned and again virtually verified. The model of the whole bone-implant system is shown in Fig. 2. Note that the pelvis was also taken into consideration.

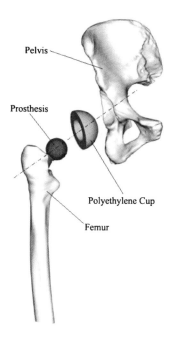

Fig. 2. Solid model of the implanted hip joint.

2.2 Finite element analysis of femur-implant system

The design process of the optimal prosthesis design was not, however, complete. The prostheses had to be checked to discern to what extent they would evoke undesirable bone remodeling (resorption) after implantation. This is a very important aspect, as bone resorption may lead to implant loosening and consequently to revision surgery, which is even more difficult than an initial operation. It has to be emphasized that the extent of bone resorption depends mainly on stem geometry, method of fixation and the material that the prosthesis is made of.

In order to predict the bone activity after implantation we performed simulations in a finite element method system – ADINA. It was crucial here to properly define the loads acting on the joint that would correspond to the weight of the patient. The discrete models of the prostheses and femur were generated on the basis of the geometrical models of these components (Fig. 3). The discrete models of the bone-implant systems consisted of about 7000 elements. The load conditions of the bone-implant systems corresponded to one-leg standing position. The reaction force acting on the prosthesis head R and the force of abductor muscles M were considered. The vectors and values of these forces are shown in Fig. 3. Perfect bonding between the prostheses and femur was assumed.

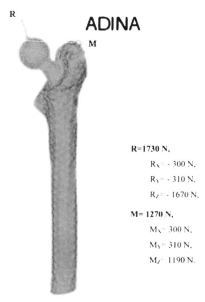

R

ADINA

M

R=1730 N.
$R_X = -300$ N.
$R_Y = -310$ N.
$R_Z = -1670$ N.

M= 1270 N.
$M_X = 300$ N.
$M_Y = 310$ N.
$M_Z = 1190$ N.

Fig. 3. Discrete model of the femur-implant system subjected to the one-leg-stance load.

In our FEM calculations of the bone-implant systems, the bone remodeling phenomenon was taken into consideration. Moreover, since it has been confirmed that bone tissue has rheological properties (7) we decided to simulate the femur as a visco-elastic material to make the results of our simulations more realistic (8). In addition to this the bone remodeling phenomenon was taken into consideration.

Mathematically speaking, the bone remodeling phenomenon may be described by one of the known kinetics equations of bone density change. These are formulated such that bone density rate is considered, thus, the whole history of the deformation process is taken into account. We simulated only internal remodeling since, as mentioned above, this adaptive activity of bone is mainly responsible for implant loosening. The mathematical model that we adopted assumes that the bone density rate is, in general, a linear function of strain energy density U. Thus, the kinetics equation that we utilised had the following form (9, 10):

$$\frac{d\rho_a}{dt} = \begin{cases} C[S - (1-s)S_0], & S \le (1-s)S_0 & \text{bone resorption} \\ 0, & (1-s)S_0 \langle S \langle (1+s)S_0 & \text{dead zone} \\ C[S - (1+s)S_0], & S \ge (1+s)S_0 & \text{bone apposition} \end{cases} \qquad (1)$$

A003/066/2003 © With Authors 2003

where: $S = \dfrac{U}{\rho_a}$, $S_0 = \left(\dfrac{U}{\rho_a}\right)_0$.

The realisation of the whole analysis consisted in integrating the kinetics equation by utilising the Runge-Kutta method and defining Young modulus E by means of the bone density, calculated in a particular time step, using the following equation:

$$E(t) = a\left(\rho_a(t)\right)^b,$$

(2)

where a, b – constants.

The visco-elastic moduli, i.e. shear modulus G, and bulk modulus K, were then determined by using the well known equations: $G = \dfrac{E}{2\cdot(1+v)}$, $K = \dfrac{E}{3\cdot(1-2v)}$, where v - Poisson's ratio. In this way the visco-elastic properties have been related to the change of bone density (11). Then for these particular material properties the stress and strain distributions were calculated.

2.3 Finite element analysis of pelvis-cup system

As the interaction between the femur and pelvis might play a significant role in the functioning analysis of the implanted hip joint we included also the pelvis in the numerical analysis of the stress, strain fields in the joint. For further calculations, solid 3D complicated model of pelvis was replaced by the shell model (the left half of the pelvis), which consisted of 2345 four-node elements, with different thickness (3 – 22) mm, for the group of elements mapping geometry of anatomical shape of pelvis bone. Basing on previous considerations (12) the numerical model was prepared and calculated applying PATRAN-NASTRAN system. The geometrical model is oriented in 3D XYZ Cartesian system, which is fixed in pubic symphysis. The axis of this system defines the planes relatively: sagittal plane XZ, frontal plane YZ and transverse plane XY.

The material properties (Young modulus) of bone, in this model, depend on the assumed thickness of cortical and trabecular bone layers for considered groups of elements. For each of the group with the same thickness of bone layers was calculated reduced Young's modulus, hence the values of Young's modulus of this model altering between 12696 – 16381 MPa, assuming 17GPa for cortical bone and 300 MPa for trabecular bone. Isotropic, linear elastic, constitutive model of bone tissues was assumed.

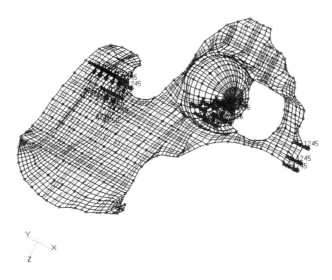

Fig. 4. Numerical FEM model of the pelvis.

Displacement boundary conditions were established in sacroiliac joint, pubic sympasis and acetabulum (12).

Force boundary conditions are determined by gravitational forces of the upper part of the body and muscle forces, which act on the surfaces of pelvis (13). The model of loading in considered case, were defined for the stance phase of locomotion cycle.

Calculations of the numerical model of pelvis bone for defined models of loading, allowed to obtain reaction forces in acetabulum, which are applied in calculations of the femur – implant system.

3. RESULTS AND DISCUSSION

The main aim of the finite element analysis of the stress and strain fields in the visco-elastic bone and elastic prosthesis was to observe the rheological effects of bone after prosthesis insertion under the applied load. The results showed that in the early stage of the load application the creep process dominates in the bone, especially in the trabecular tissue. However, the strain increase is not significant in this bone area. The low strain increase is a result of stress redistribution in the whole implant-bone system yielding decrease of effective stress in the tissue. In the following time period a part of the stresses is taken over by the cortical bone where we observed a certain increase of about 5-15%. The distributions of stress displacement and strain in the implant-bone system at the final time of calculations are presented in Fig. 5.

    A003/066/2003

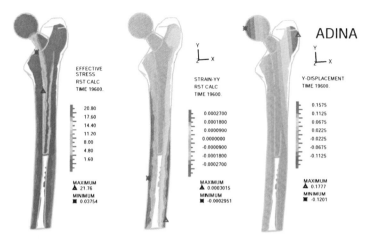

Fig.5. Distributions of effective stress (left), strain (middle) and displacement in Y-diraction (right) in the femur-implant system.

Another aim of the finite element analysis performed on the implant-bone system was to preoperatively predict the bone density distribution after implantation. In this analysis the bone was also simulated as a visco-elastic material. The initial value of bone density was 1.5 g/cm³ and was homogenous in the whole bone volume. The material properties of the bone in the first time step depended on the initial value of bone density; in the next time steps the properties changed with bone density change. The initial Youngs modulus corresponded to the initial bone density and was equal to 14000 MPa. Poisson's ratio was assumed to be equal $v = 0.4$ and was constant during the whole process of deformation. The prosthesis was defined as an elastic material of the following properties: Young modulus $E = 200000$ MPa, Poisson's ratio $v = 0.3$. The constants a and b in equation (2) where assumed as 4249 and 3, respectively. Thus, Eq. (2) takes the form $E = 4249\rho^3$. If density is expressed in [g/cm³] Young modulus is calculated in [MPa].

In Fig. 6 there are showed the bone density distribution after first five time steps (Fig. 6a) and that at the end of the calculation time. One can notice both areas of bone density increase and bone density decrease. Generally speaking, one can state that the implant insertion will cause bone resorption in the proximal part of the femur, which consequently will result in aseptic loosening of the implant. When an implant loosens in the femur the patient feels thigh pain. Such a situation should be avoided so the stem of the prosthesis must be corrected.

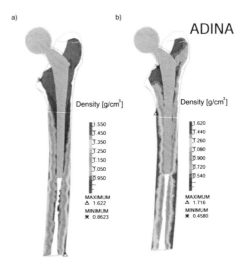

Fig. 6. Distribution of bone density in the femur after first five time steps (a) and at the end of calculation time (b).

In Fig. 7 stress distribution for the stance phase of locomotion was presented. Stiffness support in acetabulum, assumed in the numerical model influenced significantly maximal stress values. One can also observe a certain effect of the applied boundary conditions upon the stress distribution. This will be considered closer in the future investigations.

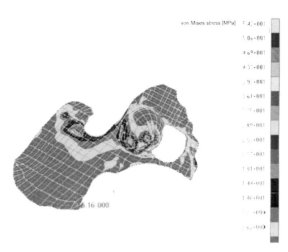

Fig. 7. Stress distribution in the pelvis.

A003/066/2003

It is also planned to develop the whole model of the implant-bone system in such a way that all the components of the hip joint will be modelled as solids and assumed stiffness support in displacement boundary conditions will be replaced by more realistic model of contact. This way more advanced analysis of the joint could be performed.

4. REFERENCES

(1)   Wall A., *Clinical Aspects of total hip arthroplasty*, Acta of Bioengineering and Biomechanics, Vol. 4, Suppl. 1, 39-46, 2002;
(2)   Bauer T.W., Schils J., *The pathology of total joint arthroplasty. II. Mechanisms of implant failure*, Skeletal Radiol., 28, 483-497, 1999;
(3)   Kim Y.Y., Kim B.J., Ko H.S., Sung Y.B., Kim S.K., Shim J.C., *Total hip reconstruction in the anatomically distorted hip. Cemented versus hybrid total hip arthroplasty*, Arch. Orthop. Trauma Surg., 117, 8–14, 1998;
(4)   Lakes R.S, Katz J.L., Sternstein S.S., *Viscoelastic properties of wet cortical bone – I., Torsional and biaxial studies*, J Biomech, **12**, 657-678, 1979;
(5)   Lakes R.S., Katz J.L., *Viscoelastic properties of wet cortical bone – II., Relaxation mechanisms*, J Biomech, **12**, 679-687, 1979;
(6)   Pawlikowski M., Skalski K., Bossak M., Piszczatowski S., *Rheological effects and bone remodelling phenomenon in the hip joint implantation*, First MIT Conference on Computational Fluid and Solid Mechanics, June 12-14 2001, Published by Elsevier Science Ltd., 399-402, 2001;
(7)   Sasaki N., Enyo A., *Viscoelastic properties of bone as a function of water content*, J Biomech, **28**, 809-815, 1995;
(8)   Piszczatowski S., Pawlikowski M., Skalski K., *Visco-elasticity and bone adaptation effects in articular joints*, International Society of Biomechanics XVIIIth Congress, July 8-13, 2001, Zurich, Switzerland, Book of Abstracts, 341, 2001;
(9)   Cowin S.C., Arramon Y.P., Luo G.M., Sadegh A.M., *Chaos in the discrete-time algorithm for bone-density remodeling rate equations*, J. Biomechanics, **26**, No. 9, 1993;
(10) Weinans H., Huiskes R., Grootenboer H.J., *The Behaviour of Adaptive Bone-Remodeling Simulation Models*, J. Biomechanics, **25**, 1425-1441, 1992;
(11) Pawlikowski M., Skalski K., Haraburda M., *Process of hip joint prosthesis design including bone remodeling phenomenon*, Computers & Structures, in press;
(12) Dąbrowska – Tkaczyk A. M., Grajek K., John A. " *Stress and strain analysis in the shell numerical model of pelvis bone"*, Proc. of Conference on Biomechanics – Modelling, Computational Methods, Experiments and Biomedical Applications, Łódź, 75-83, 1998;
(13) Dąbrowska – Tkaczyk A. M., *Analiza zmian stanu naprężenia i przemieszczeń w kości miednicy człowieka pod wpływem różnych modeli fizjologicznych obciążeń*, Mat. and Mech. Engineering, Proceedings of the Scientific Conference, Gliwice , 2000, 97-104.

On concurrent engineering and design of an intervertebral disc of lumbar spine

M DIETRICH, K KĘDZIOR, K SKALSKI, T ZAGREJEK, G KRZESIŃSKI, J SKOWORODKO, P BORKOWSKI, and P WYMYSŁOWSKI
Warsaw University of Technology, Warszawa, Poland

SYNOPSIS

In orthopaedic engineering a modern design process should comprise the issues covering not only geometrical study of a design (e. g. an implant), but also anatomical aspects of the bone tissue. Recently, such an approach has also been applied to geometrical and material modelling of an intervertebral disc of the spinal column, where the concurrent engineering (by means of strain – stress state analysis) plays very important role.

The paper aims at presenting the methodology of design of intervertebral disc (i. e. the nucleus pulpous surrounded by the annulus fibrosus). The process comprises of the following stages:

- tomographic (CT) projection of the lumbar part of spinal column,
- reconstruction of CT images and geometrical modelling of L4 – L5 of the spinal column with decease – deformed intervertebral disc,
- parametric design of an implant (artificial disc) and engineering analysis of the reconstructed region of spinal column.

Finally, the numerical results of strain – stress state simulations are taken into account in the process of the artificial disc design (i. e. metallic endplates and sliding core made of polyethylene) verification performed from both the geometrical and the material points of view.

It will be shown that the concurrent CAD/CAE data processing (based on CT) seems to create an effective tool in the design process of anatomical intervertebral disc.

A003/063/2003 © With Authors 2003

1. INTRODUCTION – VERTEBRAL JOINT AND ITS ILLNESSES

The spinal column is usually made up of thirty four vertebrae. The cervical section contains seven, the thoracial section twelve and the lumbar section – five vertebrae. The sacrum bone is made up of five joined vertebrae, whereas in the coccyx section of the spine consist of residual vertebra. Vertebrae from every section differ considerably one from another.

In humans, the transfer and intensities of static and dynamic loads on joints range considerably. Because of its segmented characteristic and the joint system, the spine burdens the load with the help of the muscles. The most frequent illnesses of the spine are kifosis, lordosis, spondylosis and scoliosis [5]. The formation of these illnesses are sometimes caused by inbred defects of the spine and are aggravated by the cumulatively increasing loads which can cause pathological change. Protection of the spine is essential as it guards the central nervous system.

During this work, questions arose regarding the vertebral joints L4, L5 as well as L5, S1. The structure of the vertebral joint is made up of two neighbouring vertebrae and an intervertebral disc. The vertebra is made up of two main parts: the frontal vertebra body and the rear of the vertebra, which is known as the lamina. From the lamina sprout seven small ribs, a pair of them pointing downwards, another pair upwards and the last pair to the sides, the last singular rib points outwards perpendicular to the face of the nucleus pulpous (Fig. 1).

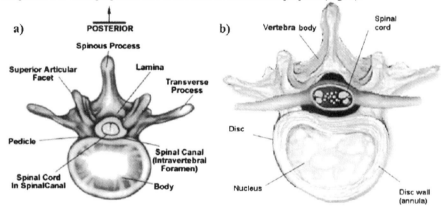

Figure 1. Vertebra (a) and intervertebral disc (b)

The body of the vertebra has approximately a cylinder-like shape. Front and side surfaces of the body are somewhat concave in the perpendicular direction and convex in the horizontal one. The rear surface of the body which limits the vertebral opening is almost flat. The upper and lower surfaces are porous and quite smooth and merge with the adjacent intervertebral cartilage.

Spongy tissue is the predominant tissue in the body of the vertebra. This tissue is covered by a thin layer of tough substance, reinforced by periosteum and oblong ligaments that follow the spine.

The intervertebral disc is a very important element of the vertebral joint for the latter to function properly. The intervertebral disc links itself closely with the surface of the vertebra body with thin layer. The discs are shaped like the surfaces of the neighbouring vertebra and are of the same size as them (as not protrude). The height of the disc is between four and twelve millimetres but also depends on the position of the vertebrae. The external part of the

A003/063/2003 © With Authors 2003

disc is tough, concentrically positioned and is called the fibrous annulus, whereas the inside of the disc is soft and flexible and forms the nucleus pulpose.

Because of the elasticity of the internal part of the disc, it absorbs any shocks that the spine might undergo. The fibrous annulus greatly inhibits the movement of the vertebrae, however the core forms a type of watery cushion, on which the vertebra stem rests during instabilities. During bending of the spinal column, the core moves in the opposite direction to the direction of bending.

The abnormal functioning of the spine may be caused by several factors. One of the most frequent reasons for this is due to traumatic damage (this then causes changes in the growths and pedicles and may even cause damage to the vertebra body), the expulsion or displacement of the intervertebral disc (the nucleus pulpous causes pressure on the spinal cord as well as on the nerve roots), vertebral dislocation (characterised by one vertebra shifting in relation to its neighbour).

Damage to the spine usually occurs in vertebrae joint L4-L5 as well as L5-S1, because this is the area where the spine undergoes the most activity and carries the largest loads..

2. CONSTRUCTION SOLUTIONS OF THE IMPLANTS IN THE LUMBAR PART OF THE SPINE

Increasing levels of pain or pathological changes in the vertebral region will usually result in surgical intervention. Due to this, in recent years, the strong development has been observed in implanting man-made intervertebral discs.

One of the earliest and typically used solutions in stabilising the spine by surgery are shown in Fig. 2. The implant - stabiliser is based on the design according to DERO [3], it is a perforated support socket that simulates the vertebra and vertebral disc (Fig 2a). This solution, as can be observed, decreases the ability for the spine to move because of the stiffening of the joint segment.

Another known solution is a cylindrical cage made of a TiNi alloy that does not deform and has unique holds that stabilise the cage in its location among the vertebra (Fig. 2b) [1].

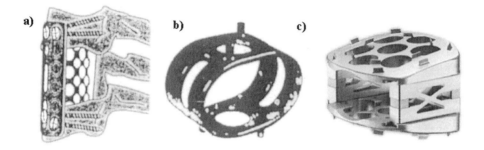

Figure 2. Implant DERO (a), cylindrical cage from TiNi alloy (b), system InFix (c)

A new construction solution has been proposed by the company Spinal Concepts Inc – Implant InFix (Fig 2c) [4]. This is an alloy construction which has the possibility of correcting the angle of arrangement of vertebrae in the implanted area. With a correct design, the load is properly distributed.

The construction is the product made by the company Link (1982), known as Charite [2, 7], the implant is made up of two metal plates with an element in between, the latter is made of plastic with the desired dimensions (Fig. 3). The fixations of these plates of the implant to the neighbouring vertebrae is done by the aid of holds on the surface of contact. In 1985 and 1987, a new and modern design was developed (Charite II and III). Presently plate sets are made and designed by PRODISC- AESCULAP [9]. These implants consist of metallic plates and plastic discs which mimic the function of the nucleus pulpous. The fixation of the plates to the neighbouring vertebra, is executed by using what is known as fixing combs. The first generation, dating back to 1990, was attached by using two of them on each plates (Fig.4a), whereas the new generation (1999) of implants only use one of these hooks (Fig.4b,c.

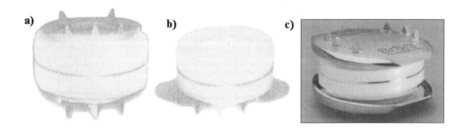

Figure 3. Artificial disc Charite I (a), Charite II (b), Charite III (c)

In this solution, the upper plate has a specific shape that allows it to rotate. The bottom plate possesses a lock that enables a secure and reliable fastening (Fig 4b). Specially prepared instruments allowed the introduction of sliding core, after the fixation of the two metal plates.

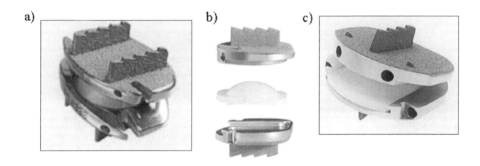

Figure 4. Set of plate PRODISC-AESCULAP: 1st generation (1990) (a), 2nd generation (1999) (b), Complete plate system (c)

          A003/063/2003 © With Authors 2003

3. GEOMETRICAL MODELLING OF VERTEBRAE AND PROCESS OF INTERVERTEBRAL DISCS DESIGN

In order to properly design the implant for an intervertebral disc, it is vital to reproduce the neighbouring geometry of the vertebrae, which is a characteristic of the surface topology. Vertebra of the spine are geometrically complicated objects [8, 10] and reproducing this geometry is a difficult process and requires a suitable approach. Creating a model is a multi-stage process as will be discussed.

The first stage is to obtain a visualisation of the internal tissues using a non-intrusive method which can be performed by the use of a CT computer scans. The gathering of the data was realised on the tomographical spiral scanner in the Central Clinical Hospital of Ministry of Interior and Administration in Warsaw. This yielded a two dimensional cross-sectional images of the spine, of the studied patient. Using the application of the specialist processing software of data, it was possible to reconstruct and segment areas of the images (producing area of equal properties for tissue and thickness). The next step of the processing of data, was the detection of edges of selected area (Fig. 5). This consists of recognising outlines, contours are attributed with appropriate two dimensional coordinate values. This data, for every scanned layer, are then transferred to a CAD system in a suitable format, then produces a cloud of points in a 3D co-ordinate system. Data in this form allows for continuous surface models to be produced.

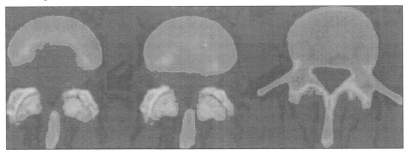

Figure 5. Images of vertebra obtained by using the contours which are an effect of reconstruction and segmentation

In Fig. 6, an example of vertebra L5 is illustrated. It should be mentioned, that the tomographic images used, belong to a person that does not suffer from joints illness.

In order to avoid repeating the ardous process of reproducing the contour of the vertebra and intervertebral disc, tests were carried out in order to develop methods of designing a parametric vertebral joint. The parametric model makes it possible to generate the vertebral joints in a faster and less-work-demanding manner.

The aim of this is, to anticipate the necessary methods used to describe the main elements of the vertebral joint layout. The assumed general co-ordinate system was (x, y, z), with one of the axis (z), being the result of the saggital plane crossing the main plane. The origine of the co-ordinate system is situated on the horizontal plane, at the point of contact with the lower part of the vertebral body, i.e. L5. Following this idea, the vertebral body was divided into n-segments (n=8), which determine the geometrical features. For every segment, a local co-ordinate system was introduced i.e. $(x_i, y_i, z_i$, for $i=$ 1, 2, ...8) (Fig. 7). Every segment is defined in its own co-ordinate system due to the characteristic set of parameters i.e. points, curves and surface paths. In Fig. 8a. the example of parametrization of the vertebral body is presented. Here parametrical defining is done using point P_1, ..., P_k, (k=12) as well as the segments of curves and solid that change curvature r_1, ..., r_m, (m=12) dotted on these parts. 3D model of the vertebral body is obtained in effect by joining of several algorithmic segments. The contour of the segment are then used to segregate the surface model from the vertebral body.

a)   b)

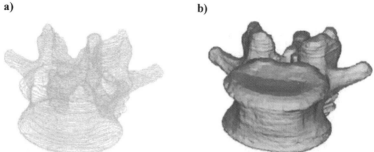

Figure 6. Reconstruction of vertebra L5: a) cloud of points, b) surfaces model

Similar methods can be applied for the process of parametric modelling of the intervertebral disc. Results of such a project are shown in Fig. 8b.

a)   b)

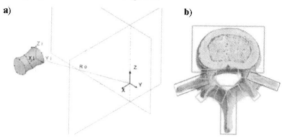

Figure 7. Coordinate systems (a), vertebra divided into segments (b)

A003/063/2003 © With Authors 2003

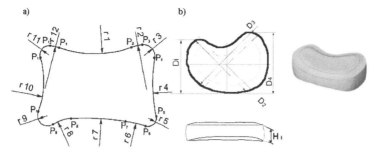

a)   b)

Figure 8. Parametrical characteristic of a section of the vertebra body (a), parametrical characteristic of a disc (b)

4. FINITE ELEMENTS ANALYSIS OF THE ARTIFICIAL DISC- SPINE SEGMENT INTERACTION

The new design of an artificial disk should be examined by performing the stress and strain analysis using finite element method (FEM).

Creating the accurate FE model is usually a laborious and time-consuming task. This may be overcome by parametric FEM modelling, recently often used - with different approaches to the problem, by engineers [11]. After a modification of the parameters describing e.g. geometrical and material properties of the considered object one can obtain a new model in a relatively short time. In a parametric model many different potential shapes should be taken into account with a limited number of parameters. Despite many advantages, the parametric model reveals also some drawbacks. The creation of such FE model is more difficult than in the case of the non-parametric one. Therefore it is reasonable to do this only in the case when the model will be used many times. In this study an attempt was made at creating of the FE model of a lumbar spine segment and then more detailed model of two vertebrae with an artificial disc. Both the vertebrae and the intervertebral disc are described parametrically. Such model may be helpful in the cases when we deal with the problem of the new design, especially custom-made type.

In the model the co-ordinates of characteristic points are assumed to be parameters representing the geometrical shape of the bones. A special attention was focused on accurate representation of a vertebral body. From cloud of points, 70 points have been used in the description of its shape. A relatively high number of the points allows the accurate representation of curvatures of upper and bottom vertebrae surfaces, which is important since those surfaces may contact with an artificial intervertebral disk. The shapes of other vertebral parts were described in a simplified way.

The points were used to create 3D spline curves over which the surfaces defining the solids were spanned. Those solids were divided into 3D solid finite elements, which represent the region of the spongy bone. External vertebral surfaces were divided into shell finite elements, which represent the cortical bone. The shell thickness of particular shell elements corresponding to the cortical shell may differ depending on the position and serve also as parameters of the numerical model. The division into finite elements is performed automatically with mesh density required by the user. In the similar way the FE model of an artificial disk is performed. The main dimensions and material properties of the disk are considered as parameters. An example of such FEM model is presented in Fig. 9.

The main goal of the analyses is the estimation of stresses within the bones and the implant. The design of the intervertebral disc should secure the relatively uniform stress distribution within the bone tissue to enable the proper bone reconstruction processes. The stresses within the implant have to remain below the assumed limits.

The examples of the results of stress analysis are presented in Fig. 10, Fig. 11 and Fig. 12. It seems that the preliminary results prove the assumed way of analysis to be effective. The important information concerning the quality of the new design may be obtained comparing the behaviour of the analysed system with a natural disc and with the artificial one. In the presented example the artificial disc leads to higher stresses within the bone (Fig. 10, Fig. 12) but the stress distribution is similar to that from 'natural case'. Much greater differences are revealed when comparing the stiffness of both considered cases (Fig. 13).

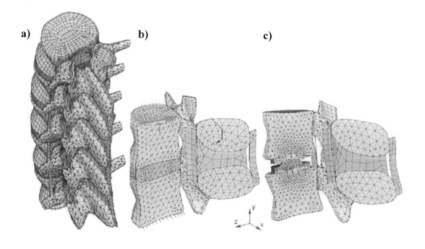

Figure 9. FE model of a spine segment (a) and the cross-sections for the cases of the natural (b) and artificial discs (c)

A003/063/2003 © With Authors 2003

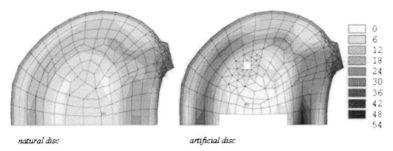

natural disc   artificial disc

Figure 10. Von Mises equivalent stress distribution on the top surface of the lower vertebra (MPa) for the compression load of 3000N.

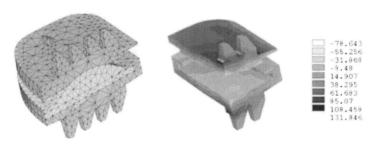

Figure 11. FE mesh of the artificial disk and stresses distribution σ_z (MPa) within it

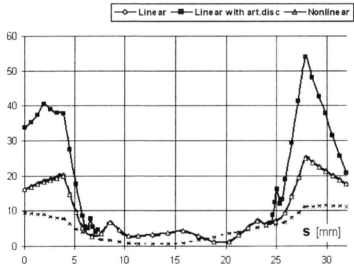

Figure 12. Von Mises equivalent stress distribution (MPa) on the top surface of the lower vertebra along the path at the symmetry plane (YZ)

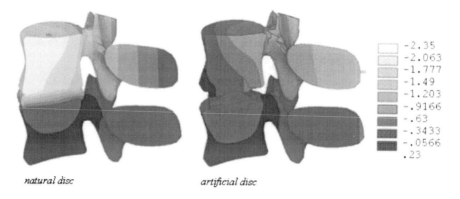

	-2.35
	-2.063
	-1.777
	-1.49
	-1.203
	-.9166
	-.63
	-.3433
	-.0566
	.23

natural disc   *artificial disc*

Figure 13. The vertical displacement (mm) distribution caused by the 3000N compression

CONCLUSION

The discussed example illustrates the thesis that only by integrating the advanced (professional) computer aided systems, such as CT, CAD, CAE and CAM, it is possible to design and manufacture modern implants. In view of the fact that the average lifetime of people has increased and that generally the society has become older the improvement of this process is an urgent task of the contemporary techniques. The progress of the engineer disciplines also stimulates the progress of orthopaedic engineering and consequently that of the human joint replacements. The role of engineers has in effect increased in this very important for the modern society.

REFERENCES

1   **Będziński R.** Engineering Biomechanics – Selected Problems, Oficyna Wydawnicza Politechniki Wrocławskiej, Wrocław 1997, (in Polish)

2   **Buttner–Jantz K., Schellnach K., Zippel H.** Biomechanics of the SB Charite lumbar intervertebral disc prothesis. Int. Orthoapeadics v. 13, 1989, pp. 173–176

3   **DERO Universal Spinal System.** Lfc. Ltd. Zielona Góra

4   **INFIX Interbody Fusion Device.** Spinal Concepts Inc. Austin, USA

5   **Kuklo T. R., Polly D. W., Owens B.D., Zeidman S. M., Chang A. S., Klemme W. R.** Measurement of Thoracicic and Lumbar Fracture Kyphosis, Spine v. 26, No 1/2001, pp. 61 – 66

6   **Kopf – Meier P.** Atlas of Human Anatomy. Wydawnictwa Lekarskie PZWL, Warszawa, 2002

7   **Link Zwischenwirbel Endoprothese.** Model SB Charite; Waldemar Link GmbH & Co. Katalog nr 241, Hamburg; Germany

A003/063/2003 © With Authors 2003

8   **Matyjewski M.** Modeling and invetigation of the spine segment taking into account poro-elasticity of fissues. Ph.D. thesis, Warsaw University of Technology, 1995 (in Polish)

9   **Prodisc. Aesculap AG & KO.KG**, Tuttlingen, Deutschland

10   **Shao Z., Rompe G., Schiltenwolf M.** Radiograhpic Changes in the Lumbar Intervertebral Discs and Lumbar Vertebrae With Age, Spine v. 27, No 3/2002, pp. 263 – 268

11   **Dietrich M., Kędzior K., Krzesiński G., Zagrajek T., Zielińska B.,** Parametric finite element models of human bones, Acta of Bioengineering and Biomechanics,Vol.4,2002,pp.135-136

Production and Control

Application of workflow technology for workshop scheduling

W ZHOU, J ZHU, and Z WEI
College of Mechanical and Electrical Engineering, Nanjing University of Aeronautics and Astronautics, People's Republic of China

ABSTRACT: To solve the complexity of scheduling problem and meet the requirement of ever-changing manufacturing environment. In this paper, a new Workflow-Based Scheduling System (WBSS) is proposed. The integration of Workflow Management System (WfMS) and rule-based scheduler provides us an effective way of generating task-sheet according to the states of system and the scheduled objects. First, the definition of workflow model for scheduling is proposed, and then follows the architecture and mechanism of the proposed WBSS; At last, an application is given to show how the established system works.

1. INTRODUCTION

Market competition in the information age requires that manufacturing should make a rapid response to the customer demands and conduct an agile and lean manufacturing as well as just-in-time delivery. However, in most job shop environment, task-oriented management and information support system, high WIP levels and long circle time caused by traditional MRP systems, prevent all these objects from realizing. As an information management system that supports process automation and process integration, workflow management technology has the potential to overcome the rigidity of current shop floor control systems，it can provide us an efficient way to model，control and monitor this complex production process，In this article, workflow technology is used to support shop floor control. Herein, an order is considered as a process, and subtasks consisting of the order are taken as activities. The relation between those activities is described by workflow. This paper describes the integration of workflow technology to the shop floor control system and discusses some design guidelines to develop such a system.

The reminder of the paper is organized as follows. In Section 2, Literature on solving related scheduling problems is reviewed; In Section 3, architecture of Workflow-Based Scheduling System (WBSS) is presented and explained; Section 4 focuses on the mechanisms of WBSS; In section 5, the application the WBSS is given to show how this system works. Finally Section 6 concludes the paper.

A003/008/2003 © With Authors 2003

2. SOLUTION APPROACHES DEVELOPD FOR SOLVING THE SCHEDULING PROBLEMS

During the past several decades, researchers have done a lot of research work in the field of scheduling and great results have been achieved (1). In general, ssolution methods developed for solving the workshop scheduling can be classified into four categories: (a) operations research(OR) approaches; (b) stochastic search methods such as genetic algorithm, taboo search and simulated annealing; (c) Artificial Intelligence (AI), such as Neural Networks, Multi-agent System, Distributed Intelligence etc(2); (d) a combination of (a),(b),(c).

However, due to the complexity of the problem, traditional scheduling problems cannot fully meet the requirements of dynamically-changed manufacturing environment. The main reason is that manufacturing is a system involving production control and monitoring, information integration, and process integration. In literature, much research has been focused on manufacturing control and monitoring, overlooking the importance of information integration and process integration. So effort should be emphasized on the information integration and process integration. Recently, Workflow Management System (WfMS) has been appeared and applied to a lot of Business Process Reengineering (BPR) successfully (3). WfMS provides us an effective way to model, control and monitor complex production process, which can achieve an integration of information and process easily.

3. PRONCIPLE OF PROCESS-CENTERED WORKFLOW MANAGEMENT

In this section, the methods developed for effective solving scheduling problems in manufacturing are presented. The method development consists of two parts. The first part involves development of workflow modeling of the scheduling problem. The second part is to develop a workflow management system based scheduling framework.

3.1 Set up workshop's workflow model

The workflow idea is defined by Workflow Management Coalition (**WfMC**) as the automation of a business process, in whole or part, during which documents, information or tasks are passed from one participant to another for action according to a set of procedure rules (**4**). A workflow management system is a commercial implementation of this idea, and is frequently deployed in business process, where high volumes of work items requiring rapid turnaround are handled. Workflow management systems aim to help business goals to be achieved with high efficiency by means of sequencing work activities and invoking appreciate human and /or information resources associated with these activities. So the kernel of workflow technology is the process. The definition of the process is the basic component of workflow management system, which is also the core of workflow management system. When apply workflow management technology to workshop control, there must be a workflow model built first. In this paper, a workflow model is established to describe the constraints of the scheduling problems, such as resources constraint, process constraint and time constraint etc. The goal of the scheduling is treated as the performance of workflow instances. The workflow model is composed of three components: process model, resource model and workflow activity model. The relationship between these three components is shown in Fig.1.

A003/008/2003 © With Authors 2003

- Process model

Process model is made up of one or several process. A process describes the operation needed for one job, and relationship between the operations (such as operation A must completed before operation B can be started. We call this kind of relation process constraints). While several processes describe operation routes for different jobs, process model is built using activity-based modeling method, in which an operation is modeled as an activity. According to the definition of the workflow coalition, there can be several instances of different process running at the same time. Therefore, the relation of these instances is the competition for resources.

- Resource model

Resource model defines the resource that can be used during manufacturing. Two kinds of resource entities are described in the model, the individual resource and resource pool. The former refers to the real production equipment, while the latter is in fact a classification of the former according to their functions. Thus the individual resource in the same resource pool has the same function.

- Workflow Control Model

Workflow Control Model describes the state changing condition of process instance and the relations between activities. According to WfMC's reference definition, the normal relations between activities include parallel relation, And-split relation, OR-split relation, AND-join relation, OR-join relation and Iteration relation (5).

Based on the above model, we can define the workflow model for scheduling as follows:
WAM = { P_1, P_2, . . .,P_m, R_1,R_2,. . ., R_m, C_1,C_2,C_m}; where P_1, P_2, . . .,P_m is the process model, R_1,R_2,. . ., R_m presents the resource model and C refers to the workflow control model. Thus, the scheduling problem is to find proper sequence of activity, which is constrained by process constraint and resource constraint. In fact, two sub-problems are involved to achieve the scheduling goal: the resource allocations and activity sequencing in logical and time matter.

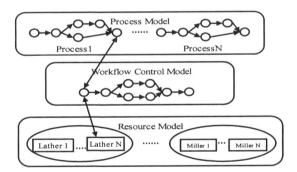

Fig.1. Architecture of Workflow Scheduling Model

3.2 The Architecture of WBSS (Workflow-Based Scheduling System)
We present a new approach to control and trigger all of the activities in the workshop scheduling system, which is called Workflow-Based Scheduling System (WBSS). From

process perspective, a task consists if all of activities. The completion of a task means its relative process stats and ends, i.e. all of activities in the process are accomplished. Only when these activities are managed effectively can the scheduling be smoothed and satisfied.

After building the workflow model, workflow engineer can execute workflow instances according to the predefined model. Being integrated with scheduler, WBSS provide an efficient way to realize the real-time dynamic workshop scheduling.

The proposed WBSS is discussed as follow. Suppose the operation of the system is based on message queue. As shown in Figure 2, the system is composed of workflow modeling tool, workflow management system, scheduler of activity, and task sheet lists.

The architecture of WSS is composed of several components as follows:

1) Workflow Modeling Tool: It consists of description of rule set and activity definition model. In description of a rule set, the precondition of each type of activity is defined. Activity definition model defines each task class is composed of what types of activities. Thus, when an instantiated task is submitted to Task Decomposer, decomposer will easily divide it into concrete and dependent activities according to rule sets and activities definition model.
2) Workflow Management System: It mainly manages activities states, such as whether these activities have completed or are in progress, how many requested tasks there are. The Workflow Administrator also collects production index and production progress from Local Database and Global Database, because these two databases store a mass of comprehensive and accurate information of production progress, task planning and states.
3) Acquisition of Knowledge Interface: It extracts main information from Local Database and Global Database. The information includes scheduling each activity and role/organization that will process the activity.
4) Workflow Engine: It is the core of the architecture: Workflow Engine will reschedule these activities according to activity states, resources and relevant knowledge provided by Acquisition of Knowledge. There are two lists of table: namely a schedulable queue and an executable queue. Because of enormous tasks or activities in schedulable queue, these tasks need to be rescheduled and optimized. Activities favorable for execution are selected out and put into executable queues waiting the assignment.
5) Scheduler of activity: It delivers reasoning results generated by Workflow Engine to relevant nodes to steer all of activities with reference to resource pool and dispatching rules. In the meanwhile, it also receives activity progress and task requisition from real-time activity progress.
6) Task list sheets: It receives the task sheets generated by scheduler of activity, and receive the feedback information about the work and returned the information to the workflow engine. A typical task sheet list is listed below, which includes activity_id, activity_name, resource used, start_time, end_time and activity status, etc.

   A003/008/2003 © With Authors 2003

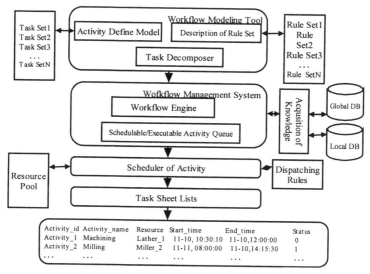

Fig.2. Architecture of Workflow-Based Scheduling System (WBSS)

4. THE MECHANISM OF WBSS

4.1 Rule set of each activity class

When a concrete task is requested or a user decides to inquire about the progress of it, the task will automatically divided into several activities according to the activity classes. The finer the granularity of an activity is, the more accurate and comprehensive its information is and the more difficult to manage it.

Object of activity class is as follows:
<meta_activity>:: <task_id, task_type,activity,role>
<Activity>::=<activity_id, name, type>;
<Role>::=<role_id, name, org_entity>;
WSS create a comprehensive and accurate database of activity states that describe, instruct, certify and record what will happen at every step of activities, including before, during and after events for each activity. From logic perspective, it needs to describe the precondition of each activity class. The representation of rules is described as follows:
<rule_activity>:: =<activity_id, pre_condition,produced_data>;
<pre_condition>::=<activity_index, data_name, data_path, data_type>;
<produced_data>::=<activity_id, data_table>;
The object model of activity is referred to figure 3.

According to object model of activity, the operation chain of each task class is easily formed.

4.2 Activity progress of database

The basis and the precondition of the WBSS are to monitor and control each activity status. The complete extent if activity states in progress library will directly affect the reliability and efficiency of the whole system. Therefore, progress library of activities must be created, and each activity progress at every stage must be effectively stored in it.

The structure of activity progress is as follows:
Activity_progress_table{
Activity_class_id, / * class identifier * /
Activity_id, / * activity identifier * /
Start_time, / * when it will start * /
End_time, / * when it will end * /
Act_Stime / * the actual starting time of the process * /
Act_Etime / * the actual ending time of the process * /
A_status / * activity status: 0-completed; 1-processig; 2-waiting; 3-suspended * /

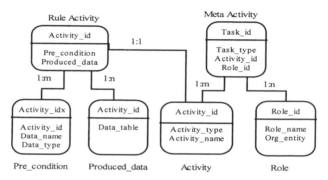

Fig.3 The object model of activity

4.3 Control strategy

Two types of trigger are defined and established, namely event trigger and timer trigger. Here an event is defined as an independent activity, The WBSS will track and control the parent event progress through the scheduling process and guide all the child events (activities) involved in the parent event. Simultaneously, in the process of guiding activities, each activity will be timely and is not on executive at any time. It means there are time triggers in the system. Time trigger will conduct these activities according to the planned time. If the activity falls below the threshold of processing time, WBSS will prompt it. Thus the whole scheduling process will be managed and controlled through these two types of trigger.

The operation step of WBSS can be stated as follows:
Step 1: Establish workflow model and store the model to the database. After the workflow instance is initiated, the instance is stored to instance database.
Step 2: Workflow engine reads the process instance from the database according to the task sets and rule sets.
Step 3: Workflow determines the activity queues, and send the activity instance to the scheduler of activity.
Step 4: Scheduler of activity chooses the right resource allocation rules and dispatching rules.
Step 5: According to the rules, the task lists can be achieved.

Step 6: After receiving the feedback information, the workflow engine will navigate the process instance to the next activity.

5. APPLICATION

In order to test performance of this approach developed in this paper. A prototype – workflow management system based scheduling system was conducted. We use CIMFlow Simulating tool to test our idea. First a workflow model of job1 and job2 is built by CIMFlow modeling tool. Then the process model is shown in Fig 4. Suppose that one job comes to process 1 and process 2 at the same time every ten minuets. After simulation, we can get a list of task sheet shown in Fig 5.

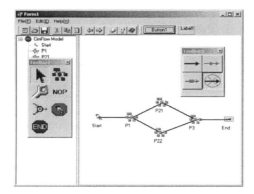

Fig. 4 Process model of the CIMFlow

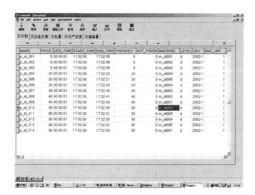

Fig. 5 An example of Task sheet lists

6. CONCLUSIONS

WBSS has been developed under Windows NT. It has the ability to collect real-time data and steer all of the activities operated smoothly. It also offers historical and real-time views of information and creates trend charts. Implementing WSS not only information integration has been achieved, but also has process integration achieved. The result shows that our model provides an effective way to solve the dynamic scheduling problem.

ACKNOWLEDGEMENT

This research has been funded by the National Scientific Foundation of China.

REFERENCE

[1] **C M Harmonosky,** and **S F Robohn**. (1991) Real-time Scheduling in Computer Integrated Manufacturing: A Review of the Recent Research. Int.J.Computer Integrated Manufacturing.Vol.4 No.6.1991, pp.331-340.
[2] **Biewirth Christian**. (1999) Production Scheduling and Rescheduling. Evolutionary Computation. Volume 7 No.1, 1999, pp.1-7.
[3] **SME BLUE BOOK SERIES,(1997)** Business Process Transformation using Workflow Management: A Case Study., Published by Computer and Automated Association of the Society of Manufacturing Engineers,1997,12, 4-24.
[4] **WfMC.** (1999) Workflow Management Coalition Terminology & Glossary, WFMC-TC-1011, Feb., 1999.56-62.
[5] **Yu Shun Fan,and Haibin Luo**,(2001) Foundation of Workflow Management Technology, Tsinghua University, Springer-Verlag Publisher, April, 2001,36-42.

A knowledge-based sequencing strategy for multiple product lines

A KHAN and **A DAY**
School of Engineering, Design, and Technology, University of Bradford, UK

ABSTRACT

Balancing is considered as the major design issue on any type of assembly line application and therefore has been widely addressed by past researchers. In many studies, an economical number of stations are claimed for the selected type of manufacturing system. This paper advances a post-balancing design strategy that relates a new balancing methodology through a knowledge based approach to a sequencing strategy for the multiple product lines. The methodology can be effectively applied to the design of multi (products processed in batches) and mixed (products processed in mixed orders) assembly lines. The methodology generates a recommendation for economical batch sizes, an appropriate batch launching sequence for multi-product lines, and a suitable launching sequence for products on the mixed product lines. This paper describes one module of the complete knowledge based design methodology, developed in a systematic step-by-step approach for manufacturing assembly lines. An economical number of stations, product cycle time(s), individual product holding cost, and the line changeover costs are the key inputs required for the knowledge based analysis. The methodology has been developed in the Windows based expert system shell AM (Application Manager). An example is presented for the analysis of the knowledge based methodology developed.

1 INTRODUCTION

Manufacturing assembly line research can broadly be divided into single and multiple product lines. A multiple product assembly line can be multi-product (flow of different products in batches), or mixed-product (flow of different products simultaneously) (1, 2). Each type can be subdivided into the specific case (one in which the line is designed for the simplest conditions) and the general case (in which a number of constraints are considered for balancing the line) (3, 4). The specific case is more widely addressed compared with the general line balancing problem (1). Researchers have also tackled various aspects of the design, balancing and performance measures for both types of assembly lines (2, 5).

Assembly system design, economics and balancing have been addressed in many researches, but still, it has been observed that various assembly system activities have very little inter-communication with each other, and because of this there is a shortage of general knowledge in manufacturing assembly line design (6). This is why the majority of manufacturing assembly lines do not appear to follow the published optimal design techniques (1, 7 & 8).

This paper links preliminary assembly system selection and appropriate balancing techniques with the post balancing activities possessed in the multiple product line by the application of a knowledge based (KB) design methodology. Initially the system gathers some factual information through the user-computer interface which is stored for the appropriate assembly system application (9). It is therefore assumed that the system already contains information related to the number of products, production demand and the suitable cycle time for each individual product in the system.

The main post balancing activity on multiple lines is the line sequencing strategy, which is the core topic for this paper. The methodology gathers factual information (e.g. individual product holding cost and the line changeover cost) from the user and recommends an economical batch sizes and an appropriate batch launching sequence for the multi-product lines and a suitable launching sequence among products for the mixed product lines. The paper covers only one module of the complete knowledge based design methodology, developed in a systematic step-by-step approach. It is developed in the Windows-based expert system shell called AM (Application Manager). An example is presented for the analysis of the developed knowledge based methodology. The line sequencing strategy is described, followed by a worked example and conclusions.

2 NOMENCLATURE

γ = Fixed launching rate for the system (sec)
$(C_i)_{Min}$ = Minimum changeover cost associated with product i (£)
$(C_i)_{Min+1}$ = Next minimum changeover cost associated with product i (£)
$(C_j)_{Min}$ = Minimum changeover cost associated with product j (£)
$(C_z)_{Min}$ = Minimum changeover cost associated with product z (£)
$(C_z)_{Min+1}$ = Next minimum changeover cost associated with product z (£)
C_1 = Stock holding cost / item / unit of time cost (£)
C_{Min} = Minimum line changeover cost associated with every product on line (£)
C_s = Line changeover cost (£)
C_y = Changeover cost associated with product y (£)
i,j,x,y,m,n = All integers between 1 and N
N = Number of products in the system
N_i = Product i demand for the balancing duration option
P_{ij} = Line changeover cost for product j following product i (£)
P_{im} = Line changeover cost for product m following product i (£)
P_{xy} = Line changeover cost for product y following product x (£)
P_{zj} = Line changeover cost for product j following product z (£)

P_{zn} = Line changeover cost for product n following product z (£)
Q^{*} = Optimum product batch size
r = Production demand

3 LINE SEQUENCING STRATEGY

As sequencing strategy is one module of the knowledge based assembly lines design methodology, the type of line (single, multi or mixed), product cycle time(s), number of stations, product holding cost and line changeover cost are predefined and are considered as prerequisite for this analysis (9). The sequencing strategy is required only for the design and analysis of multi-product and mixed-product assembly lines.

On multi product lines, the analysis is conducted for the economical batch size required for each individual product and for the economical launching sequence among the product batches. The analysis requires basic information about the individual product holding cost and the line changeover cost for each individual product.

However, the sequencing on a mixed-product line is a complex problem. An economical sequence among the products and a corresponding fixed launching rate for the products onto the line needs to be determined. The sequencing strategy for both types of assembly lines is explained in the following sections.

3.1 Line sequencing strategy for multi-product lines

A multi-product line is first arranged for the production of the first product batch in the product sequence. On completion, the line is shut off and then the preparation is started for the next product in the batch sequence. When production of one product is completed and the line is prepared for the next product in sequence, the line changeover cost or line preparation cost must be included in the total production cost. This cost includes the cost of tooling, the cost of machine / tool preparation, the cost of machine / labour idle time and is mainly dependent upon the batches of preceding and succeeding products.

The next batch production should be the one which requires the minimum possible line changeover cost for the whole production. Generally more than one product is available for launching next onto the line, therefore a methodology is necessary which analyses all possible product combinations for the selection of preceding and succeeding product batches to minimize the overall line changeover cost for the whole production (2). In the majority of cases the selection requires expert consideration. In essence the knowledge based methodology analyses all product combinations for the succeeding product batches and sometimes a higher changeover cost product is selected due to the minimum cost impact on overall production.

Initially the methodology starts its analysis by choosing the succeeding product on a minimum changeover cost basis and ensures that the selection of any succeeding product should not be repeated for any other product in the whole production. Also the methodology checks that the preceding and succeeding products should not be repeated for each other in any other pair. The product batch sequence generated from such information always results in the repetition of products in the sequence. The problem becomes more complicated when the number of products increases in the system.

An economical decision is performed through the knowledge based production rules when the same product repeats more than one times in the product batch sequence. The repeated pair products are then selected for analysis and the next minimum value cost is checked for justification. Only that pair is selected which has the minimum changeover cost impact on overall production. For the remaining pair another suitable alternative product pair is then analysed. The required knowledge based decision is described next.

Changeover cost for the product y associated with the pair of products x and y is defined such that the batch for the product x is completed and the line is prepared for the production of product y as:

$$C_y = P_{xy} \qquad (1 \leq x, y \leq N \ \& \ x \neq y) \qquad\qquad \text{Equation 1}$$

Then the minimum changeover cost assigned for the product j associated with the pair of products i and j where product i is followed by product j by the following relation:

$$(C_j)_{Min} = P_{ij} \qquad\qquad \text{Equation 2}$$

which must satisfy the following relation for all products x and y in the system

$$P_{ij} \leq P_{xy} \quad (1 \leq x, y, i, j \leq N, x \neq y \ \& \ i \neq j) \qquad\qquad \text{Equation 3}$$

Now the minimum line changeover cost is determined for the whole production by verifying the relation for every product j as:

$$P_{ij} \neq P_{zj} \qquad (1 \leq i, j, z \leq N, i \neq j, z \neq i \ \& \ z \neq j) \qquad\qquad \text{Equation 4}$$

Then the minimum line changeover cost associated with every product is determined on the line by the following relation:

$$C_{Min} = \sum_{j=1}^{j=N} (C_j)_{Min} \ (1 \leq j \leq N) \qquad\qquad \text{Equation 5}$$

(Equation 5) gives the total minimum changeover cost for the line when complete batches of every product are produced on the line and the product batch sequence is prepared such that every product appears only once in the sequence. When the same product follows more than one product, the product batch sequence will not be accepted. Therefore, a knowledge based decision is performed whenever, for any product j, (Equation 4) is not satisfied, which means that the same product j follows product i and product z in the product batch sequence. Equation 4 then becomes:

$$P_{ij} = P_{zj} \qquad (1 \leq i, j, z \leq N, \quad i \neq j, z \neq i \ \& \ z \neq j) \qquad\qquad \text{Equation 6}$$

The next minimum valued changeover cost is defined for the product i and the product z being followed by the product m and product n respectively as follows:

The next minimum valued line changeover cost for product i followed by product m is:

$$(C_i)_{Min+1} = P_{im} \ (P_{im} \geq P_{ij})$$ Equation 7

and for all the remaining products y, the following relation must be true

$$P_{im} \leq P_{iy} \quad (1 \leq i, j, m, x, y \leq N, \ i \neq y, \ i \neq m \ \& \ j \neq m)$$ Equation 8

Similarly, the next minimum value line changeover cost for product z followed by product n is:

$$(C_z)_{Min+1} = P_{zn} \ (P_{zn} \geq P_{zj})$$ Equation 9

and for all the remaining products y, the following relation must be true

$$P_{zn} \leq P_{zy} \quad (1 \leq z, j, n, x, y \leq N, \ z \neq y, \ z \neq S \ \& \ j \neq S)$$ Equation 10

As the products i and z are followed by the same product j (Equation 6), this is not possible for the product batch sequence. Therefore, a knowledge based decision is made which will select either product i or product z based on overall minimum line changeover cost. As for both products i and z, the next minimum changeover cost products are analysed using (Equation 7 & 9) respectively. The minimum line changeover cost product is selected by a knowledge based decision as follows:

Comparing the values of (Equation 7) and (Equation 9) and if the following relation is true:

$$(C_i)_{Min+1} > (C_z)_{Min+1}$$ Equation 11

Then the selection will be:

$$(C_z)_{Min} = (C_z)_{Min+1}$$ Equation 12

which subsequently decides that

$$(C_z)_{Min} = P_{zn}$$ Equation 13
and $$(C_i)_{Min} = P_{ij}$$ Equation 14

If (Equation 11) is not true, then the decision is:

$$(C_z)_{Min} = P_{zj}$$ Equation 15
and $$(C_i)_{Min} = P_{im}$$ Equation 16

The same knowledge based decision is performed for any number of products which are being followed by the same products in the product batch sequence. The methodology also checks the impact of minimum changeover cost of one product on the overall production cost. With these knowledge based decisions, the optimal product batch sequence is developed which results in the minimum changeover cost for the overall production.

In multi-product assembly lines, the product batch sizes also play a vital role in overall production cost for the system under consideration. When the product batch sizes are too large, then the stock level will rise and excessive cost will be tied up in the system. On the other hand, when the product batch sizes are too small, then a large line changeover cost will be required to complete the required demand. The selection of an optimum product batch size is therefore a compromise between the item holding cost and the line changeover cost which is diagrammatically shown in Figure 1.

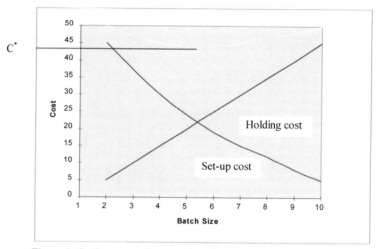

**Figure 1: Optimum cost between item holding cost and changeover cost
(Wild, 1995)**

The x-axis represents the batch size and the y-axis gives the overall cost for the system. It shows that by increasing the product batch size, the item holding cost increases while the line changeover cost decreases and vice versa. The optimum value between the line changeover cost and the item holding cost is given by C^* which eventually results in economical batch sizes for products in the system. The following knowledge based procedure was developed for selection of economical batch size for the products under consideration.

As the product batch size depends upon the item holding cost and the line changeover cost, therefore the economical product batch size can be obtained by differentiating the sum of these two costs with respect to product batch size and then equating to zero. The optimum product batch sizes are therefore given as (2):

$$Q^* = (2C_s\, r\, /\, C_1)^{1/2}$$

Equation 17

Through its knowledge based input displays, the necessary cost information is collected for this module. Necessary guidance is provided to the system user through the 'HELP' display. If the user has not provided the costs information, then random data are generated for all cost inputs and then the economical batch sizes and the product batch sequence are determined for the system under the test conditions.

3.2 Line sequencing strategy for mixed-product lines

On mixed product lines, the launching of products is completely mixed. The line configuration is not altered for launching different products onto the line. The common operations are assigned to one set of stations, while the non-common operations are performed on another set of stations. Buffer storage must be provided for the in-coming and the out-going semi-finished products (2, 10).

It is necessary to ensure that products are launched onto the line in a completely mixed fashion, otherwise various stations related to different products may not be justified for balance load on the line. Stations related to certain products may be over-loaded during the irregular launching of the products concerned onto the line and are under-loaded during the launching of non-related products onto the line. It is therefore necessary to launch every product onto the line in some predetermined frequency (launching rate) so as to monitor the overall production and their related activities accordingly.

Generally two types of launching rates are applied in mixed-product assembly lines; variable launching rate and fixed launching rate. During the variable launching rate, the products are launched onto the line by spacing the products based on the leading product cycle time. It is observed that the variable launching rate can incur high item waiting time in the system and is also difficult to monitor other related activities for such type of lines. The fixed launching rate always provides reasonable conditions for the system and is also easier to monitor other related activities at stations in the system (2).

The sequencing strategy is developed for the mixed-product assembly lines which is based on the fixed launching rate for all products in the system under consideration. The relation used for the fixed launching rate is given as follows (2):

$$\gamma = \sum_{i=1}^{i=N} N_i C_i / \sum_{i=1}^{i=N} N_i \ (1 \le i \le N)$$

Equation 18

The methodology also concludes the product launching sequence onto the line. The largest cycle time product is selected first followed by the smallest cycle time product follows and then after that the remaining products are launched one by one onto the line. If each product has different demand in the balancing duration option (e.g. a shift or a 24-hrs duration), then the product launching sequence is developed in such a way to take care of the proportion of demand of each individual product. The number of times a product is repeated within the sequence depends upon the ratio of demand for that product with the total production demand. This technique is applied when the number of products is more than 3, while for 3-product line, the longest cycle time product is launched first and then suitability of the middle cycle time product is tested. If the middle product can be accommodated in the available time, then the middle product is selected for launching as second in the product sequence onto the line, otherwise, the minimum cycle time product is selected for launching second onto the line (2).

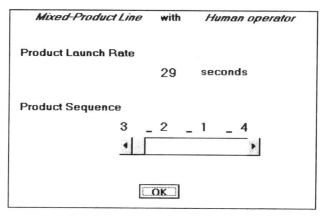

Figure 2: A knowledge based display for line sequencing strategy

Figure 2 shows the product launching rate and the product launching sequence for a mixed-product assembly line. A horizontal scrolling facility is provided for viewing the sequence for more than 4 products in the system.

As the information required for the calculation of fixed launching rate and the product launching sequence onto the line is already available in the system (product demand and cycle times), the input information is not required for the module in mixed-product assembly lines design. The fixed launching rate and the suitable product launching sequence onto the line is shown for a test example in Figure 2.

4 WORKED EXAMPLE

The performance and the effectiveness of the developed KB sequencing strategy has been verified using information provided by industrial experts. The experts were selected from UK based companies using assembly line design principles. Two examples were required for this analysis, one for a multi-product and one for a mixed product line (10).

4.1 Multi-product line
The KB information for the sequencing of a multi-product line is:

Table 1: Individual Product Requirements

Product	1	2	3
Yearly Demand	800000	700000	500000
Work Content	56	43	45
Number of stations	7	7	7
Cycle time	9	7	7
No of workers	7	7	7
Item Holding Cost	9	5	7

A003/010/2003 © IMechE 2003

Table 2: Line changeover cost for production combination

Preceding product	Succeeding Product		
	1	2	3
1	0	104	63
2	83	0	67
3	95	67	0

The multi-product line required input information (provided in Tables 1 & 2), was analysed for determination of the economical product batch sizes and the product batch sequence. The appropriate knowledge based decision for the economical batch sizes of product 1, 2 and 3 is 1108, 1250 and 866 respectively and product batch sequence was predicted as 1-3-2.

4.2 Mixed-product line

The KB information for the sequencing of a mixed product line is:

Table 3: Input information for a mixed product line

Product	1	2	3	4	5	6
Product demand / 24-hrs	1607	714	179	286	321	214
Product cycle time	24	31	23	23	25	27

The mixed-product line did not require further information and the sequencing strategy was designed on the basis of existing information of demand (for 24 hours duration) and individual product cycle times (Table 3). The knowledge based decision for the fixed launching rate for the products with a required product launching sequence onto the line is presented as follows:

Product launching rate = 26 sec
Product launching sequence = 2-2-2-2-3-1-1-1-1-1-1-1-1-1-4-4-5-5-6

5 CONCLUSIONS

The KB sequencing methodology developed and presented here has made the complete design procedure for multiple product lines easier, more efficient and more suitable for real situations. It has achieved this by linking various design stages (e.g. balancing related issues) with the post balancing sequencing strategy for the multiple product lines.

For multi-product lines the methodology determines the product sequence which incurs the minimum possible changeover cost for the system. A KB decision is used to review any alternative combinations which may result in smaller changeover cost for overall production. For such lines, the optimal batch sizes can also be determined by analyzing the line changeover cost and item holding cost.

Mixed-product lines require a fixed launching rate for products onto the line. The product sequence is best determined by launching the largest cycle time product first, the smallest cycle time product second and remaining product in the system next. Higher demand products can be repeated in the product sequence depending upon the share of product in the total demand. The conclusion is that the KB sequencing strategy is a viable technique for such applications.

REFERENCES

1.   Ghosh, S. and Gagnon, R. J. (1989) A Comprehensive Literature Review and Analysis of the Design, Balancing and Scheduling of Assembly Systems, Int. J. Prod. Res., Vol.27, No.4, 637-670.

2.   Wild, R. (1995) Production and Operations Management, 4th Edition, Great Britain: Cassell.

3.   Arcus, L. A. (1966) A Computer Method of Sequencing Operations for Assembly Lines, Int. J. Prod. Res., Vol.4, No.4, 259-277

4.   Schofield, N. A. (1979) Assembly Line Balancing and the Application of Computer Techniques. Computer and Industrial Engineering, Vol.3, 53-69.

5.   Shtub, A. and Dar- El, E. M. (1989) A Methodology for the Selection of Assembly Systems, Int. J. Prod. Res., Vol.27, No.1, 175-186.

6.   Bhattacharjee, T. K. and Sahu, S. (1987) A Critique of Some Current Assembly Line Balancing Techniques, International Journal of Operations and Production Management, Vol.7, 32-43.

7.   Chase, R. B. (1974) Survey of Paced Assembly Lines, Industrial Engineering, Vol.6, 14-18.

8.   Ma, X. and Liu, H. W. (1993) An Effective Computer Based Assembly Line Balancing Method, Proceedings of the 30th International MATADOR Conference, 31 March-1 April, Manchester, 57-64.

9.   Khan, A. and Day, A. J. (2002) A Knowledge Based Design Methodology for Manufacturing Assembly Lines, Computers & Industrial Engineering, Vol. 41, 441-467.

10.   Khan, A. (1998) A Knowledge Based System for the Design and Analysis of Assembly Lines, PhD, Thesis, University of Bradford, UK.

Performance measurement based co-ordination of enterprise resource planning systems in a virtual enterprise

Y ZHANG and A K KOCHHAR
School of Engineering and Applied Science, Aston University, Birmingham, UK

ABSTRACT

As the backbone of e-business, Enterprise Resource Planning (ERP) system plays an important role in today's competitive business environment. Few publications discuss the application of ERP systems in a virtual enterprise (VE). A VE is defined as a dynamic partnership among enterprises that can bring together complementary core competencies needed to achieve a business task. Since VE strongly emphasises partner cooperation, specific issues exist relative to the implementation of ERP systems in a VE.

This paper discusses the use of VE Performance Measurement System (VEPMS) to coordinate ERP systems of VE partners. It also defines the framework of a 'Virtual Enterprise Resource Planning (VERP) system', and identifies research avenues in this field.

1   ENTERPRISE RESOURCE PLANNING (ERP) SYSTEM

The concept of ERP can be traced back to and has evolved from Materials Requirement Planning (MRP) and Manufacturing Resource Planning (MRP II) systems. Yen, Chou, and Chang [1] described the evolution of ERP as follows: 'before creating ERP, inventory control system was the software designed to handle traditional inventory processes ... the early stage of ERP was carried out through Materials Requirements Planning (MRP), a software which focused on time requirements for sub-assemblies, components, materials planning, and procurement. Manufacturing Resource Planning (MRP-II) was developed in 1970–1980, which was the software package focused on extending MRP to the shop floor and distribution management activities ... the next stage of ERP evolution was Just-in-time (JIT) methodology ... the maturity stage of ERP occurred in mid-1990. ERP was the software package focused on extending MRP-II to cover additional areas such as finance, engineering, human resources, project management, etc ... the current ERP development intends to utilise ERP to realise and sustain a competitive advantage'. Wight [2], along with many other authors, was not in favour of viewing MRP-II as a software package, but as the way in which people effectively use techniques to run a manufacturing business: 'Technically, it's not much different from closed loop MRP … The technical differences are small compared to the real significant difference. And that is the way management uses the system'.

Gulla, and Brasethvik [3] postulate that there is a drive towards performance or value orientation: 'To measure the performance of the organization, key indicators are defined on the basis of data collected in the ERP system. Even though many of these key indicators can be computed directly from the data in the ERP system, it has become increasingly popular to implement data warehouse solutions and strategic enterprise management (SEM) components on top of the ERP transaction systems.'

2   VIRTUAL ENTERPRISE (VE)

A virtual enterprise (VE) is a dynamic partnership among companies that can bring together complementary core competencies needed to achieve a business task. Kochhar and Zhang [4] indicated three reasons why partners seek to constitute VE as a business form to achieve a business task: (a) VE can bring other partners' core competences, thus saving investment, (b) VE can make an individual company more agile, and (c) VE offers small companies opportunities to compete with large companies.

Typical VE characteristics differentiating VE from conventional enterprise are: (a) operation across single company boundaries, (b) complementary core competences, (c) membership flexibility taking account of the required competences, (d) partner equality [5], (e) partner dependence, (f) cultural differences, (g) relative temporality, (h) reduced geographical restriction, and (i) increased requirement for using electronic communication.

A virtual enterprise normally goes through five stages: **identification** stage, **formation** stage, **design** stage, **operation** stage, and **dissolution** stage.

3   COORDINATION OF ERP SYSTEMS IN A VIRTUAL ENTERPRISE

A VE is different from a single enterprise in that the VE strongly emphasises the cooperation among partners. When enterprises come together to form a VE, their ERP systems should incorporate the following additional functions:

- Identifying partners who are responsible for unusual VE performance.
- Indicating partners' core competence performance.
- Assessing the effect of the performance of a partner on overall VE performance, and on other partners' performance.
- Facilitating information sharing, which is very important for VE partner cooperation.

To achieve the above functions effectively, the ERP systems in a VE have to be coordinated. The question remains 'how to coordinate'. This paper discusses one of the possible answers: use VE Performance Measurement System (VEPMS) to achieve this coordination.

A003/050/2003 © IMechE 2003

4   VEPMS FRAMEWORK

The VEPMS framework as detailed in a previous paper [4] consists of four dimensions:

- **The 1st dimension is a set of performance measures for assessing the top-level performance of a VE.** The nature of a task carried out in a VE and the same task carried out in a single enterprise remains unchanged. However, the VE form is more complicated than the single enterprise form, because it involves not only cooperation at departmental level, but also cooperation at enterprise (VE partner) level. Thus the performance measures used in a single enterprise can all be used in a VE, except that some of them need to be systemised to measure the holistic performance of all VE partners. For example, stock turnover can be systemised to measure the whole VE.
- **The 2nd dimension is a set of performance measures assessing the co-operation between the VE partners.** Since a VE strongly emphasises partner cooperation, a specific set of performance measures needs to be established to assess the VE partner cooperation performance. An example is inter-partner satisfaction.
- **The 3rd dimension is a specification of the core competences needed for the VE business.** An important aspect of a VE is the pooling of resources and core competences of different partners. This specification indicates 'what' core competences are needed for the VE business, 'who' (i.e. which partner) will contribute which core competence, 'why' this partner contributes this core competence, and 'how' this core competence can contribute to the achievement of the VE objective.
- **The 4th dimension is a set of performance measures assessing the differences between the VEPMS and any VE partner's own VEPMS (OVEPMS).** Each VE partner enters a VE with its predefined view of what the VEPMS should be. This predefined view can be referred to as the VE partner's own VE performance measurement system (OVEPMS). A partner's OVEPMS is designed to maximize the contribution to the partner's own business from the VE. Because different partners have different OVEPMSes, the VEPMS must be the result of the negotiation among these different OVEPMSes, and finally agreed by all partners. Thus the VEPMS will probably be different from a partner's OVEPMS. In order to minimise the influence of such differences on the VE business, each partner's OVEPMS should be adjusted, as much as possible, to be same as the VEPMS. The performance measures in this dimension are used to measure such differences.

5   THE ERP COORDINATION MODEL

The underlying concepts of the coordination model can be specified as follows:

- Each VE partner is responsible for one or more VE functions.
- Each VE function can have one or more VE partners who are responsible for it.
- Each VE function has ERP functional module(s) associated with it.
- A set of performance measures can be identified to assess the performance of a VE function.
- The VE function's corresponding ERP functional module(s) can generate the values of these performance measures.
- There are identifiable relationships between these performance measures.
- There are identifiable relationships between different VE functions' performance measures.

- Each core competence brought into the VE by VE partners can be assessed via appropriate performance measures, and these performance measures are among the measures generated by ERP functional modules.

Figure 1 illustrates these underlying concepts (note: to keep this figure as clear as possible, the last underlying concept, about core competence is not included in this figure). The functions listed in section 3 can be realised as follows:

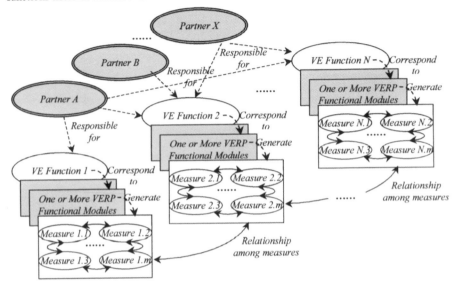

Figure 1. The underlying concepts of the coordination model

- **Identifying partners responsible for unusual VE performance:** Since each partner is responsible for one/more VE functions, and each function can be assessed via a set of performance measures, any unusual performance of a VE function can be indicated by up to date values of the related measures, and then traced back to the partners responsible for it. For example, partner A is responsible for delivery, and delivery can be assessed by a set of performance measures, such as delivery reliability. If delivery reliability is below the required level, the responsibility can easily be traced back to partner A.
- **Indicating partners' core competence performance:** The performance of each core competence can be assessed via appropriate performance measures. For example, partner A brings into the VE a core competence of a reliable distribution network. This core competence can be assessed by performance measures such as distribution cost, reliability, lead time, and damage rate.
- **Assessing the effect of the performance of a partner on overall VE performance, and other partners' VE performance:** Since relationships between measures are identifiable, upon knowing the value of a measure, the values of other related measures could be estimated via the relationships. For example, partner A is responsible for delivering material to partner B who is responsible for manufacturing and delivering products to retailers. Based on partner A's delivery lead time, partner B's manufacturing lead time can

    A003/050/2003 © IMechE 2003

be estimated. Similarly based on knowing partner A's delivery reliability, partner B's delivery reliability to retailers can be estimated.

- **Facilitating information sharing:** Up to date values of measures, and other information about these measures, such as why to use these measures, how to achieve the required levels of these measures, and how the achievement of a measure can contribute to the achievement of other measures (i.e. the relationship among measures), can be stored in a central database. All VE partners can access the central database, and share the information.

6   THE IMPLEMENTATION OF THE COORDINATION MODEL

6.1   Implement the 1st and 2nd dimensions of the VEPMS & the relationships among measures

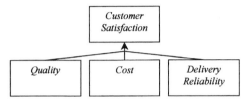

Figure 2. Example of dimension 1 measure tree for customer satisfaction

The performance measures associated with the **1st** and **2nd** dimensions, and the relationships between them can be implemented via **measure trees**. For example, a VE has the following **1st** dimension measures to assess the systemised performance of the VE: customer satisfaction, quality, cost, and delivery reliability. These four measures can be reflected in the tree illustrated by figure 2. In this situation, customer satisfaction is the **up-measure** of quality, cost, and delivery reliability. Consequently, quality, cost, and delivery reliability are the **down-measures** of customer satisfaction.

6.2   Implement the 3rd dimension of the VEPMS

The 3rd dimension 'core competence specification' can be implemented by (1) connecting each measure with the partners who are responsible for the measure, and (2) connecting each VE core competence with the measures used to assess the core competence. The relationship between a partner and a measure (i.e. **'why'** the partner can contribute to the measure) should also be specified. Each core competence should have a 'core competence description' indicating **'how'** this core competence can contribute to the achievement of the overall objectives of the VE.

6.3   Implement the 4th dimension of the VEPMS

To implement the 4th dimension of the VEPMS, it is necessary to carry out the followings.

6.3.1   *Implement the connection between ERP Functional Modules and measures*

This implementation can be carried out from two perspectives. First, the identification of each functional module application (i.e. application ID) should be connected with the related measures generated by the application. Such connection can facilitate checks upon the happening of unusual situations. Second, each application should be able to generate the up to date values of its related measures.

6.3.2 The VEPMS database

The VEPMS can be implemented as a central database consisting of ten tables.

Table 1. The VEPMS database

Table Name	Field Name
Measures	**Measure-ID**, Measure-Description, Measure-Requirement
Measure-Documents	**Measure-Document-ID**, [Measure-ID]
Measure-Bridges	**[Up-Measure-ID], [Down-Measure-ID]**
Bridge-Documents	**Bridge-Document ID**, [Up-Measure-ID], [Down-Measure-ID]
Measure-Partners	**Measure-ID, [Partner-ID]**, Why-This-Partner
Partners	**Partner-ID**, Description
Measure-Applications	**[Measure-ID], [Application-ID]**
Applications	**Application-ID**, Description
Core-Competence-Measures	**[Core-Competence-ID], [Measure-ID]**
Core-Competence	**Core-Competence-ID**, Core-Competence-Description

6.3.3 Implement the OVEPMS database

To implement the **4th** dimension, another database is needed which is a simplified version of the VEPMS database. This database resides on the site of each partner, and is used to store each partner's OVEPMS, which is then compared with the corresponding records in the central VEPMS database. In other words, copies of this database with different contents (i.e. different OVEPMS) reside on the site of each partner, and each copy needs to be compared with the central database. This database requires six tables whose structures are the same as those of the tables Measures, Measure-Bridges, Measure-Partners, Partners, Core-Competence-Measures, and Core-Competence in the central VEPMS database.

7 DESIGN OF ERP FUNCTIONAL MODULES FOR THE VE

It should be clear from the discussion in section 6.3 that the ERP functional modules related to the VE should first be identified to implement the connection between ERP functional modules and measures.

Assume that a VE has three partners A, B and C. The VE needs to use functional modules of Marketing, Product Design, Manufacturing, Supply Chain, and so on. Before entering the VE, each partner already has its own ERP system that also consists of these functional modules. Partner A is responsible for Marketing, partner B is responsible for Product Design, and partner C is responsible for Manufacturing. Note that the real situation will not be so 'simple'. A partner may well have more or less functional modules before entering the VE. But such an assumption makes the explanation more understandable. There are three situations in which the ERP systems in a VE can be set up:

(a) A separate ERP system is set up for the VE as if the VE is a single enterprise. Figure 3 illustrates this scenario. The grey-shadowed circles represent the modules related to the

VE business. In this situation, the ERP functional modules related to the VE are all included in the separate ERP system that is especially set up for the VE business.

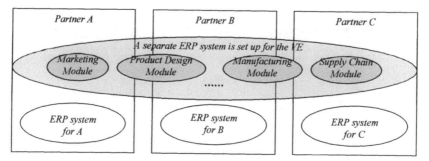

Figure 3. Illustration of a separate ERP system for the VE

(b) No separate ERP system is set up for the VE. The VE uses the relevant functional modules of each partner's existing ERP system. Figure 4 illustrates this situation.

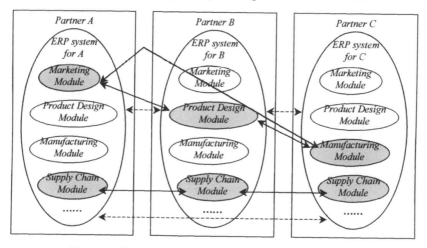

Figure 4: Illustration of no separate ERP system for the VE

- In partner A's existing ERP system, Product Design and Manufacturing modules are **not** related to the VE, since partners B and C are responsible for them. Other modules should be included in the ERP system for the VE.
- In partner B's existing ERP system, Marketing and Manufacturing modules are **not** related to the VE, since partners A and C are responsible for them. Other modules should be included in the ERP system for the VE.
- In partner C's existing ERP system, Marketing and Product Design modules are **not** related to the VE, since partners A and B are responsible for them.

(c) A combination of the scenarios 1 and 2 discussed earlier, and illustrated by figures 3 and 4. Figure 5 illustrates this hybrid scenario. In this situation, the ERP functional modules related to the VE can be identified in the following way:

- All functional modules in the separate ERP system (the Supply Chain module, and so on) are related to the VE.
- Since partner A is responsible for Marketing for the whole virtual enterprise, the Marketing module is related to the VE.
- Since partner B is responsible for Product Design for the whole virtual enterprise, the Product Design module is related to the VE.
- Since partner C is responsible for Manufacturing for the whole virtual enterprise, the Manufacturing module is related to the VE.

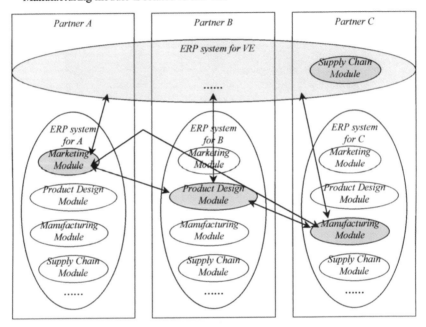

Figure 5: The hybrid situation

8   VIRTUAL ENTERPRISE RESOURCE PLANNING (VERP) SYSTEM

ERP systems evolved from Materials Requirement Planning (MRP) system and Manufacturing Resource Planning (MRP II) system, as discussed in section 1. Perhaps the next evolution of ERP system is the Virtual Enterprise Resource Planning (VERP) system, a combination of the concept of ERP system and the concept Virtual Enterprise which is a rapidly emerging business form. Technically, VERP system can be defined as a set of applications, which can be used to manage all the business functions within a virtual enterprise. This set of applications should be more compatible and portable than its predecessor ERP applications, and should be able to integrate with each other more flexibly, since VE is dynamical partnership with relative temporality. From the management point of view, VERP system can be defined as a new way in which people can more effectively use techniques to manage the VE business.

8.1   Realisation of the VERP system

The three main scenarios in which a VERP system can be set up in a VE have already been discussed in section 7, as illustrated in figures 3, 4 and 5. The situation shown in figure 3 is the least technically challenging, but requires high level of investment, money as well as human resources, since a new set of VERP applications needs to be purchased especially for the VE that may exist for only a short period of time. The situation shown in figure 4 requires the least amount of investment, since no new applications need to be purchased, but has an extremely high requirement on the compatibility and extensibility of VE partners' existing ERP functional modules. Considerable technical effort would be required to implement such a system. The situation shown in figure 5 is a hybrid of the above two situations. It requires compatibility and extensibility of VE partners' existing ERP functional modules, and a smaller amount of investment to implement new, enterprise wide, applications.

8.2   Differences between VERP system and ERP system

Since VERP system is an evolution of ERP system, it inherits all the features of ERP system. The main differences between VERP system and ERP system are as follows:

- A VERP system crosses the boundary of a single enterprise, whereas an ERP system sits within this boundary, although some modules (such as supply chain module) may communicate with some modules in other ERP systems.
- If the scenarios shown in figures 4 and 5 are used, a functional module belonging to both the VERP system and a partner's existing ERP system needs to be integrated with other modules in both systems. This poses a higher requirement on the compatibility of partners' existing ERP modules than that for the ERP system.
- If separate VERP functional modules are used for the VE, there can be serious problems. Firstly, a partner may have ERP systems in operation, one for the activities associated with the VE and a second one, for the non-VE related business. Secondly, the ownership of these modules after VE dissolution may cause disputes. There are no such problems in a single enterprise.
- Since VE places high level of emphasis on the cooperation between partners, there is a significant requirement on the coordination of VERP functional modules that reside on the sites of different partners.

9   RESEARCH DIRECTIONS FOR VERP

The research directions for VERP systems can be summarised as 'compatibility', 'extensibility', 'cooperation', and 'ownership'.

- 'Compatibility' means how functional modules can be integrated seamlessly, which is not new topic in computer-technology area. This is largely a technical issue, and requires the establishment of standards for the ERP systems.
- 'Extensibility' requires that new functions can be easily added to the functional modules to accommodate enterprises' specific requirements, such as generating values of

performance measures, and connecting different measures, as discussed in this paper. This is not only a technical issue, but also a business issue.

- Based on 'compatibility' and 'extensibility', **'cooperation'** is largely a business issue. A number of issues have been discussed in this paper. It is important to emphasise the fact that 'cooperation' does not mean the cooperation between functional modules, but the cooperation between VE partners. This is why it is mainly a business issue.
- **'Ownership'** is also a business issue. It requires negotiation between VE partners, and a well-designed VE partnership agreement.

10   CONCLUSION

The rapid growth of ERP systems during the past few years is due to the recognition that ERP systems can help enterprises increase efficiency and reduce costs. But the enterprise boundary of normal ERP systems is not suitable for the rapidly growing business form Virtual Enterprise (VE), since VE highly emphasises partner cooperation, and the agility achieved by combining the core competences of partners. A new ERP model is needed to facilitate such cooperation and agility, while still providing all the functions of normal ERP systems. Based on VE Performance Measurement System (VEPMS), this paper has outlined a coordination model of the ERP systems in a virtual enterprise. Further this paper has provided a framework for the next generation of ERP systems – the VERP system, in the belief that future ERP systems must accommodate the new VE business form.

11   ACKNOWLEDGEMENT

The authors are grateful to the Engineering and Physical Sciences Research Council (EPSRC) for the award of a research grant which facilitated the research described in this paper.

REFERENCES

1. **Yen, D. C., Chou, D. C.** and **Chang, J.** (2002) A synergic analysis for web-based enterprise resources planning systems. Computer Standards & Interfaces Vol. 24, Issue 4, September, pp. 337-346.

2. **Wight, O.W.** (1984) Manufacturing Resource Planning: MRP II. Oliver Wight Limited Publications, Inc.

3. **Gulla, J. A.** and **Brasethvik, T.** (2002) A model-driven ERP environment with search facilities. Data & Knowledge Engineering, Vol. 42, Issue 3, September, pp. 327-341.

4. **Kochhar, A. K.** and **Zhang, Y.** (2002) A Framework for performance measurement in virtual enterprises. Proceedings of the 2nd International Workshop on Performance Measurement, University of Hanover, Germany, 6-7 June.

5. **Jägers, H., Jansen, W.** and **Steenbakkers, W.** (1998) Characteristics of Virtual Organizations. Pascal Sieber and Joachim Griese (Eds.), Organizational Virtualness, Simowa Verlag Bern.

AI Applications in
Manufacturing

AI-based optimization method for the analysis of co-ordinate measurements within integrated CAD/CAM/CAE systems

S ZIĘTARSKI
Institute of Aeronautics and Applied Mechanics, Warsaw University of Technology, Poland

ABSTRACT

To achieve higher geometric accuracy of complex shapes and overall increase in surface quality it is necessary to take full advantage of capabilities available in integrated CAD/CAM/CAE systems as well as in a new generation of CNC production equipment and coordinate measuring machines. Knowledge-based automation of product design, manufacture and coordinate measurements, using artificial intelligence methods, are of great importance for effective determination of global deviations between the virtual product model and the part machined. Due to a large number of points (500 to 100 000 and more), obtained from scanning measurements, it is only recently possible to run the appropriate, specialized software. To find a global optimum in the location and orientation of the cloud of points in relation to the geometric model, a new approach has been applied. A combinatorial-cyclic iteration method is, in fact, an reliable multivariable optimization method based on either deterministic models or selected neural network techniques. The method enables : (a) to analyse all errors in manufacturing processes, even on the most complex surfaces; (b) to effectively apply reverse engineering techniques; (c) to identify all sources of errors in CAD/CAM technology; (d) to eliminate the necessity of a precise set-up of the measured part on CMMs or other measuring systems; (e) to identify a global optimum, among many local optima.

1 INTRODUCTION

Integration between manufacturing and design modules with CMM scanning and digitizing capabilities has been achieved by developed, specialized programs covering various tasks of a virtual product development (geometric model) and manufacture processes, and then by connection of the model with point sets obtained from either coordinate measuring machines or probing systems of CNC machine tools, or laser interferometric measuring systems. A knowledge-based approach to CMM measurements enables to develop a method of comprehensive, surface tolerancing, where surfaces are reconstructed using thousands of points obtained from CMM scanning techniques. In much the same way, either CL data points from CAM part programming or points from a postprocessor, can be analyzed in order to determine either the accuracy of interpolation scheme or the reliability of postprocessor algorithms. Optimum variable values are searched according to the established objective function, which can be defined depending on the specific task: geometric analysis of parts machined or reverse engineering on the basis of existing physical model. The results are not distorted by local extremum problems occurring in well-known iteration methods. The method has been made feasible, because of a new generation of computers (512 MB RAM, 1.5 GHz processors) and programming tools within CAD/CAM/CAE systems (e.g. Unigraphics, Catia).

Sources of CNC machining errors are:
- inaccuracies of CNC machine tool systems (machine tool - cutting tool – workpiece);
- errors or limitations of postprocessor algorithms;
- errors in the concept of CAM part programming;
- errors of interpolation schemes inherent in applied CAD/CAM modules;
- last but not least: errors arising from deformation of a workpiece, regardless of machine tool accuracy (except of the concept of CAM part programming) , e.g. springback errors in sheet metal forming, deformation errors in thin-walled structures, etc.

A geometric model of surface concerned is a base for any distance calculation along normals between points and the surface. It should be noted, that even in the same CAD module using the same point set for available various curve and surface definitions (a cubic spline, Bezier curve, B-spline, NURBS curves and surfaces), the different surfaces may be built. The differences arising from applying the different curves are not negligible and for inspecting the surfaces by a CMM, the geometric model used for CNC part programming should be known.

Due to a large number of points (500 to 30 000), obtained from scanning measurements, it is only recently possible to run the appropriate, specialized software. Still, it can take several hours, even when workstations or PCs have more than 512 MB RAM and 1.5 GHz. Therefore, the important feature of the programs is the capability of selecting the points from a large data file, keeping almost the same accuracy of the analysis. This is a first layer in an artificial network approach. Also, the number of combinations can be reduced and it is a second layer in this approach. 10% percent of all points and combinations can enable to obtain results with the difference below 4% in relation to results from the analysis based on the full point set and a deterministic model. Nevertheless, the most important feature of the programs is an effective, convergent iterative method, searching for optimum values of selected variables with given accuracy. These variables are either coordinates for an origin and angles for an axis orientation (in an analysis of surface errors) or dimensions of the geometric model (in reverse engineering). Optimum variable values are searched according to the established objective function, which, usually but not necessarily, is a sum of squares of distances between points and surfaces. It must be pointed out, that for an reverse engineering applications of the method, the prerequisite is to develop knowledge-based subsystem for an automated design of the product geometric model. For the other applications this subsystem is preferable to an CAD interactive model, but not necessary.

Developed procedures, based on neural network techniques, used in the optimization model of both deterministic and probabilistic modes, are mainly defined by three important values:

n – the number of points from coordinate measurements (the first layer in a neural network);

m – the space dimension (m –the number of variables, optimal values of which are searched);

k – the number of levels for each variable (k=3,5,7,9,11,...), i.e. dimensionless coded values of variables, e.g. k=3 gives coded values:-1,0,1 ; k=5 gives coded values:,-2-1,0,1,2 ; etc.

$N_c = k^m$ – the number of combinations, e.g.. $11^2 = 121$, $3^{10} = 59049$, $3^5 = 243$ (mostly used in a practical analysis of surface errors), $11^5 = 161051$, $11^{10} = 25937424601$ (a difficult problem even for supercomputers). N_c is the size of the second layer in a neural network.

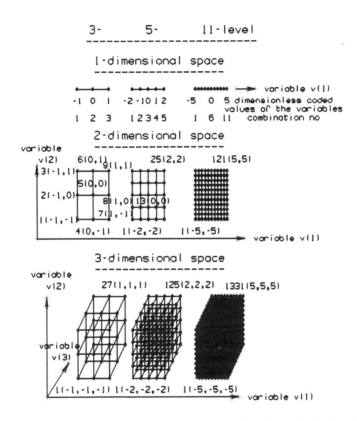

Figure 1. k-level m-dimensional spaces for the optimization method. Graphic presentation of m-dimensional spaces, where m>3, requires a non-Cartesian projective technique, but these spaces do not preclude using the same formulas for combination numbering and other calculations. It is analogous to multi-dimensional matrices.7, 9, 11 levels are not clearly visible in the same scale as 3 levels.

There are needed some additional explanations to the fig.1.The center of the space is defined by the combination no. $n_c=(N_c+1)/2$, e.g. $n_c=(9+1)/2=5$, where values of coded variables: $v(1)=v(2)=..=v(m)=0$; in general $v(j)_0=0$, $j=1..m$. Increments $dv(j)$ and code values i define boundaries (ranges) of spaces and real values of variables: $v(j)_i = v(j)_0+2$ i $dv(j)/(k-1)$; e.g. if $v(1)_0= 100$ mm, $dv(1)=2$ mm, $k=3$, then $v(1)_{-1}= 100 -2 =98$ mm, $v(1)_1= 100 +2 =102$ mm; if $k=5$, then $v(1)_{-2}= 100 -2 =98$ mm, $v(1)_{-1}= 100 -1 =99$ mm; $v(1)_1= 100 +1 =101$ mm, $v(1)_2= 100 +2 =102$ mm.

Typical problems defined in the 1-dimensional space: (a) a circular cylinder, where errors in a location and orientation of the axis are negligible and a searched variable $v(1)$=radius of the cylinder; (b) a sphere, where errors in a location of the center are negligible and a searched variable $v(1)$=radius of the sphere; etc.

Typical problems defined in the 2-dimensional space: (a) a surface roughness (Rq, Rt) from a point set of the profile , where searched variables: $v(1)$=distance of the line from the x axis, $v(2)$= angle of the line to the x axis; etc.

Typical problems defined in the 3-dimensional space: (a) a circular cylinder, where errors in an orientation of the axis are negligible and searched variables $v(1)$=radius of the cylinder, $v(2)$= x coordinate of the center, $v(3)$= y coordinate of the center; (b) a circular or elliptic cylinder, where errors in a location and orientation of the cylinder axis are negligible and a searched variables $v(1)$=semi-major axis of an ellipse ; $v(2)$=semi-minor axis of an ellipse, $v(3)$= angle of the semi-major to the x axis; there is the circle, when $| v(1)-v(2) | < \varepsilon$ (permissible error); etc.

Typical problems defined in the 4-dimensional space: (a) a sphere, where searched variables $v(1)$= x coordinate of the center, $v(2)$= y coordinate of the center , $v(3)$= z coordinate of the center, $v(4)$=radius of the sphere, etc.

Typical problems defined in the 5-dimensional space: (a) all NURBS and analytical surfaces, which are transformed to the optimal location and orientation in relation to the points from coordinate measurements; searched values $v(1)$= x coordinate of the system origin, $v(2)$= y coordinate of the system origin, $v(3)$= z coordinate of the system origin, $v(4)$= a first angle of the axis, $v(5)$= a second angle of the axis. The third angle: $\cos^2 \gamma=1- \cos^2\alpha-\cos^2\beta$. After achieving a global extremum, e.g. assuming a standard deviation as an objective function, all kinds of errors can be visualized and filed; it enables to develop a method of comprehensive, geometric and surface tolerancing.

6-,7-,8-,9-,10-dimensional space is applied to problems, where a reverse engineering (searching for geometric variables of the model) are interconnected with a transformation (searching for optimal location and orientation of the model). For example, a geometric analysis of a rotational paraboloid (satellite antennas, reflectors, etc.) requires 5-dimensional space (as above), but a reverse engineering problem requires adding one variable more and it means the 6-dimensional space. The added variable is the focal length p ($y^2=2px$), and it must be remembered that transformation problems can be eliminated by a precise set up of the measurement base what means 5 variables less. The method can be extended to m-dimensional space, where m>10, but these cases have been neither verified nor tested.

The units used for variables: m, mm, μm, nm, degrees, 1/1000 of degree.

The presented method has been validated, verified and tested on analytical shapes and NURBS surfaces as car body dies (including elimination of springback errors in sheet metals), a glider wing (the case described below in 3), TV tube molds, a parabolic antenna (including elimination of springback errors in sheet metals), an airplane fuselage, an airscrew, an yacht hull, hip joint prostheses, gears, etc. The dedicated software has been developed within the frame of the integrated CAD/CAM/CAE system UNIGRAPHICS (GRIP).

2 COMBINATORIAL-CYCLIC OPTIMIZATON METHOD

Iterative methods of optimization are applied in many areas of engineering activities. The developed AI-based optimization method has a great potential for solving engineering problems, but in the paper it is limited to identification of geometric quality of the product, and, of course, reverse engineering is in line with a quality improvements. Generally, the objective function of the optimization method is to minimize geometric deviations between a virtual product, represented by a CAD model of the product, and a real product after manufacturing, represented by points from measuring systems. It must again be pointed out, that tasks of reverse engineering require software-based automatic redefinitions of the CAD model, according to changing design variables. Usually, transformation problems can be separated from reverse engineering problems and it makes the difficult objective as a feasible one. There are criteria of an accuracy estimation of curves and surfaces, which can be used as the objective function (mostly a standard deviation) :

$$\sigma = \sqrt{\frac{\sum_{i=1}^{i=n}(d(i))^2}{n}} \; ; \quad t_{out} = \frac{\sum_{i=1}^{i=nout}+d(i)}{nout} \; ; \quad t_{in} = \frac{\sum_{i=1}^{ii=nin}-d(i)}{nin}$$

σ = standard deviation ; t_{out}, t_{in} = average upper and lower deviation; $+d(i)$, $-d(i)$ = outside and inside of nominal curve; $nout+nin = n$ (the number of points)

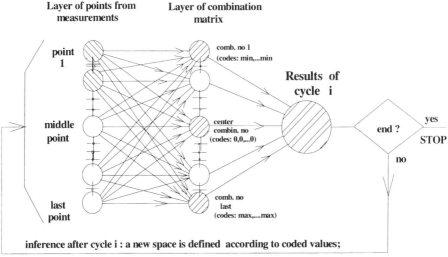

Figure 2. Combinatorial-cyclic optimization method presented in artificial neural network concept. One-directional two layers supported by evolutionary optimization algorithms; hatched elements of points and combinations are activated even in the probabilistic mode, not in the deterministic mode only.

Speaking from experience gathered in the area of iterative method applications in design and manufacture, there are two main points in optimization methods: a convergence of iterations and globality of the optimal solution. The difficulties at these points seem to be unsurmountable in terms of general mathematical formulas, but it is worthy to notice, that human brain is usually able to find an optimal solution after considering all aspects of the problem. Therefore, the combinatorial-cyclic method of optimization is embedded in a artificial intelligence techniques, including a neural network approach, where logical inference is based on designer's and manufacturer's knowledge. It is necessary, because in some cases a CAD model or CAM-NC programming are interrelated with optimization algorithms. As a result, a convergence and globality problems has been overcome, but it has been proven in the same way as in other reliable engineering software; a Coordinate Measuring Machine has been simulated and received points from the CAD model can be treated as an input to the optimization method. Of course, the point set (cloud of points) has a predetermined standard deviation, x-, y-, z-, α-, β-errors, points with maximal (+) and minimal (-) errors, and incorrect geometric parameters. If these data will be repeated as results from the method, we can assume that the method is correct. And it has been tested in 25-30 difficult cases. Absolute proof would be, when we divide m-dimensional space on elements equal to required accuracy, and a computer power is sufficient (fig.1). Still it is a long way for computers. As seen from fig.1, the simplest element in the m-dimensional space is a 3-level element and the entire space can be divided on $((k-1)/2)^m$ subspaces, all of which have the same sizes as space concerned, (in dimensionless codes), but with dislocated subspace centers. In deterministic mode, one cycle includes calculation of the objective function for all points from measurements with all combinations of the variable values, and it means that in 5-dimensional space the 11-level calculation requires 660 times more computational power than the 3-level calculation.

Table 1. Fragment of the ranking table after each cycle. X1=v(1),...., X10=v(10).

```
            C Y C L E    2
------------------------------------------------------------------
        COMBINATORIAL  ANALYSIS  FOR  OPTIMIZATION
        The best 20 combinations after  2 cycle(s)
==================================================================
:comb.no|st.dev.|  %  |   X1        X2        X3        X4       X5
        |       |     | (X1 code) (X2 code) (X3 code) (X4 code)(X5 code)
        |       |     |   X6        X7        X8        X9       X10
        |       |     | (X6 code) (X7 code) (X8 code) (X9 code) (X10 code)
==================================================================
 1:  127| .1802| 3.07|  1.024    -1.024   102.464    99.536    1.536
     |       |     |  ( 0 )     ( 0 )     ( -1 )    ( 1 )     ( 1 )
------------------------------------------------------------------
 2:  128| .1829| 1.49|  1.024    -1.024   102.464    99.536    1.024
     |       |     |  ( 0 )     ( 0 )     ( -1 )    ( 1 )     ( 0 )
.........................................................  etc.

Increment ranges of X( 1) for the cycle  3 : + .5120,- .5120
Increment ranges (accuracy) of X( 1) not changed
(coded levels of X( 1) in cycle  1 and cycle  2 :  1,   0)
-------------- etc.
```

The computing time for 1 cycle is controlled in three ways: (1) discriminated or limited random selection of points, from either cross-sections or a cloud of points; (2) discriminated or limited random selection of combination no's; (3) break of a cycle calculation after exceeding the assumed parameters of normal distribution (previous cycle is treated as a training cycle). If (1) or (3) applied, it is still deterministic mode, but (2) means probabilistic, neural network mode. Applying (1)/(2) brings decrease of an accuracy, so often we start from (1) and (2) , and finally we apply the deterministic mode, without (1) or (2).

An assumed m-dimensional space and a way of searching for a global optimum on the example of the 2-dimensional space is presented in fig.3. Typical, simplified algorithms are:

IF the space bounded by 1-7-9-3-1 is the **i**-th cycle **THEN** after analyzing the ranking table (tab.1.), the inference subprogram moves the **(i+1)**-th cycle space in the illustrated direction .

A: **IF** the objective function for combination no. **5** is in the ranking table on **1**
   THEN the space **A** is defined with decreased ranges: $dv(j)_{i+1} = dv(j)_i /2$
B: **IF** the objective function for combination no. **9** is in the ranking table on **1**
   AND the objective function for combination no. **3 OR 2 OR 1 OR 4 OR 7**
                        IS NOT in the ranking table on **2**
   THEN the space **B** is defined with decreased ranges: $dv(j)_{i+1} = dv(j)_i /2$, **ELSE** as in F:.
C: **IF** the objective function for combination no. **4** is in the ranking table on **1**

AND the objective function for combination no. **3 OR 6 OR 9**
IS NOT   in the ranking table on **2**
THEN the space **C** is defined with decreased ranges: $dv(j)_{i+1} = dv(j)_i /2$, **ELSE** as in F:
D: **IF** the objective function for combination no. **2** is in the ranking table on **1**
   AND in the **(i-1)**-th cycle combination no. **2** was also in the ranking table on **1**
   THEN the space **D** is defined with unchanged ranges and the dislocated center.
E: it is analogues to D: but two variables are concerned
F:(= F1,=F2) **IF** the objective function for combination no. **3** is in the ranking table on **1**
       AND the objective function for combination no. **7** **IS**   in the ranking table on **2**
THEN two spaces **F1 and F2** are defined with decreased ranges: $dv(j)_{i+1} = dv(j)_i /2$
AND being calculated parallel as two local optima, until one of them drops below **2**.

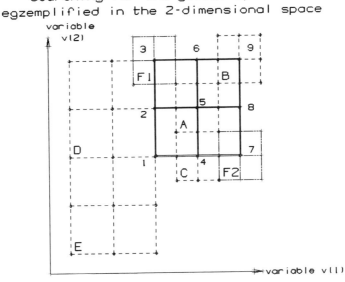

Figure 3. Graphic presentation of a simplified algorithm for searching the extremum of objective function in the 2-dimensional space.

All other directions of moves in the space are similar to A,B,C,D,E,F respectively and all of them are generalized for m-dimensional spaces. It is possible to express mathematical relationships between an objective function and variables by moving and changing a range of spaces, like 'zooming' and 'panning' on a computer screen: it substitutes derivative calculations. Dimensionless codes for variables make possible a generalization of inference logic after each cycle, particularly, when variables v(j) or increments dv(j) are measured in different units, e.g. mm and degrees, or dv(1)= 0.5 mm and dv(2)=2 mm. The method eliminates a difficulties involved in assigning initial values for variables. The incorrect initial

                                                    A003/037/2003 © IMechE 2003

value usually precluded finding an global, optimal solution, but in the presented method it only means the increased number of cycles, provided that appropriate constraints are included in the program generating a new geometric model.

3 GEOMETRIC ANALYSIS AND TOLERANCING OF SURFACES FOR A CASE STUDY: GLIDER WING

Classes of surfaces and solid bodies, which can be relatively easy defined and optimized by the method are: (a) planes, circles, ellipses, parabolas, hyperbolas, etc., (b) solid blocks and simple extruded bodies, (c) quadric surfaces as in the equation:

$$Ax^2 + By^2 + Cz^2 + D + 2Fyz + 2Gzx + 2Hxy + 2Px + 2Qy + 2Rz = 0,$$

(d) surfaces of revolution by revolving a conic curve or a series of curves, e.g. a cone, sphere, torus, rotational ellipsoids or paraboloids, etc. In these instances a geometric modeling and optimization constraining takes a few code lines in the subprogram, so often a transformation problem (error analysis) can be solved together with varying values of design parameters (reverse engineering). Also, algorithms for distances between points and surfaces are less complex, therefore, the results – what is important- are precisely the same in the top CAD/CAM systems. NURBS-defined surfaces are more and more significant among classes of surfaces, not only that it enables to define all shapes from engineering practice, but also that analytical curves can be joined and then precisely approximated by the NURBS curve as one entity.

A fuselage, wings and vertical/horizontal stabilizers of a glider are modeled as NURBS surfaces or solid bodies. Usually, both are required because, for some modules, surfaces are preferred to solid bodies (e.g. manufacturing - CAM). Achievements of the traditional design of gliders, with only size tolerancing for surfaces, are worth to be considered as a reliable knowledge base. For example, the wings of gliders PW-5 and PW-6 (PW5 – Polish World Class Glider) have been modeled , firstly, for CNC machining of molding tools needed for forming wings made of composites (glass-fiber reinforced plastic), and secondly, for a geometric and error analysis of wings manufactured. The wing, 7700 mm long, has been measured by an optical measuring system, equipped with scanning capabilities (approx.13000 points in 1 m of length, accuracy 0.05 mm). After careful consideration of various traditional approaches to a comprehensive geometric analysis of complex surfaces it has been concluded that the developed method within the frame of the integrated CAD/CAM/CAE system has the greatest potentialities.

A base curve for a CAD geometric model of the wing is the selected, after aerodynamic investigations, airfoil profile (NN 18-17, seen in fig.5) defined by points in the plane, where x, y are measured as a percent of the chordal length (according to standards approx. 22 points); chordal length of the wing 250-700-1300 mm. Points, defining airfoils for a surface, play an important as well as difficult role in a process of searching for the best design version. The upper and lower curve of the airfoil are tangential joined with circles on the leading and trailing edges (see fig.5). Tangential joining second degree curves with third degree curves was in the past a difficult task, when high accuracy had been required. There are more than 10 curve definitions (cubic splines, Bezier curves, B- splines, NURBS), which can be used for geometric modeling from points in advanced, integrated CAD/CAM/CAE systems. For example, the same 22 points used for definition of the cubic spline and NURBS (2.degree), give the theoretical maximum distance between these two curves as large as 0.35 mm. In order to select the most suitable curve definition, the specialized procedure has been developed. It has been shown in fig. 4 and in formulas below, describing an orthogonal approximation with dividing the point set in two subsets (it can also be more subsets).

$$\sigma 1 = \sqrt{\frac{\sum_{i=2}^{i=n-1}(d_1(i))^2}{n-2}} \quad ; \quad \sigma 2 = \sqrt{\frac{\sum_{i=2}^{i=n-1}(d_2(i))^2}{n-2}}$$

where: $d_1(i)$ = distance along normal between point(i) and curve 1; $d_2(i)$ = distance along normal between point(i) and curve 2; n=number of points for the curve definition.

Objective function for the most design oriented definition: $c_{do} = (\sigma 1 + \sigma 2) / 2 = minimum$.

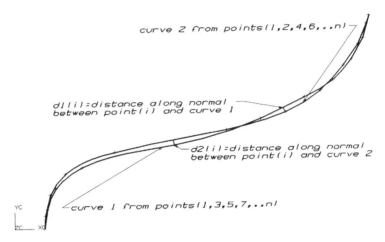

Figure 4. Curves used for finding the most design oriented curve definition from among cubic splines, Bezier curves, B-splines, NURBS.

Computing procedures for curves and surfaces are not the same in various CAD modules, though derived from the general NURBS formulas. The applied formulas for curves and surfaces are as follows: $C(u) = \sum_{i=1}^{i=r} M_i^m(u) w_i Q_i \quad ; \quad S(u,v) = \sum_{i=1}^{i=r}\sum_{j=1}^{j=s} M_i^m(u) \, N_j^n(v) w_{ij} Q_{ij} ;$

where: $C(u)$- NURBS curve , if all weights $w_i=1$; r - number of points ; m - a degree of the curve; M_i^m - B-spline basis functions of degree m (for u) ;$Q_i= (x_i , y_i , z_i)$ - points ; $S(u,v)$- a NURBS surface , if all weights $w_{ij}=1$; N_j^n - B-spline basis functions of degree n (for v).

Crucial points for applications of the method are algorithms for accurate calculations of the normal vector from the point out- or in-side of the geometric model, and the distances arising from that. It is necessary to have a specific program detecting and correcting errors in these

A003/037/2003 © IMechE 2003

calculations. Starting points are formulas below. Computing procedures for normal vectors of the surface in the point $N(u,v)$ and partial derivatives, which can also be used for building the offset surface, are based on general formulas:

$$N(u,v) = \frac{\dfrac{\partial S(u,v)}{\partial u} \times \dfrac{\partial S(u,v)}{\partial v}}{\left| \dfrac{\partial S(u,v)}{\partial u} \times \dfrac{\partial S(u,v)}{\partial v} \right|} \qquad \text{where:}$$

$$\frac{\partial S(u,v)}{\partial u} \times \frac{\partial S(u,v)}{\partial v} = \left(\frac{\partial y}{\partial u} \cdot \frac{\partial z}{\partial v} - \frac{\partial z}{\partial u} \cdot \frac{\partial y}{\partial v} \right) i + \left(\frac{\partial x}{\partial u} \cdot \frac{\partial z}{\partial v} - \frac{\partial z}{\partial u} \cdot \frac{\partial x}{\partial v} \right) j + \left(\frac{\partial x}{\partial u} \cdot \frac{\partial y}{\partial v} - \frac{\partial y}{\partial u} \cdot \frac{\partial x}{\partial v} \right) k$$

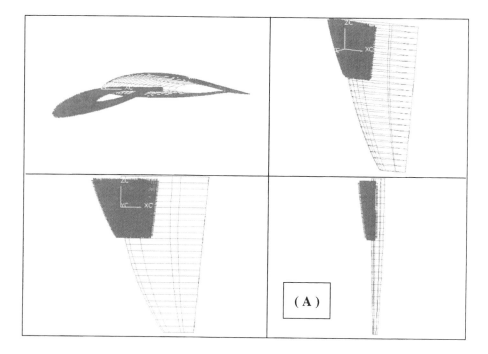

(A)

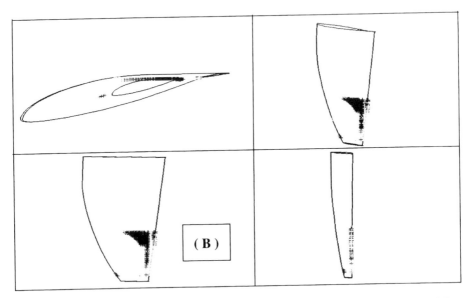

Figure 5. Graphic presentation of the program outputs. In a monitor picture of the geometric model the error areas are differentiated by colors for ±3σ, ±2σ, ±σ. points from measurements in relation to the coordinate system of the glider wing as the initial position before cycle 1; (B) points on the wing surface exceeding ±3σ.

The graphic presentation of initial positions of points from measurements and the graphic presentation of errors, exceeding ±3σ, is given in fig.5 (A), (B); a fragment of final results in tab.2. Main input data to the CAD virtual wing are: (1) span of the wing, (2) chordal lengths defining the change of outline, (3) airfoil profiles with the predefined density e.g. at step 25 mm , (4) curves along the wing, (5) set of various airfoils with standardized definitions. As seen in fig.5 (A), the surface and solid body of the wing have been defined as a mesh of curves. The CAD subsystem generating variants of the wing consists of approx. 350 code lines. It must be pointed out, that geometric models from the subsystem and from interactive mode are the same, and may be used by the method as equivalent, when the error analysis is concerned (not reverse engineering or software- based searching for the optimal design variant).

Table 2. Fragments of the final results after error analysis of the glider wing.

FINAL RESULTS OF OPTIMIZATION
OPTIMAL ITERATION after the last cycle
Accuracy ranges (increments) for optimal values:
delt(1)= .0100 (2)= .0100 (3)= .0100 (4)= .0100 (5)= .0100
Variables of the geometric model (combination no.= 1, cycle= 2):
vsrch(1)= -413.2700 (2)= -45.5000 (3)= 538.0300 (4)= 14.5800
*vsrch(5)= 4.8000 (6)=*******.**** (7)=*******.**** (8)=*******.*****

A003/037/2003 © IMechE 2003

The number of points out of the required tolerances: 3670 (30.7 % of 11969)
The number of points outside (above + 0.931 mm): 2002 (16.7 % of 11969)
The number of points inside (below -0.931 mm): 1668 (13.9 % of 11969)

------------ ANALYSIS OF SURFACE ERRORS ------------
The standard deviation: 0.93057 mm
The mean upper deviation (outside distances): 0.746 mm (6976 points)
The mean lower deviation (inside distances): -0.727 mm (4993 points)
The number of errors exceeding 3*st. dev.: 73 (.61 % of 11969)
The number of uncorrected errors, exceeding 3*st.dev.: 54 (.45 % of 11969)

OUTSIDE AND INSIDE ERRORS
Ten maximum outside distances:
ptm(11969): x=   207.655 y=   5.555 z= -261.656 dist=   3.7244 mm
ptm(11902): x=   209.265 y=   5.572 z= -256.656 dist=   3.7226 mm
..................etc..
Ten minimum inside distances:
ptm(10675): x=   -189.800 y=   -89.907 z= -201.656 dist= -3.7150 mm
ptm(11440): x=   -175.574 y=   -84.102 z= -236.656 dist= -3.7062 mm
.................etc
.

Explanations to the computer output from tab.2.:
delt(1),..... ,(5)= dv(1),...dv(5)= increments for the variables
vsrch(1),....,vsrch(5)=v(1),.....,v(5)= x(1),....x(5)
ptm(1..11969)= points from measurements.
vsrch(1)= x coordinate of translation for all points ptm in the wing coordinate system,
vsrch(2)= y coordinate of translation for all points ptm in the wing coordinate system,
vsrch(3)= z coordinate of translation for all points ptm in the wing coordinate system,
vsrch(4)= angle of rotation in relation to the x axis of the wing coordinate system
for all points ptm,
vsrch(5)= angle of rotation in relation to the y axis of the wing coordinate system
for all points ptm.

4 CONCLUSIONS

The presented method of the error analysis for complex surfaces by using CMM and other measuring systems has a great potential for applications in the airplane, automotive, and machine building industry. A software based approach to coordinate measurements enables to develop a method of comprehensive surface tolerancing and helps to identify and eliminate all errors. It can be used for reverse engineering as well. As a result, the accuracy of a new CMM generation (0.001 mm) as well as the accuracy of CNC machine tools (0.02 mm) can be fully utilized. The developed knowledge-based software added to integrated CAD/CAM/CAE systems, covering design, engineering and manufacturing, can be a meaningful factor in speeding-up the whole process toward quality products. Rapidly increasing computational power of PC and availability of integrated CAD/CAM/CAE systems make the presented method more and more effective.

5 REFERENCES

1 **Gadalla M.A., ElMaraghy W. H.:** Comparison between Fitting and Surface Offsetting Techniques to Improve the Accuracy of Sculptured Surfaces. Proceedings of the 31st CIRP International Seminar on Manufacturing Systems: Networked Manufacturing- Integrated Design, Prototyping and Rapid Fabrication. May 26-28, 1998, University of California, Berkeley, USA

2 **UG/Open GRIP Reference.** Vol. 1, 2, 3. Electronic Data Systems Corporation, 1998.

3 **Piegl L., Tiller W. :** The NURBS Book . Springer, 1997

4 **Farin G.:** Curves and Surfaces for Computer-Aided Geometric Design. Academic Press, 1988

5 **Chen Y., Johnson G., Simon D.:** Fast Complex Surface Inspection and Verification Using Optical Laser Line Scanner. Transactions of the North American Manufacturing Research Institution of SME. Vol. XXII, 1994

6 **Tyler G. Hicks S., Hicks D., Leto J.:** Standard Handbook of Engineering Calculations. McGraw Hill Inc., 1995

7 **Żurada J.:** Introduction to Artificial Neural Systems. West Publishing Co, 1992

8 **Osowski S.:** Sieci neuronowe w ujęciu algorytmicznym.(Neural Networks in Algorithmic Approach). WNT, 1996

Manufacturing and Supply
Chain Management

The design and implementation of manufacturing cells – an alternative methodology

N BOUGHTON and **H SHARIFI**
University of Liverpool Management School, UK

SYNOPSIS

This paper describes an alternative methodology to support the design and implementation of lean manufacturing cells within the aerospace sector. The methodology, which includes best practice lean principles, addresses the key phases of the cell/module design and implementation problem. The six phases of the methodology include the identification of the improvement opportunity, concept and detailed design, design implementation and the phased transition to operational status. Consideration is not only given to the design of the physical system but also the control system(s) and their interfaces, both internal and those customer/supplier facing. This paper will outline the methodology and include some initial findings.

1 INTRODUCTION

The earliest work in the field of Group Technology (GT) has been credited to Mitrofanov (1) in the 1950s. Burbidge (2) and Edwards (3) completed pioneering work in the UK in the 1960s, and since that time the volume of literature concerning the subject of GT/cellular manufacturing has grown extensively. For the purpose of this discussion cellular manufacturing (CM) involves the processing of a collection of similar parts on a dedicated cluster of dissimilar machines, in this way CM exploits the underlying commonality between parts and manufacturing processes (4). This contrasts with the traditional routeing of a product through the required but separate functional areas.

CM has for many years been advocated as the preferred way to arrange the shop floor resources of an organisation (5, 6, 7). CM has wide applicability and can yield benefits in reduced set-up time, throughput time, WIP, material handling distances and improved quality (8, 9). Furthermore manufacturing cells, particularly U-shaped cells, are central to JIT systems and the lean thinking outlined by, for example, Black (7) and Womack and Jones (10), and the Toyota production system described by Monden (11). Lean manufacturing cells boast additional benefits over traditional manufacturing cells in terms of material control, visual management and flexibility (10), and which are achieved through the adoption of a number of lean tools and techniques.

In recent years the topic of cellular manufacturing has been the focus of much research attention (9, 12). The literature, however, has to an overwhelming degree focused on the development of procedures to solve the cell formation problem. Whilst the knowledge about such techniques appears extensive, these research efforts have concentrated on determining optimal solutions and not considered how manufacturing organisations actually set about designing cells (9). Other important activities involved in the implementation of CM systems include the design of the physical system and the management control system(s), as well as supporting their phased implementation and integration within existing practices. At the practitioner level therefore there is an increasing need for the development of methodologies that integrate these different activities into a single holistic process of implementation (13, 14).

This paper outlines and critiques current approaches to the 'design' of manufacturing systems and introduces the early results from an active research project, undertaken in close collaboration with a leading aerospace supplier, to develop a methodology for the design and implementation of lean manufacturing cells. The methodology will be briefly described.

2 THE APPLICATION OF CELLULAR MANUFACTURE WITHIN INDUSTRY

Empirical research provides useful information regarding the extent of application of CM within industry. However, the volume of empirical research is fairly limited with major contributions being made by Wemmerlöv and Johnson (8, 9), Wemmerlöv and Hyer (4) and Choi (17). Ingersol Engineers (15, 16) have also produced comprehensive studies as to the extent of use of CM within the UK.

Various issues regarding the use of CM have been researched, including the rationale behind the adoption of a cellular structure, benefits sought and actually realised, the usage of formalised design methodologies and the major constraints to the design and implementation of a cell. The findings from such work have a direct impact on the research topics currently being developed. One notable finding, and perhaps somewhat surprising, is that the majority of organisations have no familiarity of, or have applied, any specific method developed by the researchers in the field (9). Of course this lack of knowledge has not impeded cell design and implementation in these organisations. Cell design *per se* is not viewed by industry as the key problem (9); it is the more 'softer' issues that are seen as the greater challenge.

The increasing awareness and subsequent implementation of lean principles within the manufacturing domain leads to, almost by definition, an increase in the consideration and application of cellular manufacturing. Current trends within the aerospace sector reflect such developments, see for example the UK Lean Aircraft Initiative (UKLAI) (www.bestpracticecentre.com).

A003/011/2003 © Boughton2002

3 THE DESIGN OF MANUFACTURING SYSTEMS

Whilst a number of methods have been developed for the design of specific areas of CM they are rarely combined into a complete framework that supports both the design and implementation of a manufacturing cell. A number of frameworks have been proposed, however, in an attempt to address this problem.

Parnaby (18, 6) has been one of the primary contributors to the design of manufacturing systems. His approach incorporates a five-stage methodology and includes (i) market assessment, (ii) products and manufacturing process analysis, (iii) steady state design, (iv) dynamic design, and (v) integration and control systems design. Black (7, 19) has developed a ten-step guide to the implementation of lean manufacturing cells. Preliminary stages of this method focus on people issues, such as, the gaining of top-level management support and the commitment of the workers to the change. The first stage of the process is to form U-shaped cells. The remaining stages focus on integrating a number of lean techniques, such as single-minute exchange of dies (SMED), quality control, preventative maintenance, level scheduling and supplier development. Once the factory has been restructured into a lean facility, the company will be required to restructure the rest of the company. This will require removing the functionality of the various departments and forming teams, often along product lines.

Other contributions made in this field include Wu (20) and Hitomi (21), who both address the issues involved with manufacturing production systems. A recent study (14) details a methodology for the implementation of CM. The methodology, which has been tested through an implementation in a manufacturer's facility, is comprised of three phases, preparation, definition and installation. The emphasis of this methodology remains, however, focussed on the cell formation problem, and fails to take into account, in any great detail, any of the wider implementation issues. Similar studies (22, 23) have also been developed with comparable conclusions.

None of the studies described here have produced a truly comprehensive model for the implementation of CM, incorporating timescales, optimisation processes, performance evaluations and lean principles. Compounding the problem Devereux *et al.* (24), who surveyed the use of design methodologies in manufacturing companies in the UK, found that no existing, single widespread design method currently in use. Also, of the companies surveyed only 53% were employing any formalised design methodology. These results concur with later studies by Wemmerlöv and Johnson (8). Given this situation there is clearly scope for the development and promotion of a formalised approach to manufacturing systems design, and one which incorporates the integration of suitable tools and techniques. Such a framework should also be applicable for small to medium enterprises (SMEs) to adopt.

4 LIMITATIONS TO CURRENT APPROACHES TO THE DESIGN OF MANUFACTURING SYSTEMS

There are a number of limitations to the approaches described in the CM literature. These limitations can be categorised in terms of scope and application.

4.1 Scope

It is clear that most of the research in CM has been directed at solving the cell formation problem. Consequently there has been an over-emphasis on techniques to define cell structures rather than methodologies for the design of the whole manufacturing system. Parnaby (6) emphasises this point by defining the two types of systems design problem: the macro problem (concerned with large systems, integrating machines, processes, control systems, etc.) and the micro problem (concerned with a small, distinct part of the complete manufacturing system). Manufacturing systems design has largely been treated as a micro problem and, on the whole, this remains the situation. Also, micro problems considered in isolation tend to lead to only a sub-optimal solution to the more holistic macro problem. Whilst it is acknowledged that the cell design issue is one of the major challenges associated with the implementation of a new cell, it is by no means the only challenge in the adoption of CM. Furthermore pragmatists hold the opinion that the mere existence of a cellular structure is sufficient to realise the improvements associated with CM (25). Jones (26) states that the most successful approaches to implementation have included four key elements. These four elements are organisation and people, engineering, logistics/control and accountability.

This view of the macro problem in the implementation of CM is supported by actual accounts of implementations of manufacturing cells. In these accounts, for example (22, 23), little attention is drawn to the cell design problem. It is the remaining implementation issues, as outlined above, that are seen as constraints on the process.

4.2 Applicability

Various techniques have been suggested for forming manufacturing cells. However, many of them have proved too complicated for practising managers to comprehend and use (27), either due to the complex computational analysis or simply the time constraints within which they have to complete the analysis. Additionally managers are more interested in simple cell formation procedures (27). Moreover there is no evidence to suggest that any of the complex algorithms developed for the design of a manufacturing cell have ever been used in industry (24). Wemmerlöv and Johnson (9) suggest the greater problems in the implementation of CM are not caused by the designing of the manufacturing cell, but rather in the implementation of the cell.

For a company to be successful with CM they need to have both the knowledge and ability to implement the design. The emphasis of current research seems to be on developing new techniques rather than evaluating the current contributions. As a result increasingly complex solutions are being discovered to solve the part/machine grouping problem where there is little evidence to suggest that these methods are ever used in practice or even understood (9), and thereby fail to address the real industrial requirements.

5 THE DESIGN AND IMPLEMENTATION OF CELLULAR MANUFACTURING SYSTEMS – AN ALTERNATIVE METHODOLOGY

5.1 Background

A methodology for the design and implementation of cellular manufacturing systems has been developed in close collaboration with a leading aerospace supplier who has been committed to significantly reorganising its operations in line with lean principles. The company is a first and

   A003/011/2003 © Boughton2002

second tier supplier located in the north west of England and manufactures medium complexity parts for use in the aerospace sector. Over recent years the site has undergone radical change from the traditional functional layout to one that utilises U-shaped manufacturing cells across a number of manufacturing modules.

5.2 The development of a methodology
An overview of the methodology is illustrated in figure 1 and summarised below. Although comprised of six sequential phases, the methodology deliberately incorporates the necessary inter-relationships and iterations via timely feedback.

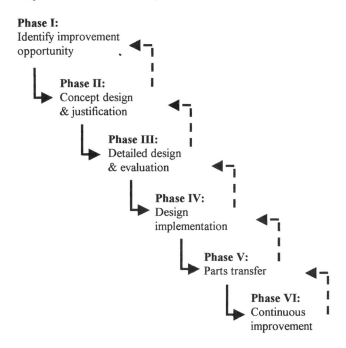

Figure 1: An overview of the cell design and implementation methodology

5.2.1 Phase I – Identification of improvement opportunity
The first phase of the methodology is the identification of an opportunity for improvement. This might be a rearrangement of existing shop floor arrangements or the identification of an additional cell to incorporate new work due to an expansion of the business. This decision should be made in line with the business strategy of the company. For this phase to be completed an understanding of the present market environment also needs to be gained. Initial data collection is required to verify the feasibility of the proposed implementation.

5.2.2 Phase II – Concept design and justification
Once an opportunity for investment or improvement has been identified the project must be defined. Definition of the cell requires a large volume of data to be collected on, for example, part routeings, machining times, part costs and quantities. At this stage of the project a

concept design (or designs) is developed. The concept design is a rough-cut design that demonstrates the material flow and levels of capacity required in the manufacturing cell. The concept design is produced through analysis of part routeings and machining requirements, and provides a first impression of how the cell will operate. Performance objectives are determined for the cell based on the improvements sought by the implementation and this data is formulated into a formal financial justification, which with the concept design aims at gaining capital expenditure approval for the project. The levels of capital required vary from small investments involved with rearranging existing operations to larger investment involved with the commissioning and implementation of a new bespoke manufacturing equipment; both options are addressed within the methodology.

5.2.3 Phase III – Detailed Design and Evaluation

The detailed design phase of the project aims to develop a robust cell and material control design. At this phase of the project best practice lean principles are introduced. The cell should be designed, where appropriate, in a U-shape to allow visibility, reduce material handling and increase operator flexibility; see for example (28). The starting point of the cell design is the concept design previously produced. From this a more detailed steady state design is developed. It is within this phase of the project where the majority of existing tools and techniques are focussed. A steady state design produces the system to meet the average requirements of the cell, in terms of resources, i.e. man, machine, and material. From this a detailed design is produced which attempts to predict more precisely how the system will behave under real operating conditions. The design is likely to require modification in order to ensure that the performance objectives are met under all operating conditions. Typically, this design (or designs) will be evaluated through dynamic/simulation modelling.

The control systems design is developed alongside the (physical) cell design. The control system should be designed to capitalise on the simplified nature of the planning and control problem as well as, where appropriate, the interface to any centralised system(s). Where applicable the use of a kanban-based pull system should be adopted to control material movements within the cell and at the cell interfaces.

5.2.4 Phase IV – Design implementation

The implementation phase of the project is the largest. Accordingly this phase is separated into three related elements: man, machine and material. The man elements concern team selection procedures, as well as the communication strategy for informing all employees of the changes and to gain feedback. Training needs are also established in, for example, lean tools and techniques as well as machine related operator training. The machine requirements are concerned with the determination of the engineering requirements for the cell. Parts may need to be re-engineered and prove-outs completed on the machines to ensure each machining process is capable, an essential requirement in the aerospace sector. For 'greenfield' implementations machinery will need to be commissioned and located in the cell; alternatively, 'brownfield' implementations will require a large number of machine moves. Material issues include supplier development and issues concerning quality. Supplier development, ongoing throughout the project, informs the (external) suppliers of the changes that are being made as well as the knock-on implications to them and expectations of them. This may take the form of direct line feeds and/or kanban delivery. Customer development is also required to develop important level schedules for production.

A003/011/2003 © Boughton2002

5.2.5 Phase V – Parts transfer

This phase of the project concerns the handover of ownership of the cell from the project (or engineering) team to the production/operations team. Until this point engineering will have managed phases II to IV. The concept of ownership is vital for the cell's future operation and, due to conflicts over the cell's resources, the handover can be a highly complex (and political) problem. The project/engineering team require capacity for the development (i.e. proving) of new parts, whereas the operations team require cell capacity for the production of parts to meet customer orders. Guidelines are therefore required to support this transfer of ownership of control of the cell's capacity as parts are proved and moved into production. The project team completes the handover of a part by developing a sealed route that ensures that the part is acceptable for production. The handover of capacity accompanies the handover of the part. In this way the resources of the cell are equally apportioned and thereby ownership of the cell is transferred smoothly and with no conflicts for resource. Although the handover is a simple concept it is difficult in practice to manage the conflicting objectives of operations and engineering; clear project management is required as well as formal agreement between the two functions.

5.2.6 Phase VI – Continuous improvement

Once the cell has been handed over to the operations team, a post implementation evaluation is required to be completed. The aim of this stage is to evaluate the success of the implementation and determine its performance relative to the initial objectives set out in phase I. Feedback will also be gained from this process that can be utilised for future cell implementations.

With concern to the lean nature of the manufacturing cells, an ongoing improvement plan is required to be constructed. The cell will have been designed to incorporate lean tools and techniques and these will need to be supported through the lifetime of the cell. Improvement activities such as 5S, visual management, and material control audits should be implemented. As the cell improves there will be a reduction in WIP, reduction in lead-time and a further improvement in quality in line with the overall strategy defined at the outset of the project.

5.3 More detailed guidelines

Further detail to each of the six phases is provided through a series of hyperlinked input-output boxes. Figure 2 illustrates the case for phase II. As indicated this phase is made up of a series activities namely, project team formation, definition of performance objectives, data collection and analysis, concept design and evaluation, and concept design justification. The key inputs and outputs to this phase are also listed. Each of the activities is hyperlinked to a detailed text-based explanation of the activity and, where appropriate, further hyperlinks to related discussion, tools/techniques, issues to consider, etc. The intention is to expand and/or update each activity based on design/implementation experiences as well as the continued development of accepted best practice.

6 SUMMARY

This paper has reviewed a number of approaches to the design of CM systems and which have focussed mainly on the underlying design techniques used to solve the cell formation problem. It has been suggested that these methods have become increasingly complex and thereby more detached from actual industrial requirements and expectation. As a consequence there is evidence to suggest that these techniques are neither being adopted nor understood by industry. Attempts to design the manufacturing system from a more holistic perspective have been described and critically evaluated, and the need for a more detailed formalised methodology for the design and implementation of CM has been proposed. The methodology outlined incorporates all of the stages required, taking into account the major limiting factors of previous techniques. This methodology has been developed through collaboration with a leading aerospace supplier. It is suggested that this methodology overcomes the major limitations identified in this paper.

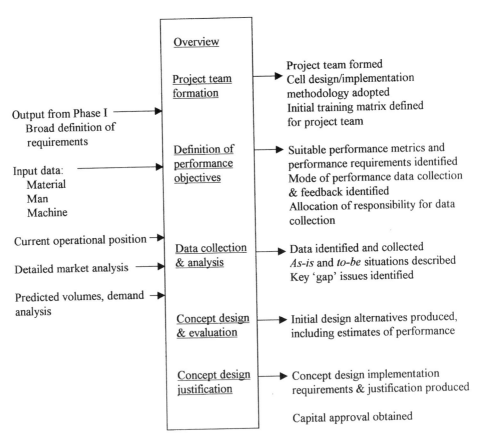

Figure 2: Phase II detail

7 ACKNOWLEDGEMENTS

The authors gratefully acknowledge receipt of EPSRC grant GR/M99545 and the support of their industrial partners.

8 REFERENCES

1 **Mitrofanov, S. P.** (1966) *Scientific Principles of Group Technology, Part 1* (National Lending Library of Science and Technology, Boston, MA).

2 **Burbidge, J. L.** (1975) *The Introduction of Group Technology* (Heinemann, London).

3 **Edwards, G. A. B.** (1971) *Readings in Group Technology* (Machinery, London).

4 **Wemmerlöv, U. & Hyer, N. L.** (1989) Cellular manufacturing in the U.S. industry: a survey of users, *International Journal of Production Research*, Vol. 27, pp. 1511-1530.

5 **Burbidge, J. L.** (1979) *Group Technology in the Engineering Industry* (Mechanical Engineering Publications Limited, London).

6 **Parnaby, J.** (1986) The design of competitive manufacturing systems, *International Journal of Technology Management*, Vol. 1, pp. 385-396.

7 **Black, J. T.** (1991) *The Design of the Factory with a Future* (McGraw-Hill, New York).

8 **Wemmerlöv, U. & Johnson, D. J.** (1997) Cellular manufacturing at 46 user plants: implementation experiences and performance improvements, *International Journal of Production Research*, Vol. 35, pp. 29-49.

9 **Wemmerlöv, U. & Johnson, D. J.** (2000) Empirical findings on manufacturing cell design, *International Journal of Production Research*, Vol. 38, pp. 481-507.

10 **Womack, J. P. & Jones, D. T.** (1996) *Lean Thinking* (Simon and Schuster, London).

11 **Monden, Y.** (1994) *Toyota Production System* (Chapman and Hall, London).

12 **Reisman, A., Kumar, A., Motwani, J. & Cheng, C. H.** (1997) Cellular manufacturing: A statistical review of the literature (1965-1995), *Operations Research*, Vol. 45, pp. 508-520

13 **Hassan, M. D.** (1995) Layout design in group technology manufacturing, *International Journal of Production Economics*, Vol. 38, pp. 173-188.

14 **Da Silveira, G.** (1999) A methodology of implementation of cellular manufacturing, *International Journal of Production Research*, Vol. 37, pp. 467-479.

15 **Ingersoll Engineers Limited** (1990) *Competitive Manufacturing: the Quiet Revolution* (Rugby).

16 **Ingersoll Engineers Limited** (1993) *The Quiet Revolution continues*, (Rugby).

17 **Choi, M. J.** (1996) An exploratory study of contingency variables that affect the conversion to cellular manufacturing systems, *International Journal of Production Research*, Vol. 34, pp.1475-1496.

18 Parnaby, J. (1979) Concept of a manufacturing system, *International Journal of Production Research*, Vol. 17, pp. 1-16.

19 Black, J. T. (1999) Design and implementation of lean manufacturing systems and cells, in Irani, S.A. (1999) *Handbook of cellular manufacturing systems*, Wiley, pp. 453-496.

20 Wu, B. (1992) *Manufacturing systems design and analysis* (Chapman and Hall).

21 Hitomi, K. (1994) Manufacturing systems: past, present and future, *International Journal of Manufacturing Systems Design*, Vol. 1, pp. 1-17.

22 Gunasekaran, A., McNeil, R., McGaughey, R. & Ajasa, T. (2001) Experience of a small to medium size enterprise in the design and implementation of manufacturing cells, *International Journal of Computer Integrated Manufacturing*, Vol. 14, pp. 212-223.

23 Afuzulpurkar, S., Huq, F. And Kurpad, M. (1993) An alternative framework for the design and implementation of cellular manufacturing, *International Journal of Operations and Production Management*, Vol. 13, pp. 4-17.

24 Devereux, S., Smith, P. & Wood, D. (1994) A survey of the use of design methodologies for implementing change in manufacturing companies in the UK, *International Journal of Manufacturing Systems Design*, Vol. 1, pp. 51-58.

25 Cantamessa, M. & Turroni, A. (1997) A pragmatic approach to machine and part grouping in cellular manufacturing systems design, *International Journal of Production Research*, Vol. 35, pp. 1031-1050.

26 Jones, D. (1994) Cell manufacture – The quiet revolution, *Colloquium Digest – IEE*, Issue 79, pp. 1-6.

27 Kerr, D. C. & Balakrishnan, J. (1996) Manufacturing cell formation using spreadsheets, *International Journal of Operations and Production Management*, Vol. 16, pp. 60-73.

28 Rother, M. & Harris, R. (2001) *Creating continuous flow: an action guide for managers, engineers & production associates*, Lean Enterprise Institute, USA.

Supply chain performance measurement

C W KHOO, A C LYONS, and D KEHOE
University of Liverpool Management School, UK

SYNOPSIS

In the Future Supply Innovations (FUSION) project at the University of Liverpool, the authors have been engaged in action research at a luxury car assembly plant and several of its key components suppliers. The study is to identify and create the necessary pre-condition to develop a prototype of an e-enabled supply chain. One of the aims of this project is to develop a method to measure supply chain performance.

In this paper, the authors raise the interaction between the components and supply chain design, as well as the measurement metrics. A case study is included in this paper to illustrate the relationship between these three aspects.

1 INTRODUCTION

Measuring supply chain performance is no longer a new topic. There are many measurement approaches suggested by different researchers. There are three measures in Beamon's (1) approach: resource, output and flexibility. Gunasekeran et al (2) classified the measurement metrics into three levels (strategic, tactical and operational level), either financial or non financial, or both. Brewer and Speh (3) suggested using the balanced scorecard to measure supply chain performance. Even though there are so many different approaches to measure supply chain performance, one common fact is there is no one perfect measurement approach that can suit all supply chains.

The criteria to judge the success of a supply chain are varied. This paper seeks to explore a supply chain measurement approach suitable for the production of high volume, customised products such as cars and computers.

However, even within a mass-customised environment, it is impossible to have one single measurement method that will suit all the supply chains in this industry. The supply mechanisms of the components that built the end product are bound to be different. For example, the supply mechanism of the key components of a vehicle (e.g. seat and instrument panel) will be tighter and more sophisticated than the non-key components (e.g. label, coin holder). A different supply mechanism often means different measurement criteria are appropriate. The selection of measurement metrics for a specific supply chain design is discussed in this paper.

2 SUPPLY CHAIN DESIGNS AND MEASUREMENTS

In the automotive industry, even though both the seat and the coin holder are components of the car, the supply chain structure for the seat is more complicated than the coin holder. In one of the UK luxury car assembly plants, the seats are supplied according to the final production sequence of the vehicle, an initiative sometimes called Sequenced In Line Supply (SILS), whereas, the coin holder is managed via a Kanban system. In SILS, the car assembly plant broadcasts the demand to the supplier, indicating the types of seats required and the sequence of fixing the seat onto the cars. There is a dedicated EDI system to broadcast the demand to the suppliers that supply sequenced components. Obviously, SILS is very different and more complicated than the Kanban system. This shows that the role of the component has to be taken into consideration when designing the supply mechanism.

Since there are different supply chain designs for different components, it is necessary to consider discrete performance measurement methods. The measurement should correspond to the supply chain structure. SILS is usually applied to key components, which have high variety and high cost. In SILS, the demand information broadcasted is to inform the suppliers about the type of component required and also the delivery sequence. Hence, it is very important to measure the OEM's production schedule adherence, which is the ability to produce according to schedule, as well as the delivery schedule adherence of the suppliers (i.e. the ability to deliver according to schedule). In order to make sure that the delivery sequence of components match with the production sequence, it is also important to measure the synchronisation level between the delivery sequence and the production sequence. The accuracy of demand information released by the OEM to the suppliers will also affect the performance of SILS. On the other hand, for the case under study, Kanban is normally applied to non-key components, with lower variety and value. Measuring schedule adherence and synchronisation accuracy are not as applicable to Kanban systems. Metrics like stockout level and delivery reliability are more suitable for measuring Kanban system performance.

However, differentiating the components according to their cost, importance and variety is a very basic classification method. There are other factors where the components can be further discriminated, such as the consumption rate, demand volatility and delivery distance. This also means that there will be more options of supply chain structure.

3 THE CASE STUDY

A case study has been carried out to examine the SILS supply mechanism of a UK-based luxury car assembly plant. As stated before, there are four elements that were considered to be important for suppliers in the performance of SILS - the production schedule adherence, the delivery schedule adherence, the synchronisation level and the material demand forecast accuracy. Measurements have been made on these metrics to assess the performance of SILS.

3.1 The SILS Supply Chain Structure

Figure 1 illustrates the information flow between the car assembly plant and its suppliers. There are two EDI (Electronic Data Interchange) information flows – forecast A and forecast B. Forecast A contains materials demand information for the next ten days in daily quantities, (followed by a longer time horizon in more tentative weekly and monthly requirements). In addition, sequenced parts suppliers receive a continuous broadcast of forecast B approximately between 8 to 12 hours prior to launch of the vehicle onto the Trim and Final assembly line. Forecast B is effectively a queue of jobs before the Trim & Final process. The disruption of the job sequence in the Trim & Final process is very rare. Hence, the delivery sequence stated in forecast B always matches the actual material requirement in the Trim & Final process. The sequenced-part suppliers deliver the materials to the Trim and Final assembly line according to forecast B, which includes the parts required and the sequence of delivery.

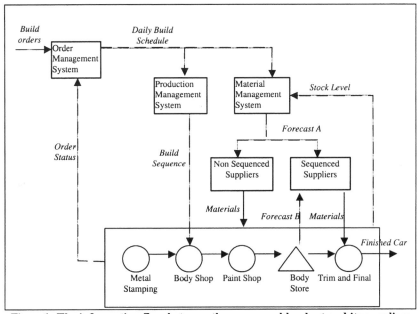

Figure1: The information flow between the car assembly plant and its suppliers

There is another information resource, which performs a similar task to forecast B (as a material demand forecast) and gives 7 days forecast of the production schedule – forecast C. It contains the specifications of the cars to be built. The receivers have to translate these specifications into material requirements, which they supply to the car manufacturer. For example, if forecast B indicates that 20 red-seat cars will be built on the 10 March 2002, then the seat supplier will be obliged to deliver 20 red seats on that day. Figure 2 illustrates how forecast B is generated. The production management system slices the WIP (Work in Progress) and committed orders within the production pipeline according to the daily build schedule. However, it is an informal material demand forecast. It is sent via email to a handful of sequenced-part suppliers, which were unable to operate with the demand uncertainty.

Table 1 provides a comparison of forecast A, B and C.

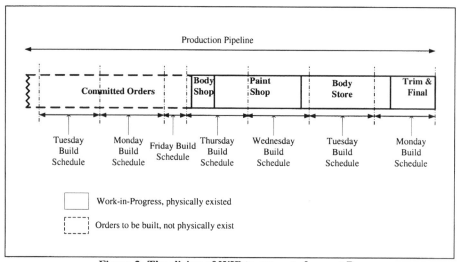

Figure 2: The slicing of WIP to generate forecast B

Features	Forecast A	Forecast B	Forecast C
Forecast Horizon	10 working days	8 – 12 hours, depends on the production rate and build schedule	5 – 7 days, depends on the WIP and committed orders
Generated by	Material Management System	Production Management System	Production Management System
Nature of information	Material demand forecast	Actual material demand in the Trim & Final process	Daily build schedule
Receiver	All suppliers	Sequenced-part suppliers	Some sequenced suppliers
Broadcast frequency	Daily (after production stops)	Continuous	Daily (after production stops)

Table 1: Comparison among the material demand forecasts

3.2 Material Demand Forecasts' Accuracy

In order to assess the accuracy of forecast A and C, graphs were constructed to compare the demand forecast of a type of front seat against the actual amount of cars fitted with that particular seat. The following are the explanations of the terms used in the graphs:

- "Actual" is the actual amount of cars that have been built with the monitored component fitted within that production day.
- "Day 1" is the forecast given 1 working day ahead. E.g. the "Day 1" forecast for 8 March is generated on 7th March evening.
- "Day 2" is the forecast given 2 working days ahead. E.g. the "Day 2" forecast for 8 March is generated on 6th March evening.
- "Day 3" is the forecast given 3 working days ahead. E.g. the "Day 3" forecast for 8 March is generated on 5th March evening.
- "Day 4" is the forecast given 4 working days ahead. E.g. the "Day 4" forecast for 8 March is generated on 4th March evening.
- "Day 5" is the forecast given 5 working days ahead. E.g. the "Day 5" forecast for 8 March is generated on 3rd March evening.

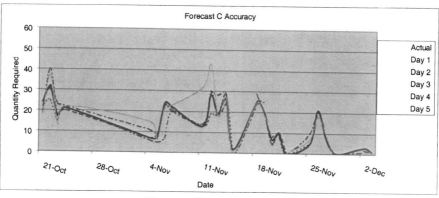

Figure 3: Forecast C analysis

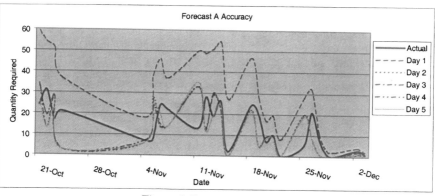

Figure 4: Forecast A analysis

By comparing the two graphs above, it can be seen that the demand forecast from C (the "Day 1" to "Day 7" lines) is closer to the actual consumption quantity. This shows that forecast C is more accurate than forecast A, even though A is the official demand forecast. The inaccuracy is due to the confusion and clashing of procedures between the material management system and the recording of production quantities.

3.3 Measuring the Production Schedule Adherence

Figure 5 show the production schedule adherence of the car assembly plant. The "Build on Day %" measures the percentage of cars that are built according to the schedule in each month. The average Build On Day percentage is about 71% (exclude the Build on Day figure from March 02). This means that on average in each month, 71% of the cars are built on the same day to the planned schedule.

3.4 The Delivery Schedule Adherence and the Synchronisation Level

The volatile line pattern and the gaps between the forecast lines and the actual line in Figure 4 show that forecast A has a very low accuracy and reliability. However, this does not affect the suppliers' delivery schedule adherence and the synchronisation level. From interviews with managers at both the car assembly plant and the seat supplier, the synchronisation level (between production in the car assembly plant and the delivery from the sequenced supplier) and the supplier's delivery schedule adherence are nearly 100% accurate. This is due to the assistance from both forecast B and forecast C.

Each sequenced supplier is required to deliver the components according to forecast B, which is derived from the WIP within the Trim and Final process. The job sequence in this process is fixed because it is a continuous flow production line. Hence, the delivery sequence matches the production sequence, as long as the suppliers are able to deliver the components according to forecast B. Meanwhile, forecast C provides the demand information 7 days ahead. This gives the suppliers 7 days to plan their production to meet the demand. Therefore, delivery schedule adherence is very good.

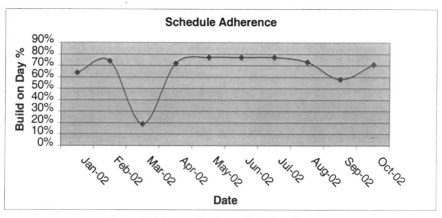

Figure 5: The production schedule adherence

3.5 The Seat's Supply Chain Performance

Figure 6 is the value stream map of the seat supply chain, from the 3rd tier supplier to the vehicle assembly plant. It shows the process to produce a seat, from the supply of foam for a headrest to the complete seat set fitted on the vehicle. The line drawn below the diagram is the throughput time for each process represented in the boxes above. The total supply chain throughput time is approximately 18 working days but the actual processing hour is about 28 hours, which is only 9.72% of the total throughput time. This is primarily due to the stock levels throughout the chain, as shown in Figure 6 and Table 2.

The suppliers are holding inventories, either in finished goods stocks or incoming material stock, as shown in Figure 3 and Table 2. Even though the vehicle assembly plant does not carry any sequenced components inventory, this does not means that the inventory has been reduced. In fact, the value stream mapping diagram and Table 2 show that the inventory has been consigned to suppliers. The seat manufacturer is able to assemble-to-order, according to forecast B. Therefore, there is no stock holding between the vehicle assembly plant and the seat manufacturer. However, there is inventory holding for the material, i.e. the headrests. The 2nd and 3rd tier suppliers also carry relatively high levels of both the material and finished goods inventories.

The demand amplification has been calculated and represented as the bullwhip coefficient (4) in Table 2. The seat manufacturer has the bullwhip coefficient of 1.0, which means there is no demand amplification. This is because the seats are assembled-to-order, according to forecast B. For the 2nd and 3rd tier suppliers, the demand has been amplified by 2.5 and 1.5 respectively. The bullwhip coefficient of 2.5 indicates that the output demand of the headrest supplier is 2.5 times more than the input demand.

Product	Supply Level	Stock Type	Stock Level	Bullwhip Coefficient
Seat	1st tier	Material - headrest	1.5 days	1.0
Headrest	2nd tier	Finished goods – headrest	2.5 days	2.5
		Material - foam	3 days	
Foam	3rd tier	Finished goods - foam	2 days	1.5
		Material	5 days	

Table 2: The inventory holding in the seat supply chain

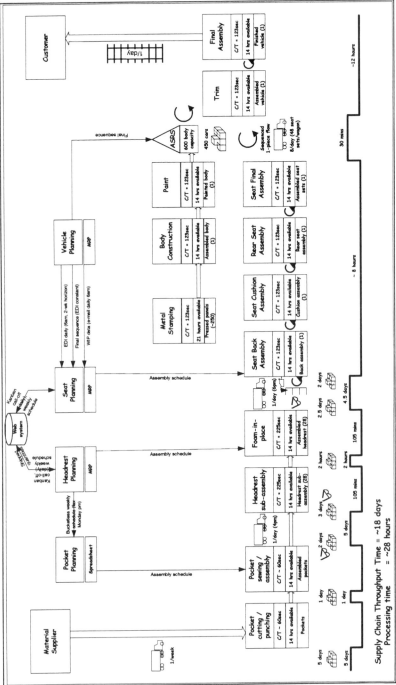

Figure 6: The seat supply chain value stream map

3.6 Case Study Conclusions

Each of the production materials in this assembly plant are categorised into two types – sequenced and non-sequenced components. The sequenced components are those key components with high variety and order winning potential. Using the SILS method, the key components' stock level can be maintained at the minimum level. For instance, there are 384 types of front seats. The inventory cost will be very high if the car assembly plant has to keep material inventory for each seat type. Therefore, the management of the supply of those key components with SILS reduces the inventory cost.

In this case study, SILS is supported by three information systems – forecast A, B and C. The system that broadcasts forecast B is purpose built, just to transmit the demand information to sequenced suppliers. There are nearly 4000 part numbers used to build a car (5). It is impossible to supply all the components with SILS. Hence, the non-key components are managed by a Kanban system, which is cheaper and easier.

In lean thinking, costs tied up in inventory are considered as a waste. In order to reduce the inventory cost, one of the methods is to reduce demand uncertainty. In this case study, even though the official demand information (i.e. forecast A) has a low forecast reliability, forecast C manages to provides an alternative source of demand forecast with higher accuracy. However, the demand amplification through the supply chain still results in a higher stock of material and finished goods inventory. Sharing the raw demand data, with an emphasis on high accuracy and reliability, from the OEM throughout the supply chain is one of the solutions to eliminate demand uncertainty and amplification.

4 CONCLUSIONS

The case study shows how the nature and characteristics of components influence the supply chain structure. For economical and efficiency reasons, the design of a supply chain has to take the component's importance, cost and variety into consideration. However, these are just the basic dimensions to classify the components. Further research is being carried out to investigate further classification dimensions and to relate them to appropriate performance measurement metrics, from the OEM to the end of the supply chain.

REFERENCES

1. Beamon, B.M. (1999) "Measuring Supply Chain Performance", International Journal of Operations and Production Management, Vol. 19, No 3, pp 275-292

2. Gunasekaran, A., Patel, C. and Tirtiroglu, E. (2001) "Performance Measures and Metrics in a SC Environment", International Journal of Operations & Production Management, Vol. 21, No. 1/2, pp 71-87

3. Brewer, P.C. and Speh, T.W. (2000) "Using Balanced Scorecard to Measure Supply Chain Performance", Journal of Business Logistics, Vol. 21, No 1, pp 75-93

4. Fransoo, J.C. and Wouters, J.F. (2000) "Measuring the bullwhip effect in the supply chain", Supply Chain Management: An International Journal, Vol. 5, No. 2, pp 78-89

5. MacDuffie, J. P. (2001) "Modularity and Build-to-Order Pull-Through: Early Trends in the Automotive Industry", Wharton Forum on e-Business

The lag between ERP software and ERP implementation – the reason of failure

M BUSI and **O J SAGEGG**
Department of Quality and Production Engineering, Norwegian University of Science and Technology (NTNU), Trondheim, Norway

SYNOPSIS

Companies that decided to implement an Enterprise Resource Planning (ERP) -system do often realize that not all the promised benefits are eventually delivered. Expenses are higher then expected, and the time schedule is delayed. This article suggests that the main reason is that enterprises are not always able to consider all the variables involved in an ERP installation and their major mistake is to consider this as a simple task instead of understanding its project nature and complexity. The two consequences studied in this paper are, first, that companies do not realize the strong relation between business process re-engineering (BPR) and ERP implementation and, second, that tasks are split in an inappropriate manner, often among a small group of people. To support the theory two case studies are briefly presented: an Italian manufacturer of steel products and a Norwegian company sheet metal and welding workshop.

INTRODUCTION

The Gartner Group (32) defines Enterprise Resource Planning (ERP) systems as:
A collection of applications that can be used to manage the whole business. ERP Systems integrate sales, manufacturing, human resources, logistics, accounting, and other enterprise functions. ERP allows all functions to share a common database and business analysis tools.

Yen et al. (36) argue that:
ERP is comprised of a commercial software package that promises the seamless integration of all the information flowing throughout the company, including financial, accounting, human resources, supply chain, and customer information.

An ERP system is an application for the execution of business processes. A common information database is the main idea behind the concept of these systems: by integrating data that used to be scattered throughout different information systems the ERP-system promises less cost of technology as well as business processes integration (23). Being ERP systems business management systems that integrate all facets of the business (36), their implementation in a company involves integrating the company computers, its pre-existing planning systems and its business processes.

There is little doubt that an ERP-system may be an important tool in gaining competitive advantages (See e.g. (3), (23)). However, several reports have been made of ERP-implementations that not had a fortunate outcome. The expenses are often higher than expected, and the time schedule is delayed. Davenport (3) reports several unfortunate implementations where large enterprises have used billions of U.S. dollars on one implementation, and thereafter abandoned the system because of changes in business relations. One of the most famous examples of an unfortunate ERP implementation is the one of FoxMeyer Drugs, where the bankruptcy trustees sued its systems' vendor and consultant company, blaming the ERP system for its business failure (3).

ERP implementation projects are the most difficult systems development projects (29) and represent a complex exercise both in technology innovation and organizational change management (19). Taking this a starting point, this paper will analyze why these projects usually end over budget, over schedule, and delivering only 30% of promised benefit (Standish Group (35)). Radding (24) states that whenever millions of dollars are invested into developing business applications and re-engineering business processes, the work is much more than any other system development project. This paper suggests that one of the causes of project failure is related to lack of management commitment and ability to understand the complexity of such a project.

This paper is structured in a theory part and two case studies. In the theory background previous proofs of strong relation between BPR and ERP are discussed, as well as previous knowledge on critical issues in ERP implementation projects. Having chosen an interpretivist approach, we use two different case studies to show the relation between: a) project team member selection and b) communication problems with project failure.

ENTERPRISE RESOURCE PLANNING –SYSTEM AND BUSINESS PROCESS RE-ENGINEERING

ERP-systems require close examinations of the affected business processes. The implementation phase may be seen as a process of making compromises between how the enterprise wants to work and how the ERP-system is allowing them to (3). This may be the reason for the common understanding of that ERP-projects impel Business Process Re-engineering (BPR).

BPR challenges the very foundation of the traditional way of doing business. The first notations of this foundation were laid in 1776 when Adam Smith (26) proclaimed the benefit of the division of labor in his book "wealth of nations". Smith discovered that by separating the production process into small, simple work tasks the productivity could be drastically increased (10). These productivity benefits were the result of the automation of human labor.

Implying that by using many different people that repeat simple tasks to produce one item is more efficient than if each of they would create the complete product from start to finish. This logic was further developed both throughout the production process and in the management- and administration functions by people like Henry Ford, Frederic W. Taylor and Alfred Solan. The thoughts from these days are strongly affecting companies in the western world even today (10)(25)(30)(31).

However, an extensive form for separation of work tasks has its shortcomings. This reveals itself through unrewarding work conditions, lack of flexibility, by generating hierarchic bureaucratic organizations and fragmenting the work processes (10)(25)(30)(31). This may lead to an absence of overview on how their specific work-tasks affect the totality among the members of an organization. This lack of overview may result in a confusion of who has responsibility of the completion of these fragmented work processes, as well as generating unnecessary operations.

BPR challenges the idea of separation of work by integrating the different business processes towards the customers' needs. Dedicated work groups have then the responsibility of the completion of the whole process from start to finish. While all works-tasks that are not contributing to the value of the customer must be either eliminated or minimised (30). In the BPR literature, a special attention is given to these effects on the work of middle management. Hammer and Champy (8) claims that a large part of the work-task of the middle management is to transfer information from the work floor to the levels above. The need for a middle management is reduced by substituting the manual information handling with electronic technology that transfers the information directly to the higher levels of the organization.

BPR may therefore be seen as a management philosophy that drives the traditional organizations and the foundation it is built on, to change through process integration project. The ERP system does integrate the information from different departments, and therefore connects traditionally separated work tasks. This indicates that the ERP-system do more than add functionality to an organizations administrative software; it will also dictate a completely new way of doing work. Or as Norris et.al put it: "What ERP really does is organize, codify, and standardizes an enterprise's business processes and data"

ERP SYSTEM IMPLEMENTATION PROJECTS

Because of their wide scope and their requirement in term of technological and process change (15), ERP systems are incredibly complex, and implementing one can be a difficult, time-consuming, and expensive project for a company (17)(22). In the near-term perspective, managers find ERP implementation projects the most difficult system development projects (29); they can take many years to complete and cost tens of millions of dollars for a medium sized company and $300 to $500 million for large international corporations (17).

Projects of such kind are usually divided in phases. Although choices at different stages are equally fundamental for the project outcome, there are certain issues in the phases that occur before the practical implementation of the ERP that must be carefully taken in consideration: these are usually about the ERP being implemented, partners selection, key-personnel choice and project plan.

When a company decides to implement an ERP system it usually carries out the following general steps (see also (15)(33)(34)):

1. a request for quotation (RFQ) is sent to different vendors
2. a comparison between various proposal is carried out
3. a selection of project partners is performed
4. a set of people in the company is appointed by the top management to be the internal project team
5. the system implementation project starts.

Companies usually tend to create a partnership with consultants or ERP vendors that offer assistance during the implementation. In a recent study (15) it resulted that 83% of an investigated sample of 20 companies created partnership with consultancy companies and 42% of the same sample chose partnering with ERP vendors. ERP market has actually grown so fast that outsourcing the needed special skills has turned out to be more reasonable than investing in developing them internally (1). It appears therefore obvious that an unfortunate choice of partners will lead to an equally unfortunate project result.

An ERP implementation project usually involve the following types of members (23):

- Users form the organization
- Consultants from the ERP-provider
- Information systems people from contracted from through agencies
- The companies own systems people

In figure 1, the main actors in this project are represented. In the middle stands the company that decided to go for an ERP system, on the left the ERP vendor that provides the system, and on the right the consultancy company selected to be partner in the system implementation.

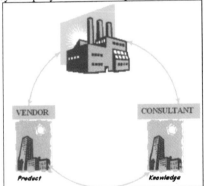

Fig. 1: ERP implementation projects: main actors.

The ERP market today us split among few very big producers like SAP, Baan, etc. plus many very small ones, as shown in the table from AMR research (17).

ERP package	Market share
SAP	32%
Oracle	13%
PeopleSoft	9%
JDE	7%

A003/035/2003 © IMechE 2003

Baan	*7%*
SSA/BPCS	*3%*
JBA International	*3%*
Intentia	*2%*
QAD	*2%*
Others	*22%*

Table 1: ERP packages' market share

The internal project team represents the "missing link" between consultants and users in the organization, and it is appointed by the organization to manage the implementation process by co-coordinating people and activities.

When the ERP package has been chosen, the internal project team has been selected and the partnership with the consultant company too, the implementation project starts. To different cases apply different approaches, but they all are quite similar. Kwon and Zmud's (16) six stage IT implementation methodology can be easily applied to ERP implementation projects. The six stages are:

1. Initiation	4. Acceptance
2. Adoption	5. Routinization
3. Adaptation	6. Infusion

Based on the same steps two other approaches are presented below, which were developed by two ERP vendors: the Implex Method by Intentia, and the ASAP roadmap method by SAP.

The Implex methodology is based on 5 different phases[1]:
1. Position the project (it groups the five steps previously presented in this paper)
2. Design business processes
3. Configure business processes
4. Implement business processes
5. Start-up

ASAP[2] is the standard methodology recommended by SAP for planning the implementation and implementing the software in the company.
1. Project start-up
2. Business blueprint
3. Realization
4. Final preparation
5. Go-live

In the course of these steps, many threats and opportunities for the outcome of the projects are hidden. In the following paragraphs, existing knowledge on management issues is discussed.

CRITICAL ISSUES IN ERP IMPLEMENTATION PROJECTS.

Analysts reported that 70% of Fortune 1000 firms had or were in the process of installing ERP systems in 1998 (11); at about the same time, the ERP market value was estimated to be $10

---

[1] For further details see the Implex Method at www.intentia.com

[2] For further details see the Accelerated SAP methodology at www.sap.com

UK billion (14) with a compound annual growth rate of 35% in 1998 (Shepherd, 1998). Despite this massive investment in ERP systems, a recent Standish Group report on ERP implementation projects reveals that these projects are, on average, 178% over budget, takes 2.5 times as long as intended and delivered only 30% of promised benefit (14). In this paragraph, possible reasons of these negative figures are analyzed.

Many ERP implementation projects have been studied in order to find the reasons of their failure or success. In literature there is common agreement that the following, among the others, are to be considered critical issues when managing an ERP implementation project (14)(15):
- Company adaptation to organizational changes
- Customization extent
- Difficulty in estimating requirements
- Difficulties in changing to new from old system
- Unavailability of skilled project people
- High cost of implementation
- Cultural barriers

Krumbholz and Maiden (14) state that there is growing evidence that inability to adapting ERP packages to fit organizational and national cultures is the main cause for project failure. This should not be surprising since ERP system installation, as we said earlier, requires changing the customer business processes to fit the ERP package best-practices processes (2).

In the study by Krumbholz and Maiden (14) three culture-related clashes influencing ERP implementation projects were identified: 1) current/new culture clash; 2) ERP/current culture clash; 3) new/old process culture clash. This view is in line with the results from Kumar et al., that in their investigation find out that companies face behavioral, organizational and management related challenges when implementing an ERP.

We take these previous studies as a starting point for opening a new window on the possible reasons of failure of such ERP implementation projects.

THE LAG BETWEEN ERP SOFTWARE AND ERP IMPLEMENTATION: THE REASON OF FAILURE.

Following an interpretivist approach, which strengths in information system research have been reported in a number of studies (13)(20)(27)(28). ERP implementation projects can be analysed by studying the meanings that emerge from the interaction of social actors, which are fluid, ambiguous and context dependent (Hochschild, 1983). Specifically, in IS research, interpretivist approaches assume that meanings are shaped and reshaped by actors through the social construction and reconstruction of information systems (Mohrman & Lawler, 1984).

As earlier introduced, the social actors in this kind of projects come from the top management, the middle management, the consultant team and the users groups. In the course of the project different phases, a sort of "Bermuda triangle" takes place between these groups (figure 2) (see also 33)

 A003/035/2003 © IMechE 2003

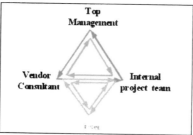

Fig. 2: Information flow between the implementation project participants.

Members in the project have different goals, different skills, access to different information and data and have different culture. The idea of implementing the ERP is therefore perceived differently by the various people, which will have a different approach to the project. Information must "fight its way through a semi permeable organizational membrane consisting of existing power networks, organizational cultures and subcultures" (5)). For this reason, IS/IT implementation projects often have unanticipated and contradictory consequences (Robey & Boudreau, 1999).

The metaphor with the Bermuda triangle is used here because it reminds what happen to important pieces of information during these projects: given the above mentioned differences, the single individuals in the project do not have that holistic vision required by the complexity of the project and therefore some details, being given for granted by the knowledge-owner, disappear along the way; it's only by working as a team that the singles knowledge/skills/vision have the chance to be complete.

In addition to facilitating communication between project actors, the internal project team have to report to the top management major milestones achievement and overall project performance: what happen here is that the middle management has a better perception of the project activities and, as opposite to the top management -that sees only main figures and facts-, it has the possibility to more closely analyze and understand eventual problem causes.

Consultants have to interact with the final users through the internal team middle management. They have an extensive knowledge of the ERP being implemented and its requirements, but have little or no pre-existing knowledge of the business processes that the ERP will support.

On the other hand the final users of the system, i.e. those that are directly involved in the processes, have no idea of what are the capabilities or requirements of the ERP, but know all about the business operation (22), which usually results in a more pessimistic attitude (14)(15).

The cases chosen in this paper show that this is a critical project phase: understanding how the processes work and re-engineering them in order to make them fit to the ERP requirement. Given the complexity of a business process –in term of involved operation and control variable-, the phase of designing and configuring the business processes can be subjected to many reviews and may lead to delays and expenses increase.

Another encountered problem in this phase is the significant resistance from the staff (15). Often the users like the "business-as-usual" and don't want to change the way they work. Depending on how well aligned are the views of the different social actors, and on how positive is their attitude (mutual determination in understanding the other party) the need of reviews diminish and the cost as well.

CASE-STUDY 1

Company A[3] is a division of an Italian manufacturer of steel products that was founded right after the Second World War with subsidiaries all over the world, more the 10.000 employees and revenue of about 1,7 billion €. Company A, that alone counts more then 800 employees was chosen in 2000 to be the pilot project for the implementation of the SAP enterprise resource planning solution.

The top management did not understand the complexity of the project. Therefore, an internal team of only two people was appointed, plus two trainees and one other person, who was not working a fix number of days per week on the project. One of them had previous knowledge on information system technology, and the other one had knowledge in information system implementation projects, the third resource had knowledge of management as well, but, as we said, he was not a fixed resource in this project.

The internal team chose one leading global business and technology integrator focused on the creation, implementation, integration and support of enterprise business solutions, from which a team of 15 consultants, with special skills in different modules of SAP, was selected to be four days per week at the company A's production site.

An additional number of workers from different departments of company A was contacted on occurrence to meet with consultants.

The internal team manager's main daily occupation was to convene the right people for the meetings with consultants and analyse the documents produced after each meeting. The internal team system expert main task was to follow the integration between the existing enterprise system and SAP. Finally, the third resource was called to solve problems that arose on the way (e.g. difficulties with information retrieval, staff, etc.)

Each consultant worked on his or her module, and the only interaction with the others took place when two different modules need to exchange pieces of information: in this case issues such as data type, format, timing etc were discussed in distinct meetings.

The business blueprint preparation lasted for more then a year, at the end of which a prototype was presented to the top management, together with main figures related to it. The result was drastic:
- Due to high inter-relationship problems between the consultants and the workers, there had been during the project the need for many reviews of the customisation parameters, because of incomplete information given to the consultants. This led to a final

---

[3] Names of companies involved in the cases presented in this paper have been masked for privacy rights.

cost for that phase about 100 % higher then planned, and a time scheduled delayed of 6 months.
- The final prototype presented would have required a massive intervention in organization, to achieve the far to extensive process re-engineering necessary for implementing the ERP.

The top management understood at that point the complexity of the project, and only because of its high economical resources, company A was able to abandon the project and start another one with the same aim –the implementation of SAP- but with a completely different approach.

CASE-STUDY 2

Company B1 is a Norwegian sheet metal and welding workshop. It used to be a family business and the grandson of the original founder is the managing director even today. The company has 46 employees and had a turnover in 2001 of 40 millions NOK (approx. 5,5 mill €). The main share of this is caused by their six large customers. Nevertheless, they do also deliver to a variety of other, considerable smaller customers. Both segments of customers are considered essential for long time survival in the marked. The company has in respect to their size and product specter remarkable modern machinery equipment. In the later years they have invested in high-tech equipment, such as automated laser cutters and automated stamping machine. These investments have, according to themselves, increased their capacity considerable. The company is focused on the workmanship for each worker, as well as they emphasis on the importance of being at the frontier in machinery equipment. This strategy has made them very flexible in product deliveries.

In 1988 they decided to buy and implement an ERP-system. Their old MRP-based systems were out of the market, and the consultant company that had delivered this system had a new agency on an integrated ERP system. They decided to go for this since this consultant company was located close to them, and they had been satisfied with their old system. The project was estimated to last approximately half a year with consultants visiting the company twice a week in order to set up tables and parameters and convert data from the old system. The project group from the company' side were consisting of one person, the office manager. This person had to support the consultants while keeping this daily work. Other involved parts had to be pulled from their daily tasks when it was needed.

After one year several different consultants with expertise on different issues like programming, hardware, economics, sales and purchase had been working on the system. If these experts needed to have something clarified, the company's representative in the project was usual not able to clarify all details of the requirements to the different departments or professional fields. This resulted in a constantly requests for people that were occupied in their daily work or that destinations were made without knowing the real effects on the daily work routines.

After one year the project costs was already more than estimated in the first place, but the system were not functioning according to the companies needs. The issue that contributed most to this unsatisfactory state was that they had to use some advanced functionality in the production module, and local consultant agency did not have expertise on this module. Therefore, they agreed to hire an expert on the production module from the consultant head office in addition to a local consultant dedicated to this task, in order to finish the configuration. These changes in the production module lasted for more than half a year and the cost that it generated were all in addition to the original project budged. In addition, the changes in the production module generated new requirements for adjustments in the software code, as well re-configuration of the other modules. This meant that the whole system configuration had to be reviewed.

After two and half years, the initial goals for the system functionality were abandoned and the new system was put into use with only marginal changes in functionality compared to the old one.

CONCLUSION

By a presentation of ERP systems and their impact on organizations, we have tried to highlight some of the challenges associated with the implementation of these. The functionality of the ERP-systems should be linked together with the difficulty of changing the physical work processes and not be treated as a matter of technology. Through our two cases we have showed that not giving an ERP project sufficient and/or right focus will have negative effect on a implementation project and the organization as a whole.

A003/035/2003 © IMechE 2003

BIBLIOGRAPHY

1. Bingi, P. Sharma, M.K. Golda, J.K. Critical issues affecting an ERP implementation, Information Systems Management (1999) 7–14.
2. Curran, T.A. Ladd, A. SAP R/3 Business Blueprint, Prentice-Hall, Englewood Cliffs, NJ, 1998.
3. Davenport, T. (1998). Putting the Enterprise into the Enterprise System."Harvard Business Review, July-August: pp. 121–131.
4. Davenport, T. (2000). Mission critical: Realizing the promise of enterprise systems. Boston, MA: Harvard Business School Press.
5. Doorewaard, H., & van Bijsterveld, M. (2001). The osmosis of ideas: An analysis of the integrated approach to IT management from a translation theory perspective. Organization, 8, 55–76.
6. Glasson, B.C., 1994. Business process reengineering: information systems opportunity or challenge?. In: Glasson, B.C., Hawryszkiewycz, I.T., Underwood, B.A. and Weber, R.A., Editors, 1994. Business Process Re-Engineering: Information Systems Opportunities and Challenges, North Holland, Amsterdam, pp. 1–6.
7. Hammer, M (1990): "Re-engineering work - Don't automate, obliterate!", Harvard Business Review, 68 (4), p. 104-112.
8. Hammer, M., Champy, J., 1993. Reengineering the Corporation: A Manifesto for Business Revolution. Harper Business, New York.
9. Hammer, M., Stanton, S. (1999). How Process Enterprises Really Work. Harvard Business Review, November-December: 108–18.
10. Hochschild, A. (1983). The managed heart: Commercializtaion of human feeling. London: University of California Press.
11. Hoffman, D., Novak, T., Project 2000 (www document) URL: http://ecommerce.vanderbilt.edu/novak/What/sld009.htm, 1998
12. Hopp, W.J. Spearman, M.L. Factory physics : foundations of manufacturing management - 2nd ed. Boston : Irwin McGraw-Hill, c2001.
13. Klein and Myers, 1999. H., A set of principles for conducting and evaluating interpretive field studies in information systems. MIS Quarterly 23 (1999), pp. 67–94.
14. Krumbholz, M, Maiden, N., 2001,The implementation of enterprise resource planning packages in different organizational and national cultures, Information systems 26, 185-204.
15. Kumar, V., Maheshwari, B, Kumar, U., 2002, ERP systems implementation: Best practices in Canadian government organizations, Government Information Quarterly 19, 147-172.
16. Kwon, T. Zmud, R. Unifying the fragmented models of information systems implementation, in: Boland, Hirschheim (Eds.), Critical Issues in Information Systems Research, Wiley, New York, 1987.
17. Laughlin, S.P., 1999, An ERP Game Plan, Journal of Business Strategy, 20, 1, 32-37.
18. Mabert, V. A., Soni, A., Venkatarama, M.A., 2001, Enterprise Resource Planning: Myths versus Evolving Reality, Business Horizons, Mai-June, 69-76.

19. Markus, M. L., and Tanis, C., 2000, "The Enterprise System Experience: From Adoption to Success," in Framing the Domains of IT Management: Projecting the Future through the Past. Zmud R. W. (ed.) Pinnaflex Educational Resources Inc., Cincinnati, 173–207.

20. Mohrman, A. M., & Lawler, L. L. (1984). A review of theory and research. In F. W. McFarlan (Ed.), The information systems research challenge (pp. 135–164). Boston: Harvard University Press.

21. Newell, S, Huang, J. C. Galliers R. D. and Pan, S. L. Implementing enterprise resource planning and knowledge management systems in tandem: fostering efficiency and innovation complementarity, Information and Organization, Volume 13, Issue 1, January 2003, 25-52.

22. Poston, R and Grabski, S, 2000. Accounting information systems research: is it another QUERTY?. Int J Account Inf Syst 1 1, pp. 9–53 March.

23. Ptak, C. Schragenheim, E. ERP: tools, techniques, and applications for integrating the supply chain, St. Lucie Press/APICS series on resource management, Boca Raton, Fla: St Lucie Press, 1999.

24. Radding, A. (1999). ERP. More Than an Application. Information Week, 728, 1–4.

25. Skorstad, E. Produksjonsformer i det tyvende århundre : organisering, arbeidsvilkår og produktivitet. Oslo : Ad notam Gyldendal, 1999.

26. Smith, A., 1789, An inquiry into the Nature and Causes of the wealth of Nations , London .

27. Walsham, G. Interpreting information systems in organization. , Wiley, Chichester (1993).

28. Walsham, G. The emergence of interpretivism in IS research. Information Systems Research 6 (1995), pp. 376–394.

29. Wilder, E., and Davis, B. (1998). False Starts, Strong Finishes. Information Week, 711, 41–53.

30. Willoch, B E (1994): "Business process reengineering" : en praktisk innføring og veiledning, Bergen: Fagbokforlaget

31. Womack, J.P. Jones, T.J. Roos, D. The Machine That Changes the World. Macmillean Publishing Company. 1990.

32. www.gartnergroup.com

33. www.intentia.com

34. www.sap.com

35. www.standishgroup.com

36. Yen, D.C., Chou, D.C., Chang, J, 2002, A synergic analysis for Web-based enterprise resource planning systems, Computer standards and interfaces 24, 337-346.

An ICT supported operations model for supply chains

S BOLSETH
Department of Production and Quality Engineering, Norwegian University of Science and Technology (NTNU), Trondheim, Norway
J O STRANDHAGEN
Department of Economics and Logistics, SINTEF Industrial Management, Trondheim, Norway

ABSTRACT

Even though SCM has existed for years, it is just in the recent years that manufacturing companies, ICT vendors and researchers have started talking about Supply Chains Operations Models. The rationale for establishing an ICT supported Operations Model is to illustrate the design, operation and improvement of the production systems that create the companies' primary product or services. This paper will compare and discuss different supply chain Operations Models, and present an industrial case where such a model was developed.

INTRODUCTION

According to Browne et al. (1), (2) the manufacturing systems today are subject to tremendous pressure of the ever-changing marketing environment. Browne et. al (1) concludes that individual companies have to work together to form inter-enterprise networks across the total product value chain in order to survive and achieve business success.

Supply Chain Management (SCM) is based upon cooperation and coordination, and the key is the integration of processes, both up- and down-stream, in the supply chain. Information and Communication Technology (ICT) is an important enabler for this integration process, and thereby for SCM. Much of the current interest in SCM is motivated by the possibilities that are introduced by the abundance of data and the savings inherent in sophisticated analysis of these data (3). According to Huang and Nof (4) the impact of modern ICT on enterprise systems can be classified into three categories: 1) Speeding up activities, 2) Providing intelligent and autonomous decision-making processes, and 3) enabling distributed operations with collaboration. The introduction and utilization of integrated ICT for managing the supply chain will enable companies to gather vital information along the whole supply chain and quick act upon it and be in advance on market changes, and thereby gaining competitive advantages (5).

A003/036/2003 © With Authors 2003

Even though the concept of SCM has existed for almost 20 years, and ICT offers the possibilities to achieve visibility and integration in the supply chain, there have been few attempts to develop ICT supported operations model for the supply chain. This paper will discuss the concepts of operations model for supply chain, highlight 2 different operations models, and present a ICT supported operations model developed at a Norwegian 1. Tier supplier to the automotive industry.

SUPPLY CHAIN MANAGEMENT

Supply Chain can be defined as: *"a set of three or more entities directly involved in the upstream and downstream flows of products, services, finances, and/or information from source to a customer"(6)*. It is important to note that these supply chain exists whether they are managed or not.

The concept of Supply Chain Management (SCM) is well documented in literature; see (7), (8), (9), (3), (10) and (6). SCM can be defined as (7): *"The management of upstream and downstream relationships with suppliers and customers to deliver superior customer value at less cost to the supply chain as whole"*. Each company in a supply chain is dependent on each other, and, yet, paradoxically by tradition does not co-operate very closely with each other. Supply chain competitiveness can be achieved through chain integration and process re-design that decrease waste through unnecessary activity, reduction of stocks as well as faster response times. The goal is to get everyone in the supply chain onto a common platform of logistics transactions and information systems for greater interorganizational "seamlessness" or transparency resulting in faster system response time (11). The goal of these Supply chain Information Systems are to (3):

- Collect information on each product from production to delivery or purchase point, and provide complete visibility for all parties involved
- Access any data in the system from a single-point-of-contact
- Analyze, plan activities, and make trade-offs based on information from the entire supply chain.

In order to achieve these three goals, ICT systems must be able to support strategically, tactical and operational activities, both internally in a company and externally in a supply chain.

Lee and Whang (12) points out that the use of ICT (E-business) impacts supply chain integration on four critical dimensions: 1) Information integration, 2) Planning synchronization, 3) Workflow coordination, and 4) New business models (see table 1).

Table 1: Supply Chain Integration Dimensions
Source: Lee and Whang, 2001

Dimension	Elements	Benefits
Information	- Information sharing & transparency - Direct & real-time accessibility	- Reduce bullwhip effect - Early problem detection

integration		- Faster response
		- Trust building
Synchronized planning	- Collaborative planning, forecasting & replenishment - Joint design	- Reduced bullwhip effect - Lower cost - Optimized capacity utilization - Improved service
Workflow coordination	- Coordinated production planning & operations, procurement, order processing, engineering change & design - Integrated, automated business processes	- Efficiency & accuracy gains - Fast response - Improved service - Earlier time to market - Expand network
New business models	- Virtual resources - Logistics restructuring - Mass customisation - New services - Click-and-mortar models	- Better asset utilization - Higher efficiency - Penetrate new markets - Create new products

OPERATIONS MODEL

Operations Management tries to ensure that the transformation process is performed efficiently and that the output is of greater value for the customer than the sum of inputs. Thus, operations can be defined as the process that transforms inputs into outputs of greater value (13), (14), and (15).

Operations Management is the organisation and control of processes that are needed to produce a company's goods and services. Operations Management is directly responsible for the satisfaction of customers through activities that include (13):
- the design of the physical transformation processes that provides the specific value a customer desires
- the design of concepts and systems for planning and controlling the physical work, and the material and information flows within those transformation processes
- the design of systems for monitoring and improving the company's effectiveness in satisfying customers
- the effective operation of the planning and control systems to create products or services that satisfy customers

For Operations Management to be successful, it must add value during the transformation process. The value is created for a customer that receives the products or services, and a network of suppliers and customers is linked together in order to satisfy the final customer. In other words, every manufacturing plant is situated in a *value chain*, defined as the network of actors that are involved, through upstream and downstream linkages, in the different processes that produce value in the form of products and services for end consumers.

The Operations Management in the supply chain can be executed through a computerised *Operations Model*. The Operations Model is a representation of the operations in the Supply Chain, and also an ICT-tool that ensures efficient and co-ordinated control, and real-time information and communication, in Supply Chain. The Operations Model gives access to on-line and structured information regarding all processes, performance and status in the supply chain, and can be used to view current or future states of the supply chain. The Operations

Model also contains tools and directions for management, collaboration, and control in the supply chain.

The SCOR model

The fare most widespread and known Supply Chain Operations Model, is the Supply Chain Operation Reference (SCOR) model developed by the Supply Chain Council (www.supply-chain.org). The SCOR model, see figure 1, is a reference model that allows companies to introduce standards to business process re-engineering, benchmarking and process management. It is further a tool that enables users to address, improve and communicate supply chain management practices within and between actors. The SCOR model is just to some extend supported by ICT, where the e-SCOR application is the major ICT component.

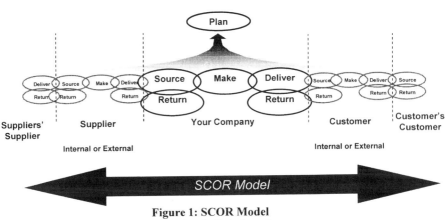

Figure 1: SCOR Model
Source: Supply Chain Council

E-SCOR provides a graphical modelling environment for the dynamic analysis of supply chains, based on the SCOR model. The e-SCOR application can simulate various configurations, test robustness of the supply chain and identify the service levels required of a company and its supply chain. It can further help to identity the weak links and areas for improvement in the supply chain, using the metrics for the five processes in the SCOR model (www.gensym.com).

The SCOR model (with the e-SCOR application) cannot be classified as a truly ICT supported Operations Model accordingly to the previously stated definition. It is mainly a reference model for performance measurement and benchmarking in the supply chain. The SCOR model do not offer real-time, on-line, access to structure information regarding all processes, performance and status in the supply chain.

Enterprise Resource Planning (ERP) Systems

An ERP-system is a standard application program, which support execution of business processes throughout the whole company. The ERP-system has functionality that makes the company able to replace many of their applications with a single seamless system with one common database.

   A003/036/2003 © With Authors 2003

The existing solutions of ERP software packages are mainly centralized, company specific customized, in-heterogeneous in the own organization an inflexible to adapt new conditions. Only a few solutions do support process-oriented structures in the own organizations, however, not to mention external organizational processes (16). Hieber and Alard (16) points out that it should be possible to extend the functionality of the ERP systems out of the enterprise and into the whole supply chain. Such an ERP system should incorporate five crucial activities of every business:

- Configuration - the ability to model and configure the supply chain to link different partners for an optimal customer solution
- Planning - the ability to anticipate the future and respond to changing situations by providing an integral end-to-end view
- Optimization - the ability to find the best solution for the whole supply chain
- Execution - the ability to standardize and automate the daily business within the supply chain
- Control - the ability to identify the weak chain and locate the possible impacts on the common business

The ERP vendors have now offered supply chain functionality, to some degree corresponding with the list above, for some years. This supply chain functionality is based on an APS (Advanced planning and scheduling) solution and offer functionality to plan for demand, supply, and production functions for the extended supply chain and also to support plant scheduling and global available-to-promise functions. The focus of these SCM applications is still mainly on the planning side rather the execution side.

ERP systems are not labelled or sold as an operations model, rather as a planning system. But an ERP system can be viewed as a very promising basis. It contain an incredible amount of information regarding the processes within the company, and in an increasingly way also about the supply chain. In a few years these systems probably will be able to talk (without any modifications and/or obstacles) with each other, and this will give on-line access to information from the whole supply chain. A major concern is that the ERP systems very complex, and will not be able to the desired functionality in and simple and visual way.

CASE - RAUFOSS CHASSIS TECHNOLGY AS

Raufoss Technology (RCT) is a part of the Raufoss group, and is developing and manufacturing aluminium alloy chassis components for automotive industry. Due to a larger contract with General Motors (GM), RCT has built a new plant at Raufoss and are building a similar plant in Canada. Start Of Production was January 2002 at the Raufoss plant, and is scheduled to be June 2003 at the Canada plant. The manufacturing in each plant is organized in two fully automated manufacturing lines: one for front control arm, and one for rear (each with a capacity of 1,4 million finished products per year). There are 14 different assembled parts from 7 different suppliers in addition to the aluminum part. Extruded aluminum profiles are delivered from two different suppliers. Suppliers are located in Europe and USA. Even though there is only one customer, there are call offs from 7 GM plants in Europe, and a similar number of plants in USA. Logistically this acts as different customers. A simplified picture of the supply chain is illustrated in figure 2.

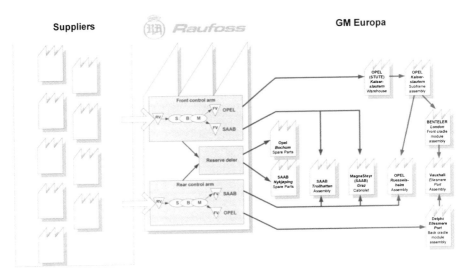

Figure 2: RCTs Supply Chain

Development of an Operations Model for RCT Supply Chain

The idea behind the Operations Model at RCT was grounded in thoughts similar to those in page 3-4, regarding ICT and E-Business in the supply chain. The rationale for establishing an Operations Model at RCT has been:

- To have a common description with all relevant information about the operation of RCTs supply chain.
- To be able to use the RCT Operations Model as a tool for future development of the supply chain as well as developing the Operation itself
- To be able to use the RCT Operations as a learning tool for new employees
- To secure that all description and information is stored once and only once, and is globally real-time available to all personnel involved.
- To create one single information pool whose flat structure with cross-linking and hypertext features is preferred to the more traditional hierarchical- and partition-based one. This makes the user's information quest process more time- and cost-efficient.
- To enable multi-media technical support, helpful in a wide range of situations (e.g. machinery brake-down or machinery upgrading design, etc.)

The Operations model is web-based, and is a part of the intranet at RCT. It will be further developed into an extranet, where every actor in the supply chain have access, and to a portal, where the interface and content are customized to the user. The Operations Model has direct links to all software applications used in operation of RCT, where it can import and export data.

A003/036/2003 © With Authors 2003

It will in the future (2003) be the single source of information on how RCT runs its Operations, as well as source of information about the current status of the Value Chain. The Figure 3 shows the main picture and Intranet access page, as well as the main elements of the Model.

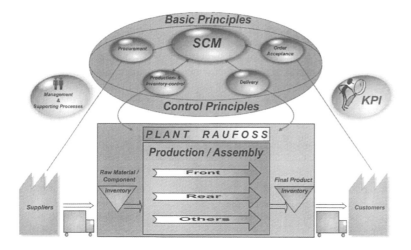

Figure 3: Operation Model at RCT

Elements of the Operations Model
The Operations Model consist out of several elements like:
- Basic and control principles
- A SCM centre, with administrative processes
- Actors, with a set of primary processes
- KPIs
- Management and supporting processes

These elements will in the following be presented.

Basic and control principles
The Operations Model is based on some basic principles influenced by concepts like Lean Manufacturing, the Extended Enterprise and the Control Model Methodology, and they are: 1) Time focus, 2) One-way flow and tasks dine once, 3) Real-time information, 4) Transparent Supply Chain, 5) Integrated processes, and 6) Visual and Simple. By applying the basic principles it has been possible to establish some overall control principles for the complete flow of goods:
- Masterplan of resources and delivery volumes based on customers year plan
- Daily delivery to customers based on call-offs
- Delivery executed from finished goods stock
- Smoothened production plans, fixed and stable for 4 weeks horizon

- Material and components purchase making use of year-orders, forecasts, and call-off based on production plans and components stock levels
- All plans updated weekly (Monday)

A SCM centre, with administrative processes

One of the most central elements of the Operations Model is the establishment of the Supply Chain Management Centre (SCM). This centre co-ordinates all transactions regarding the RCT node in the value chain, and consists of a dedicated group of people who have a complete and shared responsibility for all supply chain processes related to information and flow of goods.

The SCM is responsible for the yearly planning process and four daily administrative processes:
- Order acceptance (receive Call-off)
- Delivery scheduling and execution
- Productions and inventory control
- Procurement

Actors, with a set of primary processes

The Operations Model consist out of 5 actors:
- Suppliers
- Transporters
- Warehouses/distribution centre
- Manufacturer
- Customers

The Model can be customised into different configurations of these actors, both in amount and which is present or not. Actors can also be added and/or removed during the operation phase.

Each actor will always contain one or more out of the primary activities, where a primary processes are those concerning the physical material flow in the value chain.

KPIs

- The KPIs used in the Operations Model at RCT will show the values for the past, the present and the expected future for both traditional and more SC oriented KPIs. The indicators will be used to:
    1. To monitor performance
    2. To evaluate and analyse performance
    3. To monitor the status of the flows and processes in the logistics value chain. The flows are flow of information and flow of goods
    4. To be used as decision support for controlling the flow of goods and information in the value chain (supply chain control)
    5. To be used in improvement projects in order to pinpoint identify problems and focus effort.

     A003/036/2003 © With Authors 2003

The same type of indicators must be measured at several steps of the value chain in such a way that they can be compared along the chain, and so that an aggregation of them gives a complete value chain picture. This is thus based on an assumption on some degree of transparency in the supply chain.

Management and supporting processes
These are processes not directly connected with the flow of materials and goods through the value chain, e.g. maintenance, finance, accounting, HR, top management, etc.

CONCLUSION
Companies today are forced to collaborate across the supply chains, and Information and Communication Technology is an important enabler for this supply chain integration. Operations Management can be executed through an ICT supported Operations Model, which is a representation of the operations in the supply chain, and also an ICT-tool that ensures efficient and coordinated control, real-time information and communication in the supply chain. The fare most widespread and known Supply Chain Operations Model, is the SCOR model. This model is mainly a reference model for performance measurement and benchmarking in the supply chain, but lack functionality to offer real-time, on-line, access to structure information regarding all processes, performance and status in the supply chain. An ERP system can also be viewed as an Operations Model, and are superior to the SCOR model in many areas. Major concerns with ERP systems are that they are to complex, and have an internal instead of an external focus. A Norwegian first tier supplier to the automotive industry, RCT, have developed an web-based operation model for the supply chain consisting out of several elements like:

- Basic and control principles
- A SCM centre, with administrative processes
- Actors, with a set of primary processes
- KPIs, and
- Management and supporting processes

The Operations Model has direct links to all software applications used in operation of RCT, where it can import and export data.

REFERENCES

(1)   Browne, J., Sackett, P. J. and Wortmann, J. C., 1995, Future manufacturing systems - towards the extended enterprise, Computers in Industry, Vol. 25, pp 235-54
(2)   Browne, J. and Zhang, J., 1999, Extended and virtual enterprises – similarities and differences, in International Journal of Agile Management Systems Vol. 1 No 1
(3)   Simchi-Levi, D., Kaminsky, P. and Simchi-Levi, E., 2000, Designing and Managing the Supply Chain – concepts, strategies and case studies, MacGraw-Hill, Boston
(4)   Huang, C.Y. and Nof, S. Y., 1999, Enterprise Agility:A View From the PRISM lab, in International Journal of Agile Management Systems Vol. 1 No 1
(5)   Narasimhan, R. and Kim, S. W., 2001, Information system utilazation strategy for supply chain integration, in Journal of Business Logistics, Vol. 22, No.2

(6) Mentzer, J. T., DeWitt, W., Keebler, J. S., Min, S., Nix, N. W., Smith, C. D. and Zacharia, Z. G, 2001, Defining Supply Chain Management, In Journal of Business Logistics, Vol. 22, No.2

(7) Christopher, M., 1998, Logistics and Supply Chain Management: Strategies for reducing costs and improving service, Pitman Publishing, London

(8) Bowersox, D. J, and Closs, D. J., 1996, Logistical management: the integrated supply chain process, McGraw-Hill, New York

(9) Schary, P. B. and Skjøtt-Larsen, T., 2001, Managing the Global Supply Chain, Copenhagen Business School Press, Copenhagen

(10) Jagdev, H. S. and Thoben, K. D., 2001, Anatomy of Enterprise Collaboration, in Journal of Production Planning & Control, Vol. 12, No. 5, 437-451

(11) Boyson, S., Corsi, T.M., Dresner, M.E. and Harrington, L.H, 1999, Logistics and the Extended Enterprise Benchmarks and Best Practices for the Manufacturing Professional, John Wiley & Sons, inc., New York

(12) Lee, H. L. and Whang S., 2001, E-Business and Supply Chain Integration, White paper, www.standford.edu/group/scforum

(13) Hanna, M.D. & Newman, W.R., (2001), "Integrated Operations Management – adding value for customers", Prentice hall

(14) Rusell, R & Taylor III, B, (1998), "Operations Management – Focusing on quality and competiveness" Prentice Hall

(15) Meredith, J.R. & Shafer, S.M, (2002), "Operations Management for MBAs" John Wiley & Sons

(16) Hieber, R. and Alard, R., 1999, Next generation of information system for the extended enterprise, Proceeding form the IFIPWG5.7 Conference in Berlin

A003/036/2003 © With Authors 2003

Collaborative supply chain management – an implementation stage

Z M UDIN and **M K KHAN**
School of Engineering, Design, and Technology, University of Bradford, UK

ABSTRACT

The importance of Supply Chain Management (SCM) in organisations specifically to manufacturers is to play a vital role in managing the flow of material and information along the chain from suppliers to customers. With the emergence of collaborative computing by utilising the power of information and communication technology facilities, the concept of the Collaborative Supply Chain Management (CSCM) is being accepted by many organisations. Apart of these technologies, people factor plays an important aspect in developing CSCM. In this research, the development of CSCM, which is supported by the knowledge-based approach, is divided into three stages. These stages are planning, design and implementation. As discusses in this paper, the implementation stage of this model delivers an action plan that could be followed by suppliers, focal organisations and customers in implementing CSCM. These action plans cover three critical aspects of organisations, which are human developments, process improvements and technology enhancements.

1 INTRODUCTION

Collaborative Supply Chain Management (CSCM) is a management concept that is believed to dominate the industry's Supply Chain Management (SCM) in this 21st century. In order to face the challenge in this century, organisations should prepare themselves with new strategy in SCM. The competition in the industry is shifting from between organisations to competition among the supply chain (1-3). The integration is needed between and within organisation and should be able to deliver a win-win situation among suppliers, focal organisations and customers. This smart partnership would not only develop a competitive advantage to organisations but also could improve the organisation's operation and lead into profit improvement. Furthermore, the advancement in information and communication technology should be utilised by organisations in order to improve the effectiveness and efficiency of business processes. Organisations should plan strategically their direction towards CSCM, which is believed could become a strategic solution in facing challenges in this competitive environment.

In developing a CSCM environment, a proper plan and design are needed by organisations in order to reduce the complexity that exists in the current environment. These complexities such as human, process, technological and financial factors could reduce the efficiency and effectiveness of organisations (4). In addition, organisations should also need to evaluate their capabilities in order to sustain their competitiveness in the CSCM environment. This paper discusses the implementation stage of CSCM model as shown in the framework in Figure 1 below. This model, which is supported by a Knowledge-Base (KB) mechanism, is used to support organisations in planning, designing and implementing a CSCM.

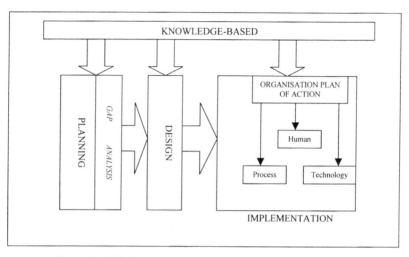

Figure 1: Collaborative Supply Chain Management (CSCM) Model

In implementation stage, based on the information that gathered in planning and design stage, the model will provide a suggestion to users. The suggestion is developed from the rules and knowledge that stored in the system which is currently in the developing process. These rules and knowledge are only valid to a group of organisations (supplier, focal organisation and customer) that involved in one particular CSCM. The organisation plan of action component is used to assist supplier, focal organisation and customer on what action they should undertake in order to success in implementing CSCM.

2 CSCM KNOWLEDGE-BASED SYSTEMS

A knowledge-based approach is used in developing a CSCM environment. This selection is due to some reasons. Based on the survey of literatures, none of the earlier studies had used a Knowledge-Based System (KBS) as a decision-making tool or approach in redesigning supply chain and developing a CSCM. However, some applications that are based on knowledge and human expert are used in certain activities in SCM such as in supplier selection (4), purchasing (5), process planning (6), job scheduling (7) and process of make or buy decision (8). KBS has the capability to diagnosis problems and supplying alternatives to users in a fast reaction time compared to humans. In addition, KBS could also provide explanations and suggestions to users based on the problem situation (9). KBS is effectively

A003/045/2003 © With Authors 2003

used in solving an open-ended problem where no suitable approach to solve them. Furthermore, KBS is independent and not influence by emotion or bias in suggesting solutions to users (10).

The CSCM knowledge-based system utilising a production rules technique in retrieving and inferencing knowledge from the knowledge base. This system consists of three main modules, which relates to planning, design and implementation. The first level flowchart of the implementation module is illustrated in Figure 4.2 in the final page. In this module, users could interactively communicate with system before an action plan that relates to human development, process improvement and technological enhancement could be provided by system. The input for this module is supplied from the previous modules in planning a nd designing phase, which is stored in the information and knowledge base. The rules are used in mining and extracting knowledge and try to reach a conclusion before solution could be provided. The simple example of rules that used in this system is provided as below.

Example 1

IF *supplier is dependable*
　　AND *supplier integrity is high*
　　AND *supplier financial is stable and promising*
THEN *potential to be a partner in collaboration is high*

Example 2

IF *staff training is provided*
　　AND *staff incentives is offered*
THEN *proceed to staff availability sub-module*

In the implementation module, there are three sub-modules will be developed in order to provide an action plan for organisations in implementing CSCM. These three modules are relating to human developments, process improvements and technology enhancements.

3 HUMAN DEVELOPMENTS

Human developments relate to the plan that organisation should do in developing its human resources in the CSCM environment. Human developments program is an important aspect that organisation should emphasise in implementing CSCM. Most of the factors that enable the success of supply chain collaboration are relate to human aspects (11), (12). Among these factors are including commitment, training, leadership, openness, responsibility and knowledge. In implementing collaborative supply chain, the attitude of 'give and take' should exist in organisations environment. Every staff in supplier, focal organisation and customer that collaborate with each other should aware about this situation. In addition, in order to make CSCM working smoothly and success, it is believed that the attitude such as dishonest, betrayal and closeness should be avoided. Among the factors that should be considered in human development is commitment from top management to shop-floor, appropriate training, transformation of working culture and value, establish leadership, trust and openness development and developing a clear expectation. Based on information from organisation capability component in the design stage such as skills a nd e xpertise o f s taff,

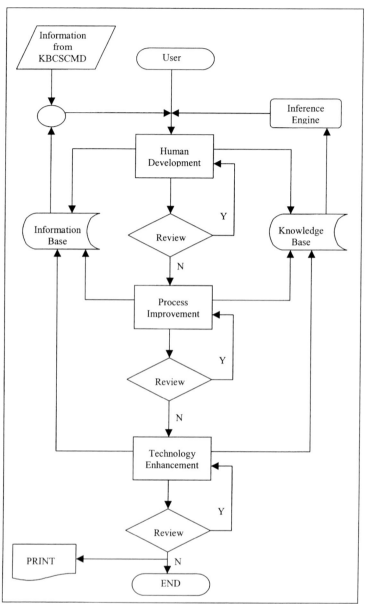

Figure 4.2: The Flowchart of Implementation Module

A003/045/2003 © With Authors 2003

degree of support and commitment, relationship between management and lower level staff and staff willingness in supporting organisation objectives, this module would analyse and suggest to organisation a plan of action on what should be done in human development in order to implement CSCM. This plan of action would cover all the human resources aspects that significantly contribute in executing a CSCM.

4 PROCESS IMPROVEMENTS

The second aspect in this module relates to process improvements program. Process improvement is needed with the purpose to improve organisation efficiency, eliminate waste process, reduce process lead-time, add value to existing process, eliminate process redundancy and automate the process. In addition, through process improvement the flow of material, service and information within CSCM become more transparent and easy to monitor by supplier, focal organisation or customer. Furthermore, organisations could develop their competitive advantage and would influence the CSCM competitiveness. According to (13) in their research about relationship between Japanese automakers and the U.S suppliers, process improvement that done in the U.S suppliers give an impact to their performance and significantly improve the relationship between these two organisations. Information from the organisation capability and organisation alignment in the design stage is a base for KB mechanism in this system to provide a suggestion to the supplier, focal organisation and customer.

The model (system) would suggest a plan of action relating to strategies that needed to align business processes or areas in order to make CSCM working effectively. This plan of action would consist strategies that enable organisations to collaboratively work in streamlining the business processes among them. Furthermore, suppliers, focal organisations and customers could make a preparation to integrate the appropriate business process within their own organisation.

5 TECHNOLOGY ENHANCEMENTS

Finally, in technological enhancements, a suggestion would be given to organisations on how technology could be used in improving the organisation competitiveness in the CSCM. Apart of human development factor, technology is one of important enabler in CSCM success and among the challenges in new economic model in the Internet era (14). The utilisation of technology not only could transform the traditional operation into automate operation but also contribute in improving the effectiveness of the supply chain (12). Along with advancement in the information and communication technology, the supplier, focal organisations and customer should take advantage of using the collaborative computing technology in order to improve the communication among them. In this module, based on the information from design stage where assessment on the technological capability of organisation is done, the KB provides an action plan for suppliers, focal organisations and customers. This action plan could consist strategies in utilising an existing technology in the new environment, which would include technology used in product design and production. In addition, the utilisation of information and communication technology also be emphasised in this action plan.

6 CONCLUSIONS

This paper outlines the implementation module, which is the final module in the research framework. T he objective of this module is to utilise all information and knowledge that gathered in planning and design phase in order to deliver an action plan that could be used by organisations in implementing CSCM. This action plan would cover three organisational aspects that are critical in implementing CSCM environment.

This integrated module is supported by the knowledge-based mechanism that used to inference knowledge and information before reaching a conclusion. In this module, the production rules are used for this purpose. In order to support organisation in implementing CSCM, action plan that are delivered by the system would cover human development, process improvement and technology enhancement.

Human development action plan would emphasise on strategies that relates in preparing organisation human resources for CSCM environment. This strategies including on how organisation could utilise these human resources in order to make CSCM works effectively. Organisations also need to improve their business processes in order to make the integration among them work smoothly and effectively. In addition, doing the process streamlined should eliminate the possibility of process redundancy that resulted by the integration. Since technology aspect is one of important enabler in implementing CSCM, an action plan for organisation to enhance the technology capability is produce by this module. These action plans are important for organisations to follow in order to assure the success of CSCM implementation.

REFERENCES

1   **Helms, M.M., L.P. Ettkin, and S. Chapman**, *Supply chain forecasting: collaborative forecasting supports supply chain management.* Business Process Management Journal, 2000. 6(5): p. 392-407.

2   **Horvath, L.**, *Collaboration: the key to value creation in supply chain management.* Supply Chain Management: An International Journal, 2001. 6(5): p. 205-207.

3   **Cokins, G.**, *A collaboration enabler: sharing profit and cost data.* www.ascet.com, 2002. 4.

4   **Vokurka, R.J., J. Choobineh, and L. Vadi**, *A prototype expert system for the evaluation and selection of potential suppliers.* International Journal of Operations & Production Management, 1996. 16(12): p. 106-127.

5   **McIvor, R.T., M.D. Mulvenna, and P.K. Humphreys**, *A hybrid knowledge-based system for strategic purchasing.* Expert Systems with Applications, 1997. 12(4): p. 497-512.

6   **Ming, X.G., K.L. Mak, and J.Q. Yan**, *A hybrid intelligent inference model for computer aided process planning.* Integrated Manufacturing Systems, 1999. 10/6: p. 343-353.

7   **Zhang, Y. and H. Chen**, *A knowledge-based dynamic job-scheduling in low-volume/high-variety manufacturing.* Artificial Intelligent in Engineering, 1999. 13(3): p. 241-239.

8   **Humphreys, P., G.Huang, and R.McIvor**, *An expert system for evaluating the make or buy decision.* Computer and Industrial Engineering, 2002. in press.

9   **Mockler, R.J.**, *Knowledge-based systems for management decisions.* 1989: Prentice-Hall International Inc.

10   **Turban, E. and J.E. Aronson**, *Decision support systems and intelligent systems.* 1998, New Jersey: Prentice-Hall Inc.

11   **Mentzer, J.T., J.H. Foggin, and S.L. Golicic**, *Collaboration: the enablers, impediments and benefits.* Supply Chain Management Review, 2000. September/October: p. 52-58.

12   **Akintoye, A., G. McIntosh, and E. Fitzgerald**, *A survey of supply chain collaboration and management in the UK construction industry.* European Journal of Purchasing & Supply Management, 2000. 6: p. 159-168.

13   **Liker, J.K. and Y.-C. Wu**, *Japanese automakers, U.S. suppliers and supply-chain superiority.* Sloan Management Review, 2000(Fall): p. 81-93.

14   **Anthony, T.**, *Supply chain collaboration: success in the new internet economy.* www.ascet.com, 2000. 2.

Note: * lecturer of School of Information Technology,
        University Utara Malaysia, MALAYSIA.

Using simulation to evaluate e-business process implementations

P D BALL, P ALBORES, and **J MACBRYDE**
Centre for Strategic Manufacturing, University of Strathclyde, Glasgow, UK

ABSTRACT

E-business adoption within companies appears to be driven by high level justification and "me too" attitudes rather than from detailed investigation and assessment of the operational benefits. This paper argues the need to perform this detailed investigation and proposes the use of discrete event simulation as a tool for this analysis. Case studies from two small to medium sized enterprises illustrate features of simulation tools required that could improve the evaluation of e-business process implementations. Results from the case studies show the effect of e-business process implementations on the dynamics of the business whilst demonstrating the greater understanding of the dynamic performance that can be gained by using simulation.

1 INTRODUCTION

The business process approach to organising and managing a company's activities is well established. The approach focuses on how to deliver value to the customer, an example being the order fulfilment process, containing all the activities to transform a customer order into a delivered product or service including order receipt, check availability, credit check, source materials, produce, pick, dispatch and invoice. Sweet (1) has identified the realignment of business processes as one of the main worries of companies when implementing e-business. Companies perceive the benefit of adopting of e-business as a means of simplifying their business processes (1).

Despite the high general awareness of e-business, within the UK the manufacturing sector has the lower rate of adoption of e-business strategies and the highest rate of "e-laggards" (2). Interestingly many companies see the move towards e-business not as whether to proceed but when to proceed (3). It is important that the improvements offered by business process change and e-business are valid and sustained. The move to e-business must be evaluated not just from an apparent structural improvement but one of actual performance improvement.

This paper addresses the need to assess impact of e-business and the techniques with potential to assist. Modelling techniques are reviewed and the relevance of modelling techniques in assessing the impact of e-business is illustrated through two case studies. The case studies are discussed and the challenges for evaluating e-business process change identified.

2 ASSESSING E-BUSINESS PROCESS CHANGE

In the course of improving business processes to improve performance companies must assess the specific impact the changes will have on them. Such assessment will typically be presented as local benefit with costs to demonstrate benefit to the business. Whilst this type of evaluation is important it lacks the evaluation of the effect on the wider business on operational performance and financial, not just cost, benefit (4).

McFarlene et al (5) state "...remarkably little work has addressed the full impact of internet developments at the operational level of manufacturing companies". The use of a business process approach to structure and document the operation of a business could lead a wider view of the impact of changes. Work on developing generic business processes for e-business includes the MIT e-Business repository in the US (6) and that conducted by the Open University in the UK (7). This work can guide analysts to focus on the area of e-business change as well as the overall effect on the process, e.g. the whole of the order fulfilment process. It should be noted, however, that the changes observed will be static and structural changes, not changes in operational or financial performance.

One form of assessment is through the use of benchmarking. Benchmarking ranges from informal comparison with local companies to formal models. Some benchmark schemes contain suggestions for specific applications that can be used to close gaps with those best in class and are therefore extremely useful for suggesting overall change. One formal approach at supply chain level makes use of the Supply Chain Operations Reference (SCOR) model (8). By using a standard process reference model and common metrics benchmarking an audit of a company can be carried out, compared against best practice and indications of the areas that the company should focus on derived. The gap identified against the benchmark data is quantitative and will indicate the extent of improvement necessary. Whilst the change can be identified, the impact of the change at local level, such as e-business implementation, is not assessed. Of interest here is Gammelgaard and Andreassen's (9) discussion of the current lack of insight of the differing burdens and benefits at different points in the supply chain and subsequent insight at company level.

Modelling techniques can be applied to assess impact within a specific company and are therefore compatible with benchmarking techniques. Benchmarking can highlight areas of relatively weak performance and suggest solutions whilst modelling techniques have their strength in evaluating solutions. There are a variety of modelling techniques that can be deployed, each varying in the degree of quantitative analysis, insight to operation and time and skill required to use. Modelling techniques can be used to generate cost, and in some cases financial, information that in turn can be built into the business case. The next section looks at the characteristics of modelling techniques and their application.

3 MODELLING E-BUSINESS PROCESSES

3.1 Modelling techniques

Modelling techniques vary in the way a process can be represented and the degree to which they can assist in quantitative analysis. This section introduces generic properties of the techniques.

One of the key strengths of modelling techniques is to enable an abstract representation of a system, in this case a business process, to be built up. The abstract representation captures key information at a certain level of detail. There is no predetermined level of detail and modelling could be hierarchical in nature. For example, in the order fulfilment process introduced earlier the check availability stage could be further broken down into check stock, check production plan, allocate, etc.

Some modelling techniques are static in nature and provide diagrammatic representation of a process. Techniques include simple flow charting or process flow to more developed methods such as Data Flow Diagrams, IDEF modelling and GRAI grids. The use of such techniques for communicating the structure of a business process makes them an obvious candidate for assessing the impact of e-business processes. All these techniques have the ability to capture the activities and interactions present, but lack the ability to model the dynamic and random aspects. IDEF0 models are static in nature and through extensions to IDEF3 can capture time but require links with simulation to execute the dynamic behaviour.

Other techniques allow quantitative analysis. The advantage of tools such as spreadsheets is that they can be used for "what if?" type analysis whereby a model is repeatedly run under different conditions and using different parameters. For example, what if the demand varied, what if an activity was more responsive or, more significantly, what if the process was reconfigured in a different way? Whilst in practice such analysis is often quick to carry out there are difficulties in representing the passage of time and conveying the process flow and interaction for communication and validation purposes.

A more select group of techniques are able to look at dynamic and random behaviour of a process. Such simulation techniques can capture and replicate real world effects such as variation of customer orders over time, the level of inventory in the supply chain over time, etc. This extends analysis from a static, stable view of a business process to a real-life fluctuating performance over time. This is an important aspect of analysis as it could be argued that companies never settle into steady state and it is the dynamic, random aspects that have a critical impact on performance. Furthermore, Albores et al (10) suggest that in modelling the differences between 'traditional' business processes and e-business processes the most significant differences will appear in how each activity performs rather than their sequence or interaction. Such dynamic and random behaviour therefore is more topical. Inevitably techniques able to represent dynamic and random behaviour will require more time, skill and expense to use and these will have an impact on their adoption.

The degree to which simulation techniques can represent interaction varies. Techniques include system dynamics and discrete event simulation. System dynamics, using software such as Ithink (11), is typically deployed at the supply chain level rather than detailed business process level. Building more detailed models of processes and in particular representing individual transactions is extremely difficult and this technique is therefore appropriate to more high level analysis. Discrete event simulation is more common and is able to represent detailed transactions, information and physical flows over a period of time as well as incorporating randomness. Software tools can help enormously in this to reduce the time and skill and costs can be minimized with packages such as Simul8 that retail in the office software price range.

3.2 Modelling applications

Albores et al (11) identify five different groups of reported business process simulations. These will be considered in turn with references to applications.

Conceptual business process simulation. Work here deals mainly with the identification of the requirements for simulating business processes. For example, Hlupic and Robinson (12) model order taking and higher level coordination as well as physical processes. They have identified the importance of hierarchical capabilities within simulation software, the need of a good process model interface (usually linked to standards like IDEF0, IDEF3 or flowcharting) and the need to have flow objects, resources, activities and routings in order to better represent business processes.

Manufacturing systems simulations. This group include applications that are focused on the physical flow of entities through a series of workstations. An example application is (13). Although it is presented as a business process simulation, it lacks the flow of information and cross-functional activities that should characterise a business process (14).

Process Specific Simulations. Work on simulating a specific process within an organisation include resource analysis in a Police Custody Suite (15) and process for providing basic telephone service (16). Vusksic et al. (17) present a business-to-business simulation model of a procurement process and include some aspects of wider business, e.g. finance thus showing the potential to evaluate the wider impact of e-business process change.

Enterprise-wide simulations. This type of simulation refers to the interaction between different areas of an enterprise. Examples of this category are the Whole Business Simulator (4) and the application of simulation to understand and improve the co-ordination between three different plants of the same automotive company (12).

Supply-chain-wide simulations. Examples here include demand management policies (Make to Stock, Make to Order, etc) and its implications for the members of the supply chain (18) as well as the IBM Supply Chain Simulator (19) for modelling at the supply chain level, able to capture the high level process design and evaluate in more global metrics. Also at a supply chain level is the e-SCOR (20) simulator, able to model the processes set out in the SCOR model and generate results using the same metrics although there appears to be nothing specific to e-business applications.

4 CASES STUDIES OF E-BUSINESS PROCESS MODELLING

This section briefly introduces two case studies where e-business process designs have been modelling using simulation techniques. The following section will discuss the lessons learnt.

4.1 Modelling of mineral water company

An analysis of a rapidly expanding mineral water manufacturer was carried out to ascertain whether the increasing number of order transactions could be more efficiently handled using an e-procurement solution. The analysis focused on the sell from stock process, in particular receiving customer orders, stock allocation, haulier purchase order issue, haulier collection from warehouse and delivery to customer. Transactions from their business system including the issue of pick notes to warehouse and invoices to customer were included. Performance metrics were processing time, throughput, staff utilization and cost. The model was built at two levels of detail using Simprocess. Whilst this software was used because of its process orientation and ability to support multiple views of model detail, the choice of software was not considered as important as the challenge of how to represent the scenario conceptually before using the software.

The analysis showed that operational cost savings and order process time reductions could be achieved. The modelled e-business improvements allowed a faster flow of information but the physical process stages remained unchanged, see table 1 for representative figures.

	'traditional'	e-process
Processing time/unit (min)	272	215
Staff Utilisation (POH)	78%	18%
Cost/unit	£ 4.58	£ 2.90
Cost/month	£ 8755	£ 5650

Table 1. Comparison between 'traditional' and e-enabled purchasing process.

4.2 Modelling of a printing company

The second case study refers to a printing company specialising in the production of labels and labelling systems for industry and commerce. The simulation model was built to analyse the effect of introducing new e-business strategies, specifically towards customer interaction and involvement along the design and production process. Having current order-processing lead times of up to 14 days, they plan to reduce this significantly through the introduction of e-business. The process examined looked at business-to-business interfaces and workflow between areas of the company including quotation, planning and production.

The model was built in Simul8. The results of this analysis showed that implementing a workflow application would greatly reduce the time the orders spend being processed, but it will build a bottleneck in the first stages of production. Similarly, allowing the customers to have access to the different stages of the production process will have the effect of removing the slack that is currently inbuilt in the system and would put strain on the account managers and the technical planners. Once again, the level of detail of the process maps did not show the differences between the traditional and the e-business process, the differences were only picked up on simulation.

5 DISCUSSION

The cases showed that simulation can be a useful tool for the analysis of e-business implementation and allow a better understanding of the performance of an e-business processes. Interestingly in the first case, although the order processing time reduced the overall lead-time remained largely the same since the transport policies were not revised, affirming the value of simulation in understanding the wider effect of local improvements (10). The model took account of the faster flow of information but did not address the responsiveness or characteristics of an e-business environment (21) in which there is greater volatility in ordering and expectations of faster response.

It was observed that business process models created were essentially generic and structurally differed little between the 'traditional' and e-business scenarios. The behaviour of the process models did differ and therefore simulation has an important part to play in quantifying this. More detailed standard business process models in general (and e-business process models in particular) could have helped guide the model building, however, in these particular examples there was little room to further improve the company's business process structure to gain performance improvement, only in the way activities were carried out.

The simulation software was able to represent the flow of information readily as well as separate product flows, however, challenges existed in areas where information and physical product had to be joined up rather than one simply being the trigger for the other. Scope of modelling is also an issue. As well as having a richness to be able to represent different stock (physical stock as well as the stock records), models need to have the scope to enable wider evaluation to avoid optimising a specific area at the expense of the whole. This issue of scope can be extended to cover modelling of the e-element of businesses. Assuming e-business systems do have a significant impact on performance then the behaviour and the impact must be captured.

Modelling tools should not be used in isolation. Guidelines or methodologies are necessary to assist in the development of the analysis of e-business. This is very much application specific and beyond the more generic modelling methodologies for building a model of any type. Such enhancements could guide which business processes to improve, the possible process improvements that could be carried out and how to assess whether the impact is beneficial. As introduced earlier, benchmarking techniques have the potential to guide best practice and identify areas were performance is lacking in comparison to competitors. Some approaches (e.g. SCOR) guide process mapping and combine this with benchmark data to indicate which tools can be deployed.

6 CONCLUDING REMARKS

This paper has examined the use of modelling techniques, simulation in particular, to measure the impact of e-business process change. It has examined the attributes of simulation and made reference to work carried out in a variety of relevant areas.

With business process change seen as a pre-requisite for e-business there are two notable gaps. Firstly, there is a lack of detailed but generic e-business process models for specific

    A003/062/2003 © P D Ball 2003

implementations of e-business. Secondly, there is a lack of simulation tools that can directly use such generic business processes as templates to create simulation models.

As e-business systems are used more and more for the planning and control of companies they will be more responsible for the dampening or otherwise of the dynamics of the business. Guidance and tools are required on how to capture the dynamic effect of e-business systems and measure their impact.

REFERENCES

(1) **Sweet, P.**, 2001, "Time for a return", *CONSPECTUS*, Dec, pp. 2-4.
(2) **Farish, M.**, 2001, "Quiet revolution", *Engineering*, April, pp. 18.
(3) **Porter, M.**, 2001, "Strategy and the Internet", *Harvard Business Review*, Mar, pp.63-77.
(4) **Love, D. and Barton, J.**, 1996, "Evaluation of design decisions through CIM and Simulation", *Integrated Manufacturing Systems*, vol. 7, no. 6, pp. 3-11.
(5) **McFarlane, D., Gregory, M., and Thorne, A.**, 2001, "E-manufacturing in Japan: Assessing the state of the art in Internet-Based Manufacturing and its impact on practice, strategy and policy", Institute for Manufacturing, Cambridge, UK.
(6) **MIT**, 2000, *E-business process repository http://process.mit.edu/*
(7) **Barnes, D., Hinton, M., and Mieczkowska, S.**, 2001, "Towards a Framework for investigating the impact of E-commerce on Internal Business Processes", *Proceedings of the European Operations Management Association 8th International Annual Conference,* Bath, UK, pp. 798-806.
(8) **Supply Chain Council**, 2002, *Supply Chain Operations Reference (SCOR) Model.* http://www.supply-chain.org/default.htm
(9) **Gammelgaard, B. and Andreassen, M.A.**, 2002, "Strategic Benchmarking of the Supply Chain." *Proceedings 9th International Conference of the European Operations Management Association (EUROMA) Conference,* Copenhagen, Vol. 1, pp. 571-581.
(10) **Albores, P., Ball, P. D., and MacBryde, J.C.**, 2002, "Generic Business Process Models For E-Business In Manufacturing Companies: Is Simulation Useful?" *Proceedings of the 2002 IEEE International Engineering Management Conference,* Cambridge, UK, pp. 685-690.
(11) **Michaelides, Z., Ho, J.C.C., Boughton, N.J. and Kehoe, D.F.**, 2002, "The Development and Evaluation of Internet Based Supply of Non-production (MRO) Items", *Proceedings of the 2002 LRN Conference*, pp. 43-50.
(12) **Hlupic, V. and Robinson, S.**, 1998, "Business Process Modelling and analysis using discrete-event simulation", *Winter Simulation Conference Proceedings,* pp. 1364-1369.
(13) **Irani, Z., Hlupic, V., Baldwin, L., and Love, P.**, 2000, "Re-engineering manufacturing processes through simulation modelling", *Logistics Information Management*, vol. 13, no. 1, pp. 7-13.
(14) **Albores, P., Ball, P. D., and MacBryde, J. C.**, 2002, "Assessing the impact of electronic commerce in business processes: A simulation approach", *Proceedings of the 9th International Conference of the European Operations Management Association Conference,* Copenhagen, pp. 15-28.
(15) **Greasly, A.**, 2000, "Effective uses of business Process Simulation", *Winter Simulation Conference Proceedings,* pp. 2004-2009.
(16) **Dennis, S., King, B., Hind, M., and Robinson, S.**, 2000, "Applications of business

process Simulation and lean techniques in British Telecommunications PLC", *Winter Simulation Conference Proceedings,* pp. 2015-2021.

(17) **Vuksic, V., Stemberg, M., and Jaklic, J.**, 2001, "Simulation modelling towards e-business models development", *International Journal of Simulation*, vol. 2, no. 2, pp. 16-29.

(18) **Strader, T. J., Lin, F., and Shaw, M. J.**, 1998, "Simulation of Order Fulfilment in Divergent Assembly Supply Chains", *Journal of Artificial Societies and Social Simulation*, vol. 1, no. 2.

(19) **Bagchi, S., Buckley, S. J., Ettl , M., and Lin, G. Y.**, 1998, "Experience using the IBM supply chain simulator", *Winter Simulation Conference Proceedings,* Washington, D.C., pp. 1387-1394.

(20) **Barnett, M. W. and Miller, C. J.**, 2000, "Analysis of the virtual enterprise using distributed supply chain modeling and simulation: an application of e-SCOR", *Proceedings of the Winter Simulation Conference,* Orlando, Florida, pp. 352-355.

(21) **Fryer, B.**, 2001, "Competing for Supply, Conversation with Bryan Stolle", *Harvard Business Review*, Feb 2001, pp. 25-26.

Manufacturing Process

Physical and numerical simulations of a continuous extrusion process

A BARCELLONA
Department of Production Engineering, University of Palermo, Italy

SYNOPSIS

A continuous extrusion process, based on the friction actions between tool and workpiece to obtain extrusion pressure and softening of the material, is studied and analysed in this research. By means of a traditional extrusion process only one piece of metal can be worked at a time, frequent reloading is necessary and welding of the parts has to be frequently executed. In order to avoid these difficulties, a machine that produces a continuous extruded rod by means of shear actions that pull and push the material against a fixed die has been studied by means of numerical and physical simulations. The principle of the process, called CONFORM and invented in the UK, is based on the friction existing between the workpiece and a grooved tool container. Shear actions play a fundamental role, and the process has been studied with the aim to find the correct operative parameters to generate sufficient pressure and temperature to extrude the material trough a die.

1 INTRODUCTION

A sample scheme of the analysed geometry of the Conform extrusion process is shown in the figure 1. A grooved extrusion "wheel" drags and push the feed material by means of a continuous rotation, basing on the frictional force existing between wheel and input rod material. The material is pushed into holed dies of various shapes or geometries. The frictional force between the rod feed and the wheel must overcome not only the force required for the deformation but also the frictional force existing between the rod feed and the stationary dies, named "shoe" and "abutment". The material is extruded when it meets the "abutment" and the die cavity. In this way, the product is continuous and welding phase of the billets is not required. A wide range of products may be manufactured, such as wire, profiles, multi-void tubes and strips. During the process, heat is generated due to the friction between material and shoe, and due to the deformation. Some experimental and numerical results of the process, are reported in literature. Theoretical and numerical studies of the process have been conducted, in some situations by considering a plain strain deformation field, isothermal conditions and several geometric reductions, when employing two dimensional numerical codes, but also by means of full three dimensional thermal mechanical analysis, in order to evaluate the performances of the process and to study the quantity of heat generated due to the slipping and to the deformation (1), (2), (3), (4), (5), (6).

One of the most serious problem in the process is due to the eventually excessive heat generation in the material; for this reason in a previous paper (6) a thermal mechanical numerical analysis has been performed in order to evaluate and study the quantity of heat generated by friction and by the imposed high strains. Furthermore, the choice of some operative parameters, such as die geometry, velocity and cooling rate of the dies, has been performed by means of an outranking approach (7), (8), (9) in order to select the best combination of them, by comparing the numerical results in the different cases. In this research the synergic effect of physical and numerical simulations gave, as a result, the possibility of better find a threshold with regard to the feasibility of the process. Different friction conditions are furthermore studied by means of the physical model.

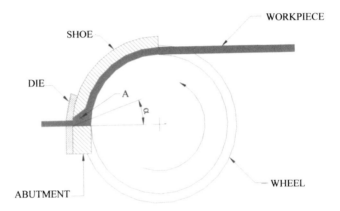

Figure 1. The Conform process

2 THE NUMERICAL MODEL

2.1 The FEA model
Basing on the results of a previous research (9), in which several numerical simulations were performed by varying the operative parameters (velocity, tool geometry, presence of coolant), it has been decided to concentrate the attention on the cases characterized by the velocity of 90 mm/sec and 10 mm/sec and the absence of coolant, and that shown high performances by employing the outranking approach discusses in the paper. Furthermore, the velocity of 500 mm/sec, and the presence of coolant would have comported several difficulties in the physical model.

The new experimental plan of experiments, numerically simulated and, some of them, physically simulated, is the one obtained by considering the values, variable and constant, in the following Table 1, for a total number of 12 cases of study. As shown, two velocity of the output material have been analysed, actually depending on the rotational velocity of the wheel.

Table 1. Operative parameters for numerical simulations

Velocity (mm/sec) - (r.p.m)	10 – 4	90 – 36	-
Hole die diameter (mm)	14	12	10
Tool geometry α (degree)	20	30	-
Feed aluminium rod (mm)	16 x 38	-	-
Wheel diameter (mm)	300	-	-
Initial temperature of aluminium (°C)	25	-	-

Workpiece and tooling have been considered initially cold; in this way, the increment of temperature only due to the slipping and to the deformation has been analysed in order to evaluate the heat generated during the process. The feed aluminium rod is 16 x 38 mm, the die diameters considered are 14, 12 and 10 mm and the rotating extrusion wheel has a diameter equal to 300 mm. Tooling for numerical simulations have been considered made of H13 steel. Not only the workpiece, but also die, wheel, abutment and shoe were meshed in order to study the variation of the temperature in the tooling; dies were modelled as rigid surfaces, thus leading to a thermal-mechanical analysis of the workpiece and to a thermal analysis of the dies. Simulations have been carried out by means of the finite element DEFORM3^{TM} code (10), that is a very powerful tool in 3D numerical analysis. More than 34000 elements and almost 9000 nodes have been employed for the analysis. Due to the tooling geometry, several automatic remeshings were required during the simulations that where executed on a high performances Compaq workstation with a mean CPU time equal to 25 hours. The plastic flow curve of the material employed for the simulation has been considered as a function of temperature and deformation rates; data have been introduced in the code by points. The Aluminium alloy is the 6062 alloy (0.6% Si, 0.2% Fe, 0.05% Cu, 0,6% Mn, 0.7 Mg, 0.05% Cr, 0.07% Zn, 0.1% Ti) in O conditions, showing a melting point equal to 650°C. Frictional effects at the interface workpiece-dies have been taken into account by means of the constant shear friction model. The friction factor was assumed very high and equal to 0.8 between the material and the rotating wheel and equal to 0.1 between the material and the others dies. The simulations have been carried out according to the values of the thermal characteristics of the materials reported in Table 2. As a previous analysis confirmed, before the beginning of the grooved wheel rotation, an upsetting of the material inside the die is necessary to generate a pressure inside the dies. For this reason, firstly, a movement of the shoe in the direction of the wheel, while the workpiece is located between them, has to be numerically simulated.

Table 2. Thermal and mechanical properties for FEA of Aluminium 6062 and H13 steel.

Workpiece material and tooling	Heat capacity N/mm^2/sec	Thermal conductivity W/m °K	Linear expansion coefficient (20-100°C) 1/°C	Flow stress curve at room temperature
ASTM 6062	1.75	220	23 * 10E-6	$\sigma = 210\,\varepsilon^{0.13}$ MPa
AISI H13	3.77	30	12 * 10E-6	-

2.2 Numerical results

By means of the numerical simulations it has been possible to find the variation of temperature due to the slipping of the rod feed against the shoe and due to the deformation, that is anyway dominant. In the figure 2(a) it is possible to see the feed material entering the tooling, in figure 2(b), the shoe and the extruded material through the die, in figure 2(c) only workpiece, die and abutment are shown, while wheel and shoe are intentionally omitted to simplify the observation. In Table 3, for each combination of velocity, die diameter and tool geometry, the rising of the temperature of the workpiece in the point named A in figure 1, is reported. As above mentioned, the initial temperature of workpiece and dies has been chosen equal to 25 °C and, as an example, the rising of temperature for V = 90 mm/sec, α = 20 is 230 °C with the die diameter = 10 mm and 220 °C with the die diameter = 14 mm.

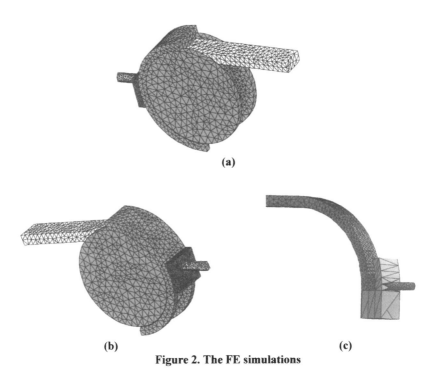

(a)

(b)   (c)

Figure 2. The FE simulations

Missing results are due to the non feasibility of the process because for some combinations of the operative parameters, slipping of the material over the rotating wheel is obtained: in this situation, temperature rises very rapidly due to friction, without any deformation or movement of the feed material. The reached temperatures indicate that locally melting of workpiece material may happen. This occurs when the reduction in area is too strong, and the required extrusion pressures are very high and the shear actions for the chosen geometry are not enough to generate pressure and temperature to extrude the material.

The feed material is not dragged by the rotating wheel and slipping between workpiece and die occurs. On the contrary, when the feasibility of the process is obtained, the increasing of temperature of the feed material inside the dies due to deformation and due to friction between itself and the stationary tools reduces the heating costs, because for this reason it may be possible to reduce the engagement of heating equipments (furnaces).

Effectively, as confirmed in previous researches, only for higher velocities of the ones analysed here it is possible to forecast the success of the extrusion process even without heating of dies and workpiece.

Table 3. Temperatures in point A (°C).

Temperatures in point A (see fig. 1)	Hole die diameter 10 mm		Hole die diameter 12 mm		Hole die diameter 14 mm	
	$\alpha = 20$	$\alpha = 30$	$\alpha = 20$	$\alpha = 30$	$\alpha = 20$	$\alpha = 30$
V= 10 mm/sec	slipping	slipping	slipping	slipping	188	220
V= 90 mm/sec	slipping	slipping	230	245	220	235

As above mentioned, simulation results have shown that some combinations of the operative lead to the non feasibility of the process, because slipping between workpiece materials and wheel happens. But the prediction about slipping occurrence may also depends from the modelled friction conditions; in fact, in the numerical models, it is necessary to specify the correct friction conditions, strongly depending from the actual interface conditions and furthermore the analyst has also to chose the correct theoretical friction model. With the aim to better investigate the occurrence of slipping, a simple physical model has been constructed and employed.

3 PHYSICAL SIMULATIONS

3.1 Physical modelling

Prediction of material flow and necessary loads in the design of a forming process are conducted today by means of several engineering methods, such as upper bound, slab, finite difference and finite element methods. Furthermore, experience and intuition of forming experts are always necessary. The use of engineering methods requires the knowledge of material properties, frictional conditions and process mechanics. When the forming process concerns full three dimensional or anyway complex geometries, these operative data are often not available; these situations have been successfully analysed by conducting physical modelling experiments. It should be noted that development of numerical methods and codes, which have been largely improved by the use of supercomputers, have not made experimental simulations obsolete; in fact a strong synergic effect may be obtained when the two approaches are together applied.

In order to predict metal flow, die filling, defect occurrence, strain distribution and forming loads, the use of highly deformable model materials represents a valid and powerful tool. One of the main advantage of physical simulations over numerical simulation by computers, is that the current three dimensional FE codes are powerful but often expensive and time consuming.

In the tool design phase, a soft material, that may be either metallic or non metallic, is used to carry out several tests by changing tool geometry because dies are made by easily to machinery and not expensive materials, such as acrylic, glass or aluminium. Costs of the trial and error procedure, very high with the actual tools, are reduced or totally cancelled. Other advantages of modelling tests are that low temperatures are generally employed and that it is possible to scale up the dimensions to allow finer details because of the smaller forming loads.

In the production phase, physical simulations are successfully employed to obtain information about the preform geometry and the necessary forming loads (12), (13), (14), (15), (16).

The non metallic model materials, such as plasticine and wax, are characterised by small forming loads, low energy and mainly inexpensive tooling and equipments. Another advantage of the non metallic model materials is the possibility of the use of multicolour layers inside of transparent dies, thus allowing an easy observation of the flow during and after the process. Disadvantages of the non metallic materials are that they are more sensitive to strain, strain rate and temperature and furthermore are less homogeneous than the metallic materials. Furthermore, plasticine is very difficult to machine and the production of the specimens is very time consuming (14). Advantages of using metallic materials are that flow stress curves are well known and quantitative predictions are possible. It should be noted that physical simulation, as a quantitative method, may be conducted in the prediction of forming loads or die pressures only if condition of "plastic similarity" exists and correct comparisons between the model and the process are only possible if linear relationships between model and effective material exist. The minimum condition that have to be held between model and actual process are geometrical identity, equal friction conditions and equal (or similar) plastic flow behaviour between model and actual materials; plastic similarity can be obtained if in the Holomon equation, $\sigma = k\, \varepsilon^n\, \varepsilon^m$, the strain hardening coefficient n and then strain rate coefficient m are the same for the model and the actual material.

3.2 The model material
In this research, the physical simulations, by means of grey plasticine, that is one of the most popular non metallic model materials, have been used in order to predict feasibility of the process. The employed grey plasticine is produced in the original shape of thin bars and has to be homogenize in order to remove entrapped gases and to obtain an homogeneous material. This has been done by means of a vacuum extruder machine. After this operation, plasticine is stored at 5°C and cut in the preformed shape before the experiments. Specimens of 19x38x500 have been employed in the physical simulations.

The model material employed in this study has been characterised at room temperature (25°C), by means of compression tests on 30x50 mm cylindrical specimens; the interface between specimens and plates were lubricated with vaseline and a constant velocity of the compressive plate has been assigned equal to 10 mm/min. In figure 3 is reported the true stress strain curve for the employed plasticine. The curve shows that the employed grey plasticine has a strain hardening behaviour. By elaborating the experimental results of the compression test, it has been possible to compute, the strain hardening coefficient and the constant of the material, so that flow stress curve may be represented by the relation:

$$\sigma = 0.46\, \varepsilon^{0.126} \quad (MPa)$$

that is valid for the strain $\varepsilon < 0.5$.

It has to be noted that plastic similarity between the strain hardening coefficients of the aluminium and of the plasticine exists, at least for the flow stresses curves obtained at room temperature.

A003/006/2003 © IMechE 2003

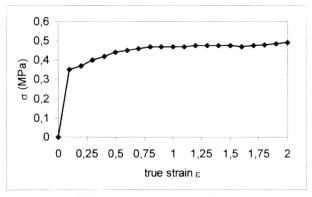

Figure 3. True stress-strain curve for plasticine

3.3 The physical simulated cases

Up to now, with regard to the Conform process, only the velocity of 10 mm/sec has been physically investigated; in fact higher velocity, as the one of 500 mm/sec that, in the actual process requires the presence of coolant, would have comported more complexity in the physical modelling. Tooling and equipments for physical simulations have been constructed in plexiglass and aluminium. In order to model the two different friction conditions between plasticine and rotating wheel and between plasticine and stationary tooling (shoe) it has been operated as follows: as far as regard the interface plasticine-rotating wheel, in order to model an high friction coefficient, such as the one employed in the numerical simulations, the surfaces of the wheel in contact with the plasticine have been shagreened by means of cutting tool; with regard to the interface plasticine-shoe, with the aim to model two different lubricant conditions, it has been interposed powder of talc or vaseline. The operative parameters employed in the physical plan of experiments are those shown in Table 3.

Table 3. Operative parameters for physical simulations.

Velocity (mm/sec) – (r.p.m.)	10 – 4	-	-
Hole die diameter (mm)	14	12	10
Tool geometry α (degree)	20	30	-
Feed plasticine rod (mm)	16 x 38	-	-
Wheel diameter (mm)	300	-	-
Initial temperature of plasticine(°C)	25	-	-
Lubricant	talc	vaseline	-

Since two different lubricants have been employed (talc or vaseline) the plan is composed of a total number of 12 physical experiments. The rotation of the wheel is given by an electrical engine, but it is also possible to give the rotation by hand.

The first set of physical simulations have been executed employing powder talc as a lubricant, at the interface plasticine-stationary die; in this way, it is tried to replay the numerical condition in which the shear actions were modelled by a friction factor m equal to 0.1. The results of these physical experiments are compared with the numerical predictions in Table 4, confirming the trend about the feasibility of the process, but reducing the threshold values: in fact, as it is possible to note in the Table 4, while the numerical prediction gave, as a result, the non feasibility of the process for the cases characterised by the values of the hole die diameter equal to 10 mm and 12 mm, physical simulations predicted slipping only when the smallest hole die diameter, equal to 10 mm, is employed.

Table 4. Comparison between numerical and physical results.

Combinations of operative parameters			Feasibility of process	
Velocity (mm/sec)	Hole die diameter (mm)	Tool geometry α (degree)	Numerical prediction	Physical prediction
10	10	20	Slipping	Slipping
10	10	30	Slipping	Slipping
10	12	20	Slipping	Extrusion
10	12	30	Slipping	Extrusion
10	14	20	Extrusion	Extrusion
10	14	30	Extrusion	Extrusion

As above mentioned, an other set of physical simulations has been executed by employing a different kind of lubricant, strewing with vaseline the interface plasticine-shoe. In this way a different lubricant condition has been realised and a comparison, in terms of feasibility of the extrusion process, has been made. The results are reported in the Table 5, in which it is possible to observe that a good lubrication of the interface workpiece-shoe allows the success of the process, in terms of extrusion feasibility, in the cases in which others combinations of operative parameters didn't consent it. This result gives a precious indication for the actual process, but it is necessary consider that friction at the interface material-shoe contributes at the heating of the workpiece and for this reason it has not always to be held as a negative factor.

Table 5. Comparison between two different lubricant conditions in physical modelling.

Combinations of operative parameters			Feasibility of process	
Velocity (mm/sec)	Hole die diameter (mm)	Tool geometry α (degree)	Lubricant: talc	Lubricant: vaseline
10	10	20	Slipping	Extrusion
10	10	30	Slipping	Extrusion
10	12	20	Extrusion	Extrusion
10	12	30	Extrusion	Extrusion
10	14	20	Extrusion	Extrusion
10	14	30	Extrusion	Extrusion

4 DISCUSSION

The thermal analysis of the extrusion process has shown that temperatures rise very rapidly and intensively, yet with not too high rotational velocities; it is also possible to observe that heating of dies and workpiece before the operation may result in the failure of the process. Effectively, with regard to the numerical results it is possible to forecast, for a chosen combination of geometries and operative parameters, the necessity or the absence of heating equipments and pre heating phase. As an example, for the feasible and analysed numerical cases, it should be noted that an initial temperature of workpice and dies should not over come

A003/006/2003 © IMechE 2003

the value of 200 °C in order to avoid an excessive heating of the parts or the possibility of local melting.

The use of the numerical code, especially in this strongly three dimensional analysed situation, represents a powerful tool but expensive and time consuming; furthermore, the numerical simulations of the actual friction conditions, that in this process play a fundamental role, may be affected by the employed numerical model and coefficients, often not exactly know, specially in new processes and technologies. The possibility to physically simulate the process, by means of highly deformable model materials, helps the analyst to better study the mechanics of the process; the employed model material utilised in this research has shown a good plastic similarity with the actual material; in fact a very good similarity between the two strain hardening exponents n has been found. The limit of the realised physical model consists in the fact that, as it has been numerically observed, while the process is strongly non isothermal, the model material, here characterised only at room temperature, is also heavily temperature dependent. Thus, for a correct employment of the model material, also plastic similarities at high temperatures have to exist. For some combinations of geometries and operative parameters physical predictions were in contrast with the numerical ones and, even if a complete comparison with the actual experimental tests has still to be done), it is possible to think that physical model may better match the actual situation, for the reasons above explained regarding the limits of the numerical friction models.

5 CONCLUSIONS

Basing of previous studies, an innovative continuous extrusion process, that takes advantage of friction between workpiece and rotational tools, has been analysed in this research. Advantages of the process are in terms of low production costs and times. Two different sets of combinations of operative parameters have been analysed by means of numerical and physical approaches. The numerical analysis has been employed with regards to the thermal management of the process, while the physical approach, based on the use of plasticine as model material, together to the numerical one, has been employed with the aim to better understand the mechanics of the process and to predict the feasibility of the process itself. Both a more careful characterisation of the model material, here characterised only at room temperature, and a decomposable version of tooling, in order to observe strain fields by means of the physical model, represent, and they are already started, the final steps of this study.

REFERENCES

1 **Reinikainen, T., Korhonen, A. S., Andersson, K., Kivivuori, S.** (1993) Computer-Aiede Modelling of a NEW Copper Extrusion Process, Annals of the CIRP, Vol.42/1/1993.

2 **Reinikainen, T., Andersson, K., Kivivuori S., Korhonen, A. S.** (1992) Finite-element Analysis of Copper Extrusion Process, Journal of Mat. Proc. Technology.

3 **Newman, R.** (1995) Precision Extrusion with Conform Technology by Holton Machinery Limited - UK, Int. Conference on Aluminium Extruded Products, Brescia, Italy.

4   **Green, D.** (1972) Continuous Extrusion-Forming of Wire Sections, Journal Institute of Metals.

5   **Tomiatu S., Fukuoka S., Ozaki M. et al.** (1999) Thermal Management in Conform Etrusion of Multi-Voids Tubes, Proc. of the 6th ICTP, Vol. III, Germany.

6   **Barcellona, A.** (1999), On The Influence of some Operative Parameters in the Conform Extrusion Process, Proc. of the IV Convegno AITEM, Brescia, Italy.

7   **Roy, B., Bertier, P.** (1973), La Méthode ELECTRE II - Une Application au Média-Planning, OR '72, M.Ross, North-Holland Pub. Company.

8   **Antonelli, D., Barcellona, A. and Cannizzaro, L.** (1999) Application of a Decision Making Method to Improve an Industrial Hot Extrusion Forging Process, Annals of the CIRP, Vol. 48/1/99.

9   **Barcellona, A., Cannizzaro, L., Riccobono, R.** (2000) The continuous extrusion process CONFORM: a full 3D thermal and mechanical analysis, Proc. of the 33rd International MATADOR Conference, Manchester, UK.

10 **Oh, S.I., Wu, W.T., Tang, J.P. and Vedhanayagam, A.** (1991) Capabilities and Applications of FEM code DEFORM: the Perspective of the Developer, Journal of Mat. Processing Technology, 27:25-42.

11 **Clode, M. P.** (1992) Hot Torsion Testing to Model the Extrusion Process, J. Aluminium Industry, 11:24-39.

12 **Maegaard, V.** (1985) The use of the model technique in the prediction of the pressure distribution over the tool surfaces in cold forging, Journal of Mechanical Working Technology, n. 12, pp. 173-191.

13 **Finer, S., Kivivuori, S., Kleemola, H.** (1985) Stress-strain relationships of wax-based model materials, Journal of Mechanical Working Technology, n. 12, pp. 269-277.

14 **Barcellona, A., Fujicawa, S., Altan, T.** (1994) Flashless forging of a connecting rod from Aluminum alloy and a metal matrix composite (MMC) material, Internal report at ERC/NSM, Ohio State University.

15 **Bay N., Hansen, A., Andersen, C.B., Oudin, J., Picart, P.** (1992) Comparison between FE and physical modelling of closed die forging. Proc. of 13th RISO Int. Symp. on Materials Science - Modelling of Plastic Deformation and its Engineering Applications, Riso, pp. 221-226.

16 **Bay, N., Wanheim, T., Arentoft, M., Andersen, C.B., Bennani, B.** (1995) An Appraisal of Numerical and Physical Modelling for Prediction of Metal Forming Processes. COMPLAS, 4th Int. Conf. on Computational Plasticity - Fundamentals and Applications. Barcelona, pp. 1343-1354.

Modern process energy sources for electrical discharge machining (EDM)

M LÄUTER, H-P SCHULZE, and G WOLLENBERG
Otto-von-Guericke University, Magdeburg, Germany

ABSTRACT

In this paper, different concepts of process energy sources for electro-erosive machining are presented. The evolution of the process energy sources and the possibilities of its control for attaining a stable process are described.

The most important high-quality feature of an EDM process control is the complete avoidance of process degeneracy and a maximum number of removal-effective pulses. The modern semiconductor technology makes possible a high variety of the pulse shape modifications.

As a result, the number of the parameters to be influenced by the pulse control is increased essentially. Thus, with one generator concept/process control many applications can be processed successfully.

Different types of process energy sources like relaxation generators, static pulse generators, and needle pulse generators can be assigned to a special application and a special machining process (sinking, cutting, micro machining, roughing, finishing,...). For several examples, the characteristic differences are shown and the limits of machining are determined.

A special group of process energy sources is designed for combined methods and, therefore, the parameters have great variability. The combination EDM/ECM represents an extreme demand.

1 MOTIVATION FOR THE DEVELOPMENT OF PROCESS ENERGY SOURCES

For the development of new process energy sources for EDM there are different reasons. On the one hand it is new knowledge about the process what makes necessary adapted energy sources, on the other hand there are newly developed process variants. The development of semiconductors allows the realization of new concepts what results, in the consequence, new findings and ideas again.

In the following the evolution of the energy sources and the motivations for the corresponding development steps are described.

The first energy sources for the spark erosion were the RC generators, which consist of a d.c. voltage source U, a charging resistor R_A, a discharging resistor R and a storage element C. The connection of these RC generators with two electrodes, composing a gap filled with dielectric fluid, results in case of continuous reduction of the gap between the electrodes in a stable removal (spark erosion). This self-regulation at the use of RC generators allows partly good machining results, so that this principle is used also today at specific applications. The removal depends on the time constants of loading and unloading-circuits and the capacitor energy (Figure 1). The attainable removal rates are not very effective. Wear (tool removal) and surface roughness are only insufficiently controllable.

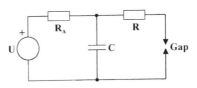

Figure 1: Simple relaxation generator (RC-Generator)

In order to be able to use the spark erosion in more applications, the pulse parameter must allow a greater adaptability, for example the constant energy content of each pulse of the pulse series independent of other parameters. Therefore, the development of the static pulse generators with which these demands can be realized, was a big jump in the development of the process energy sources.

With introduction of the static pulse generators into use a separate development of the process energy sources occurred for the sinking erosion and the wire erosion, because the parameters of both procedures are very different. The simple static pulse generators with rectangular current pulses and relatively long pulse durations satisfied the requirements of the sinking erosion, however, for the wire erosion shorter pulse durations were demanded. Circuit concepts for short pulses with an adjustable maximum current rise time were developed. In order to get a minimum pulse duration, the current fall follows to the current rise immediately. Therefore, these pulses have a triangle-shaped current (needle pulses).

A further development of these generators was the independent adaptability of current amplitude and pulse duration with the aim to allow a process optimization which permits not only a modification of the pulse break.

Also a motivation for the development of the process energy sources aimed at the elimination of the undesirable electrochemical dissolution parts. With a.c. generators producing a mean value of gap voltage of zero Volt by bipolar pulses the electrochemical effect was reduced strongly.

The specific demands of the ED micro machining initiated again a further development of the energy sources, with the aim to generate pulses with small energy and a essentially higher pulse repetition frequency.

A003/012/2003 © IMechE 2003

Figure 2 shows the most important stages of the development of the process energy sources for the EDM.

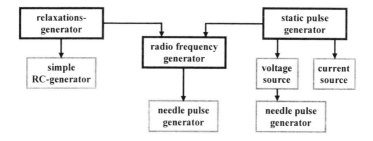

Figure 2: Stages of the development of EDM process energy sources

These development phases of the process energy sources for the EDM show many different motivations focussing in the following points:

- higher flexibility of the pulse parameters (larger optimization field),
- application-specific features (wire erosion, micromachining),
- increase of the productivity and decrease of the thermal affected zone
- decrease of the wear and higher machining accuracy.

2 PROCESS ENERGY SOURCES AND THEIR DEVELOPMENT PHASES

From the description of the motivations the individual stages of the development of the process energy sources can be derived. At the example of the main development phases (Figure 2) the special features characterizing the individual circuit concepts, and the signals usable for process control are described. Only the relevant characteristics are treated in this paper.

2.1 Relaxation generators

The first developed EDM energy source was the RC circuit in Figure 1. In Figure 3 the scheme is represented in detailed form.

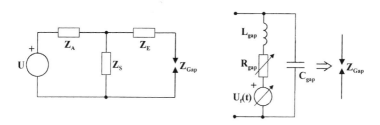

Figure 3: Scheme of relaxations generator and equivalent circuit of the working gap

The loading impedance Z_A consists of the resistance R_A and the line inductance L_A. The storage element Z_S includes the capacity having as a technical component a parasitic shunt resistor. In the case of the use of a diode as a storage element the capacity is given by the diode capacitance.

Whereas the loading-circuit limits only the charge of the storage element, the discharging circuit has influence on the pulse form. Z_E includes the line resistance, the line inductance and the line capacity of the gap interconnection. Z_{gap} contains the capacity of the electrode arrangement C_{gap}, the time-dependent resistance $R_{gap}(t)$ and the back-voltage source $u_f(t)$ (spark burning voltage). In more exact equivalent circuits an inductive part L_{gap} of the electrode arrangement has to be included. The Z_{gap} can be regarded as a circuit according to Figure 3 with parameters depending on the special machining case.

Because loading and discharging circuits represent resonant circuits, in particular the discharge pulses show the influence of these resonant circuits onto the breakdown mechanism. By changing the wiring oscillations can be amplified or damped. Both cases need attention. In the case of a too high damping the pulse repetition frequency is reduced considerably, which results in slower productivity. Because of unwanted supereelevations of the current amplitude near of the resonance frequency defaulted maximum pulse energy may be exceeded.

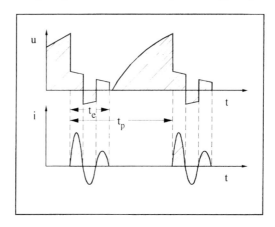

Figure 4: Ideal curves of gap voltage and gap current (relaxation generator)

The idealized current and voltage curves of a relaxation generator are represented in Figure 4. The bipolar pulse is not desired during the spark erosion because it increased the tool wear. This behavior can be reduced by an inverse diode.

Further improvements of the relaxation generators were achieved through the capacitor charge by a constant current (high-frequency pulses) or by modification of the charging resistance R_A according to a defaulted time scheme [01],[02]. For controlling of the process stability a modification of the discharging resistor R or the storage capacity C is also possible. In this case the storage element can be also an inductance.

As evaluable signals for the process control almost the mean value of the work- or gap voltage can be used. The main optimization of the process was done by the gap flushing system and the die gap control.

Nowadays relaxation generators are used for finishing, the machining of small faces and in low cost machines (start hole machines).

2.2 Static Pulse Generators

The static pulse generators were developed in order to be able to change the pulse energy independently of the pulse repetition frequency. Higher removal rates are achievable, however, the process has no more its self-regulating mechanisms. The appearing gap conditions lead to more unwanted discharges that have to be recognized and avoided.

Figure 5 shows the basic scheme of the static pulse generator and an ideal rectangle shaped pulse with a small ignition delay time. Discharges at the working gap are generated by switching on/off. A rectangle shaped pulses an idealization. At a defaulted current amplitude with this form the pulse energy a well defined. For long pulses (> 50 μs) the current seems be rectangular, at shorter pulses because of the limited current rise rate it is trapezoidal. By increasing the current rise rate of the pulses a higher removal rate results, however, the wear increases substantially.

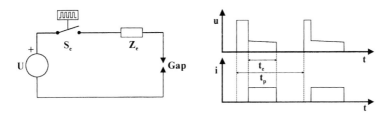

Figure 5: Static pulse generator (scheme of equivalent circuit) and pulses

The process control can be combined with an off-line technology optimization. One of the technological ratings (removal rate, relative wear, surface roughness, machining accuracy ...) is used as primary goal for optimization, while the other technological ratings restrict the allowed parameter space. For this on the way off-line optimization the pulse parameters eroding current i_e and pulse duration t_e are selected as variables.

An online process control for the process stabilization in an optimum range is possible through generators, which supply pulses with defined and reproducible parameters. The ignition delay time, the spark burning voltage and the gap voltage during the breakdown are measurable and can be used for the process control. For this purpose it was important to determine operating points in the bounded process space. Through a statistical analysis [03], [04], for example, for the ignition delay time appropriate mean and limiting values can be determined.

A modern method for the process analysis is the evaluation of the rf-part of the spark burning voltage and/or of the eroding-current [06]. This process analysis is based directly on the frequency-dependent processes in the plasma channel and allows a reliable distinction between spark and arc discharge.

For the increase of the removal rate of the EDM-process generators with a higher pulse current are needed. Boundaries had the utilization of the basic circuit (Figure 5). Because of the high power dissipation arising in the current-limiting resistance Z_E the basic circuit concept according to Fig. 5 is no more applicable efficiently for higher currents.

Therefore, the next evolution step was the separation in power units and ignition-unit. The last one generates only the low ignition-current. Additional to the decrease of the power dissipation by this step, the process variables chosen for the process control became independent of the energy parameters of the pulse. The process variables can be related to an fixed or in a small range modifiable ignition-current through which the occurring process stages and/or gap conditions are more exactly determinable. Thus, the process analysis based on reproducible conditions.
The modular construction of the power units is mainly used for high current amplitudes. However, the modular construction allows also the shaping of the pulse slopes through appropriate control of the individual power modules, which is important for a low-wear machining.

A further development of the modular circuit technology are the current sources which supply an constant current amplitude independent from the gap state and workpiece resistance. The total current for the parallel connection of several power sources results from the sum of the single currents [05],[07]. In Figure 6 it is represented an eroding-pulse generated from a current source with four power units.

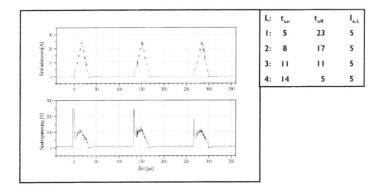

L:	t_{on}	t_{off}	$I_{e,L}$
1:	5	23	5
2:	8	17	5
3:	11	11	5
4:	14	5	5

Figure 6: Shaped current pulse for ceramic EDM [07] with modular parameters

On the other hand, when modules on basis of voltage sources are used the total current depends on the internal impedances of the individual modules and of the loads. In this case, the total current is not equal the sum of the currents of the individual modules. For pulse shaping, this fact we have to take into account.

A further advantage of the ED process energy sources on the basis of constant-current modules is that they give the opportunity to work out theoretical model for an exact optimization of the feed systems, of control algorithms and process stability and technological aims. Therefore, these generators an important precondition for fundamental research on spark erosion.

 A003/012/2003 © IMechE 2003

Needle pulse generators producing pulses with very short duration, also belong to the category of static pulse generators. The current waveform does not have any constant sections, a current fall follows the current rise immediately. (Figure 7)

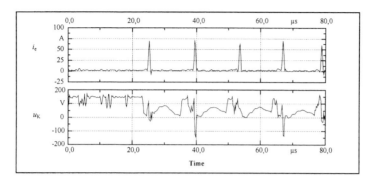

Figure 7: Erosion current i_e and terminal voltage u_k of the needle pulse generator

The current rise rate of the needle pulses has values between 0.1A/ns to 0.9A/ns. The generators are used mainly for Wire-EDM. The shortest pulse durations can amount to 200ns.
The modular construction and the separation of ignition and power units allow an ignition-delay-analysis and a breakdown-voltage-analysis. Because of the fast switching operations without compensation stage a rf-analyse of the process is complicated.

Technological investigations show that with reduced current amplitudes and higher pulse repetition frequencies improved surface qualities can be achieved. A sharp boundary to the high-frequency eroding-generators can not be indicated. For the following considerations the lower boundary for high-frequency generators should be drawn at about 1MHz.

2.3 High-frequency generators

EDM process energy sources are called high-frequency generators, when the pulse repetition frequency is greater than 1 MHz. In general, the pulse duration is shorter than 200ns, the current amplitude limited from circuit-technical reasons is smaller than 25 A. The aim for the utilization of these generators is an improvement of the surface properties with a maximally high removal rate. High-frequency eroding-generators are used mainly for micromachining. The form of the current pulses resembles a triangle or a needle.
With static pulse generators according to Figure 5 it is possible to compose required pulse trains. However, the main problem with circuits, at which the eroding-current is switched on and off, is the power dissipation in the switch S_e. At high frequencies the power losses reach critical values and require big attenuation from the designer.

At high frequencies the parasitic components in the circuit have to be taken into consideration and gain a great influence on the current-voltage-curves. Therefore, the circuits of high-frequency generators are often mixtures of static generators and relaxation generators. As consequence, the limiting parameters of the pulses depend strongly on application and installation.

If the switch S_e in Figure 5 is replaced by only parasitic components, then the circuit is a modified form of the relaxation generator (Figure 3). Then the pulse waveform at the gap depends only on the feeder impedance Z_e and the impedance of the electrode arrangement Z_{gap}.

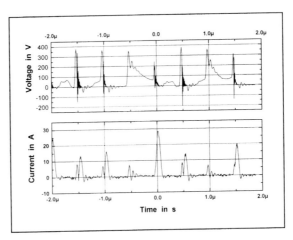

Figure 8: Current and voltage of the hf-generator (100ns/30A/2MHz) [08]

Figure 8 shows the voltage and current curves of a hf-generator for the machining of surfaces smaller than 1cm² and an attainable roughness R_a smaller than 0.3μm. These values demonstrate possibilities of relaxation generators. Because it is a voltage source, the current amplitude changes with the gap conditions Z_{gap} (t), only the maximum eroding-current is defaulted. For micromachining this change of the current amplitude is uncritical. The maximum current amplitude guarantees the required roughness, the high repetition frequency an effective removal rate. Due to the small pulse energy the thermal influences (tensile stresses, bending strength, phase change) of the machining surface decreases. This was verified by experiments.

Other circuit principles are based on an a.c.-source with a high-frequency oscillating current. However, the frequency of the discharges amounts less than 1MHz. At these generators pulse duration is in the ns-range and the pulse period is in the μs-range. The first half wave can have a pulse width of 50ns [10]. The mode with low frequency generated needle pulse series show a low current flow in the pulse break. The bipolar character of pulse series results in an elimination of the electrolytic dissolution effect .

The process control at hf-generators can only be realized by fade out of pulses or the insertion of additional pulse breaks.
High ignition voltages as in Figure 8 cause a high process stability and, as consequence, a continuous removal and a minimum number of unwanted discharges (larger gap). High ignition voltages are also favorable, when the eroding-area (active area) considerably changes during the machining process. If the active areas relatively constant the ignition voltage and the work gap can be reduced. Thus, the machining accuracy can be improved considerably .

A limit for the ED micro machining is given by the physical properties of the spark discharge. For the single discharge the minimal critical value of the current amplitude is approx. 800

 A003/012/2003 © IMechE 2003

milliamperes. This is the critical value for the transition of the glow-discharge into the arc discharge. The glow-discharges have a sufficient energy in order to evaporate the fluid working medium in the work gap. In this case, discharges occur in a mixture of air and gaseous hydrocarbon/metallic vapor. By that, the removal structure is changed considerably and results in unwanted deviations of the technological limits.

Micromachining is also possible in gaseous working media (air)[09]. During the microspark erosion in fluid working media the energy density is increased and, by that, the minimal crater structure is defined. By using additives (powder) or pulse shaping (control of the delay period between ignition and power unit) an enlargement or distribution of the plasma channel is possible so that the crater structure becomes smaller.

Fundamentally the spark discharge in the gaseous medium does not have a high energy density at the foot-point of the plasma channel so that the workpiece-sided removal occurs in very small portions per discharge.

3 SUMMARY

Process energy sources are a main part of EDM machining plants. Generally, their detailed construction is not represented in publications. However, the basic circuits are known and can be adapted for applications. Different generator principles require also a development of different process control systems (signal acquisition, process signal selection, control algorithm). Both, generator and control system, are the precondition for a high efficiency.

With the modular pulse shaping many new possibilities are given for the process influencing (breakdown/discharge process) which must be included in control strategies. Not only the productivity of the process is the decisive technological aim. More and more the surface quality (roughness, hardness, bending strength ...) and the machining accuracy gain importance. For the accuracy the tool wear is important which can be influenced also by the pulse shaping in modern generators.

Main fields in the future of the spark erosion (EDM) are micromachining and process combinations. For micromachining the productivity must be increased under retaining a machining quality attainable today.

REFERENCES:

01 **Wollenberg, G., Schulze, H.-P., Pape, Th.,** and **Läuter, M**. (1999) Moderne Generatortechnik für die funkenerosive Bearbeitung. 4. Aachener Fachtagung Funkenerosive Bearbeitung, pp. 9/1-26.

02 **Schiedung, H**. (1983) Zusammenhang zwischen dem Prozeß des funkenerosiven Schneidens und der Struktur mikrorechnersteuerbarer Impulsgeneratoren. Diss. TH Magdeburg .

03   **Weck, M.; Dehmer, J.M.** (1992) Analysis and adaptive control of EDM sinking process using the ignition delay time and fall time as parameter. Annals of the CIRP Vol. 41/1.

04   **Dehmer, J.M.** (1992) Prozessführung beim funkenerosiven Senken durch adaptive Spaltweitenregelung und Steuerung der Erosionsimpulse. Fortschritt-Berichte VDI-Reihe 2 Nr. 244. VDI Verlag Düsseldorf.

05   **Wollenberg, G., Timm, M., Pape, Th., and Schulze, H.-P.** (1995) Stepped pulse power supply and feedrate control for EDM of materials which are poor conductors. ISEM XI, Lausanne, Switzerland, pp.353-361.

06   **Rajurkar, K.P. and Wang, W.M.** (1989) A new model reference adaptive control of EDM. Annals of the CIRP Vol. 38/1, pp. 183-186.

07   **Timm, M.,** (1996) Elektronische Stromquellen für das funkenerosive Schneiden von elektrisch schlecht leitfähigen Werkstoffen. Diss. Otto-von-Guericke-Universität Magdeburg.

08   **Wollenberg, G., Schulze, H.-P. and Läuter, M.** (2001) Process energy supply for unconventional machining. ISEM XIII, Bilbao, Spain, 9-11 ,2001, pp. 283 –294.

09   **Furudate, C. and Kunieda, M.** (2000) Study on dry WEDM. 2nd International conference on machining and measurement of sculptured surfaces, Krakow, Poland, pp.325-332

10   **Magara, T., Kobayashi, K,, Yamomi, T. and Hon, K.K.B.** „Micro-Finishing by high frequency AV source in Wire-EDM".

Ultrasonic precision cutting

J A McGEOUGH
Division of Engineering, University of Edinburgh, UK
S J EBEID
Faculty of Engineering, Ain Shams University, Egypt
R T FAHMY and **S S HABIB**
Faculty of Engineering, Zagazig University, Egypt

ABSTRACT

Ultrasonic machining processes have attained a recognized status in manufacturing. USM can be used successfully for cutting of precise components. USM has solved many problems for the medical, aerospace, electronics and automotive industries. Various processes are able to be combined in rotary ultrasonic machining where, the machining efficiency could be increased. New methods for grinding ceramics and hard materials by using non-rotational tools have been developed.

The objective of the present paper is to highlight the state-of-the art of hybrid and recent USM processes. Moreover, the paper summarizes the latest technologies regarding precision cutting and ultrasonic assistance for medical applications. Ultrasonic precision cutting proves to be promising in the micro machining field as well as in biomedical applications and it looks to be a beneficial tool in the coming era of nano-technology.

1 INTRODUCTION

In USM, high frequency electrical energy is converted into mechanical vibrations, which are then transmitted through an energy focusing device. This causes the tool to vibrate along its longitudinal axis at a high frequency usually ≥ 20 kHz with an amplitude of 5-50 μm. Typical power ratings range from 50-4000 W. A controlled static load, as shown in Fig 1., is applied to the tool and an abrasive slurry is pumped into the space between the workpiece and the tool. The vibration of the tool causes the abrasive particles held in the slurry between the tool and the workpiece, to impact the workpiece surface causing a pocket in the material. The abraded material is removed along the surface perpendicular to the direction of the tool vibrations. The material is removed in minute particles and a cavity is formed in the workpiece, exactly copying the profile of the tool face (1,2).

A003/014/2003 © With Authors 2003

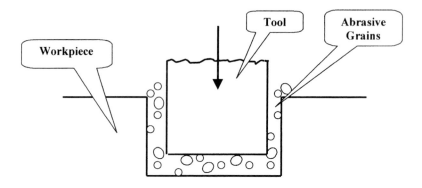

Figure 1. Ultrasonic cutting

The mechanism of metal removal Fig 2. comprises: mechanical abrasion by direct hammering of the abrasive particles against the workpiece surface, then micro chipping by impact of the free moving abrasives, and finally cavitation erosion, where the ultrasonic shock waves give rise to bubbles in the liquid carrier. The collapse of these bubbles causes very high local pressures, which generate cracks and chip pockets in the workpiece causing material removal (3,4).

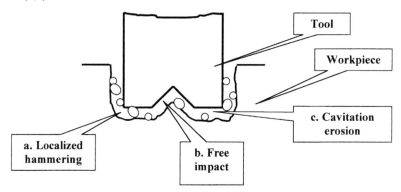

Figure 2. Mechanism of material removal

2 MAIN ULTRASONIC MACHINING PROCESSES

The use of ultrasonics for machining processes is a technology that has attained a recognized status in manufacturing as in drilling, milling, turning, threading, grinding, engraving, parting, lapping and polishing of faces. Various materials are possible to be machined as semiconductors, glass, sintered carbides, ceramics, diamonds and hardened steels.

Anantha Ramu et al (5) discussed the ultrasonic drilling of hard and brittle materials. A dynamic model of ultrasonic drilling was made by Neilson et al (6), in which they explained the mechanism of impact force generation during the process. They postulated a method for

    A003/014/2003 © With Authors 2003

calculating the material removal rate, which explains the observed fall in material removal rate at higher static loads.

The ultrasonic drilling of fine diameter holes for various applications like holes in injector nozzles and wire drawing dies of about 135 μm up to a depth of 500 μm in glass was investigated by Jana et al (7). The ultrasonic machining of blind cylindrical, taper and form holes, grooves and slots of curvilinear shape in brittle non-metallic materials has been studied by Markov et al (8). Zhang et al (9) have studied the ultrasonic machining of holes in ceramics. The concept of fracture mechanics, on which the machining mechanism is based has been analyzed. It has been concluded that any increase in the amount of energy imparted to the ceramics in terms of the amplitude of the tool tip, the static applied load and the size of the abrasive will result in an increase in the material removal rate. The results showed that ultrasonic machining is an effective method for the machining of ceramics.

Soundararajan et al (4) used mild steel with boron carbide abrasive to machine high speed steel, tungsten carbide and plate glass workpieces. It has been concluded that material removal is found to be mainly activated by the hammering mechanism. Venkatesh (10) discussed the machining of glass by two impact processes namely, abrasive jet machining and ultrasonic machining. Comparisons between both were drawn and it was found that ultrasonic machining is a much faster process than the abrasive jet machining process.

3 HYBRID ULTRASONIC MACHINING PROCESSES

Mult et al (11) studied the combination of the grinding process with ultrasonic vibrations at high frequencies of about 22 kHz and low amplitudes between 4 and 15 μm. USM assists feed grinding and provides enormously reduced normal forces at slightly increased wheel wear and surface roughness. For the production of holes and other complex contours, ultrasonic assisted face grinding proves to be very efficient compared to that of ultrasonic machining alone.

Using USM with electro-discharge machining leads to drastically reductions in machining production times and costs (12). USM increases the efficiency of EDM operations, by easing the removal of the residue a good surface quality is obtained, with higher penetration depths and feed speeds, particularly for precise and intricate shapes. Thoe et al (13) combined USM with EDM to drill small diameter holes less than 1mm diameter in an electrically non-conductive ceramic coated nickel alloy, where the removal rate was greatly increased. Guo et al (14) investigated the combined technology of ultrasonic machining and wire electrical discharge machining. Experimental results show that wire vibration induced by ultrasonic action has a significant effect on the overall performance of the process. There exists an optimum relationship between the vibration amplitude of the wire and the discharge energy, by which the highest cutting rate and the optimum machined surface quality can be obtained.

Lau et al (15) used a combined ultrasonic-laser technique to improve the quality of laser drilled holes and increase the depth of cut. A mathematical model was proposed to describe the shape of the machined hole. On the other hand Yue et al (16) developed a theoretical model using finite element method to determine the shape of ultrasonic-aided laser drilled holes and the thickness of the re-cast layer.

4 RECENT ULTRASONIC MACHINING PROCESSES

Drilling and milling processes can be combined in rotary ultrasonic machining processes (RUM). An ultrasonically activated drill bit is rotated against the workpiece. An abrasive slurry is not used in this method and is replaced by a coolant liquid. The machining efficiency is increased by RUM. Milling, drilling, threading and grinding operations can be performed with rotary USM. The tool must be either diamond impregnated or coated (17). Komaraiah et al (18) found that the material removal rates in RUM are up to four times those in conventional USM.

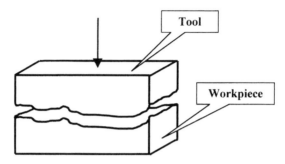

Figure 3. Three-dimensional sinking

Many ultrasonic machining applications are involved in drilling where a tool of either simple or complex cross section penetrates axially into the workpiece, to produce either a through or blind hole. Where a three dimensional cavity is required, a process analogous to die sinking is generally employed as shown in Fig 3. The problem of using tools of complex forms, is that the tools are not subject to the same machining rate over the whole of their working surface and experience differential wear rates, both of which affect the product shape. In addition, there are also greater problems in turning a complex tool to achieve maximum performance compared to more basic tools. An alternative approach is to use a simple "pencil" tool and contour the path of the complex shape with a CNC programme, as shown in Fig 4 (19).

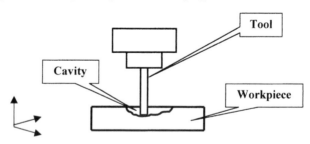

Figure 4. CNC path of a complex cavity

A new method for grinding ceramics and hard materials by using a non-rotational tool has been developed by Suzuki et al (20). A fixed abrasive type-grinding tool is vibrated by a complex ultrasonic transducer longitudinally and torsionally at the same time. Experiments showed that the complex ultrasonic vibration keeps grinding force low and stable for a long time in grinding hard ceramics even without rotating the tool. This method has a great effect on producing cavities with sharp inner corners of almost zero radii.

Moriwaki et al (21) presented a new ultrasonic cutting method named "Elliptical vibration cutting" (Fig 5). They proposed the cutting edge to vibrate in the plane including the cutting direction and the chip flow direction. Orthogonal cutting experiments of copper are carried out within a scanning electron microscope, and it is observed that applying the method proposed reduces the chip thickness and the cutting force remarkably.

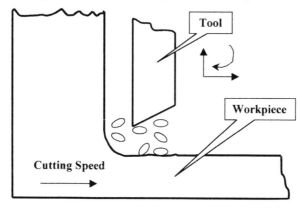

Figure 5. Elliptical vibration cutting

Zhao et al (22) studied a new ultrasonic polishing method for machining on free form or sculptured surfaces at an oblique angle (Fig 6). They developed a robotic polishing system, with an elastic polishing tool. The experimental results on free-form surfaces in 2-D verify the effectiveness of the new machining method and the robotic polishing system.

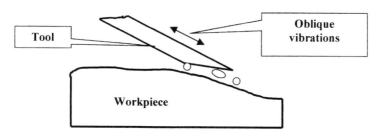

Figure 6. Oblique ultrasonic machining

5 ULTRASONIC PRECISION CUTTING

Ultrasonic cutting has been successfully applied to precision cutting to manufacture precision components with a surface roughness of a few nanometers and with a tolerance, which is in submicron range. As it is difficult to vibrate tools as small as 15 μm in diameter, it has been proposed to vibrate the part and rotate the tool (23). The rotation eccentricity is in a range smaller than 0.5 μm. Ultrasonic machining can produce micro holes as small as 5 μm in diameter, in glass and silicon. With such case very fine grains were used in the order of 0.2 μm diameter diamond abrasives. Micro drilling requires amplitudes smaller than or equal to 1 μm. However, studies carried out on micro holes as small as 5 μm failed to investigate the accuracy and circularity of such ultra-precise holes (24).

The technique of elliptical vibration cutting was used for ultra-precision diamond cutting of hardened steel (25). The performance of the elliptical vibration cutting is examined in comparison with the conventional vibration cutting. The experimental results showed that the proposed method has superior performance regarding low cutting force, high quality surface finish and long tool life.

6 ULTRASONIC ASSISTANCE FOR MEDICAL APPLICATIONS

The medical area represents a broad field having a number of specific applications as in dentistry and precision microsurgery. In dentistry, ultrasonic vibrations impart small movements that are less likely to harm and are more acceptable to a patient in comparison with conventional rotary tools. The high tool oscillation frequency and the very small amplitude allow plaque to be removed without damage to the tooth. Scalpels are now ultrasonically vibrated to aid surgeons in eye micro surgery. Another example is the use of a scalpel which is vibrated at an ultrasonic frequency of 55 kHz, 100 μm amplitude. The rapid movement causes separation of the tissue ahead of the blade and aids dissection (26).

New applications of ultrasonic systems are being proposed such as removal of cement during hip replacement prosthesis and removal of brain tumors. Recent research is proceeding to clear blocked passages in arteries. The aim is to resize the artery to a satisfactory diameter whereby required operations can then be undertaken, avoiding open thorax surgery. Results on tests made on coronary and leg arteries have shown the possibility of clearing 70% of blockage (27,28).

7 CONCLUSIONS

The present work highlights the powerfulness of the USM process in nowadays industry. A survey on the present status of hybrid and new techniques of USM is clarified. Ultrasonic precision cutting proves to be promising in the micro machining field as well as in biomedical applications and it looks to be a beneficial tool in the coming era of nano-technology.

REFERENCES

1 McGeough, J. A. (1988) Advanced methods of machining, Chapman and Hall, London.

2 Pandey, P. C. and **Singh, C. K.** (1992) Production engineering sciences, Standard Publishers Distributors, Delhi.

3 Khairy, A. B. E., (1990) Assessment of some dynamic parameters for the ultrasonic machining process, Wear, 137, 2, pp. 187-198.

4 Soundararajan, V. and **Radhakrishnan, V.** (1986) An experimental investigation on the basic mechanisms involved in ultrasonic machining, *Int. J. Mach. Des. & Res.*, **26,** 3, pp. 307-321.

5 Anantha Ramu, B. L., Krishnamurthy, R. and **Gokularathnam, C. V.** (1989) Machining performance of toughened zirconia ceramic and cold compact alumina ceramic in ultrasonic drilling, *J. Mech. Work. Tech.*, 20, pp. 365-375.

6 Neilson, R. D., Player, M. A. and **Wiercigroch, M.** (1993) A dynamic model of ultrasonic drilling, Mach. Vib., 2, pp. 136-143.

7 Jana, J. K. and **Satyanarayana, A.** (1973) Production of fine diameter holes on ultrasonic drilling machine, *J. Ins. Eng.* (India), 54, pp. 36-40.

8 Markov, A. I. et al. (1977) Ultrasonic drilling and milling of hard non-metallic materials with diamond tools, Mach. and Tool., 48, 9, pp. 45-47.

9 Zhang, Q. H. (1999) Material removal rate analysis in the ultrasonic machining of engineering ceramics, *J. Mater. Proc. Tech.*, 88, pp. 180-184.

10 Venkatesh, V. C. (1983) Machining of glass by impact processes, *J. Mech. Work. Tech.*, 8247-260.

11 Mult, H. C., Spur, G. and **Holl, S. E.** (1996) Ultrasonic assisted grinding of ceramics, *J. Mater. Proc. Tech.*, 62287-293.

12 Kremer, D. et al (1983) Ultrasonic machining improves EDM technology, Electro-machining Proc. 7th Int. Symp., Buckingham, UK, pp. 67-76.

13 Thoe, T. B., Aspinwall, D. K. and **Killey, N.** (1999) Combined ultrasonic and electrical discharge machining of ceramic coated nickel alloy, *J. Mater. Proc. Tech.*, 92-93323-328.

14 Guo, Z. N. et al. (1997) A study of ultrasonic-aided wire electrical discharge machining, *J. Mater. Proc. Tech.*,63823-828.

15 Lau, W. S., Yue, T. M. and **Wang, M.** (1994) Ultrasonic-aided laser drilling of aluminum based metal matrix composites, Ann. CIRP, 431, pp. 177-180.

16 Yue, T.M. et al. (1996) Analysis of ultrasonic-aided laser drilling using finite element method, Ann. CIRP, 451, pp. 169-172.

17 Wiercigroch, M. et al. (1993) Experimental study of rotary ultrasonic machining: dynamic aspects, Mach. Vib., 2, pp. 187-197.

18 Komaraiah, M. and **Narasimha Reddy, P.** (1993) A study on the influence of workpiece properties in ultrasonic machining, *Int. J. Mach. Tools Manuf.*, 333, 495-505.

19 Benkirane, Y., Kremer, D. and **Moisan, A.** (1999) Ultrasonic machining: an analytical and experimental study on contour machining based on neural network, Ann. CIRP, 481, pp. 135-138.

20 Suzuki, K. (1993) A new grinding method for ceramics using a biaxially vibrated non-rotational ultrasonic tool, Ann. CIRP, 421, pp. 375-378.

21 Moriwaki, T. and **Shamoto, E.** (1995) Ultrasonic elliptical vibration cutting, Ann. CIRP, 441, pp. 31-34.

22 Zhao, J. et al. (2000) An oblique ultrasonic polishing method by robot for free-form surfaces, *Int. J. Mach. Tools & Manuf.,* 40795-808.

23 Egashira, K. and **Masuzawa, T.** (1999) Micro ultrasonic machining by the application of workpiece vibration, Ann. CIRP, 481, pp. 131-134.

24 McGeough, J. A. (2002) Micromachining of engineering materials, Marcel Dekker, Inc., USA.

25 Shamoto, E. and **Moriwaki, T.** (1999) Ultraprecision diamond cutting of hardened steel by applying elliptical vibration cutting, Ann. CIRP, 481, pp. 441-444.

26 Coulson, A. (1999) The use of the harmonic scalpel in minimally invasive coronary artery bypass surgery, www.inreach.com.

27 Drobinski, D. and **Kremer, D.** (1997) Ultrasonic percussion device, Patent No. 5,649,935.

28 Drobinski, D. et al. (1993) Effects of ultrasonic energy on total peripheral artery occlusions: initial angiographic and angioscopic results, *J. Int. Cardiology,* 62, pp. 157-163.

A003/014/2003 © With Authors 2003

Investigations of the electrochemical drilling supported by the laser beam

M ZYBURA-SKRABALAK, A RUSZAJ, G SKRABALAK, and **R ZUREK**
Department of Electrochemical Machining, The Institute of Metal Cutting, Krakow, Poland

ABSTRACT

The tests have shown that for the examined machining process very important is electrode tool construction. The best results have been obtained in case, when the electrode tool with insulating face was applied. The results of investigations for ECLM process proved that using the electrode tool with insulating electrode tool in electrochemical drilling supported by the laser beam it was possible to increase six times current density by using the additional heating by laser beam in comparison to classical ECM (for the same electrochemical machining parameters) and also decrease the surface roughness R_a from 2.15 μm do about 0.7 μm.

1 INTRODUCTION

During last years the primary investigations in the field of electrochemical machining with assistance of laser beam took place [Datta and other 1989; Davydov 1994; Kozak 1998; Rajurkar and other 1999; Kozak and other 1999; McGeough and other 2000]. In this way of machining, the area of electrochemical dissolution process is under the influence of laser beam. Nowadays, this method is being investigated with high intensity, however it's practical application in industry is not very wide. It is expected that its prospective application will take place in shaping of small elements with high accuracy for microelectronic and space industry, especially when they are made of special materials (alloys, composites, ceramics) which are difficult for cutting.

The application of laser beam for electrochemical machining can give the increase of the velocity of dissolution process by:
- changing the balance potential of the machined material immersed in the electrolyte,
- the current density increase by increasing the temperature,
- improving of the dissolution product transportation out the machined area,
- increasing of the efficiency of dissolution process by eliminating the indirect reactions,

- the surface passivation elimination,
- activation energy decrease on the machined material surface, what decreases the energetic barrier for electrochemical reactions.

As a result of laser beam influence many physical and physico-chemical transformations on the machined surface and on dipper layers of the material occur. These transformations are the result of heat flow into machined material. Laser radiation can change physical properties of machined material as well as kinetics of electrochemical reactions. The temperature distribution is the main factor with high influence on phenomena occurring into machined area. The run and kinetics of the electrochemical reactions during electrochemical machining depend mainly on electrolyte and machined material constitution, temperature, current density and velocity of dissolution product evacuation out of machined area. In order to eliminate the indirect reactions and increase the velocity of the main reaction it is necessary: to increase the current density, to decrease the electrodes polarisation (by decreasing the thickness of diffussion layers adjacent to electrodes) and also by decreasing the dissolution product concentration in interelectrode gap; to increase the temperature in machining area.

During electrochemical machining, the irreversible processes are occurring. So, very important are diffusion processes, which depend mainly on reaction temperature and transportation processes.

If a diffusion gradient exists between the surface and the bulk of solution, the movement of ions in a concentration gradient is controlled by Fick's first (1) and second (2) laws:

$$\text{flux} = D\left(\frac{\partial C}{\partial x}\right) \tag{1}$$

$$\left(\frac{\partial C}{\partial t}\right) = D\left(\frac{\partial^2 C}{\partial x^2}\right) \tag{2}$$

The concentration gradient is given by Fick's first law of diffusion and is linear in the stationary state. As material cannot accumulate at $x = 0$ then the rate of electrochemical reaction must balance the rate of diffusion and the same, the rate of electrochemical dissolution will increase.

The simplified model of the process makes it possible to evaluate the relationship between constant of dissolution rate and temperature. The dissolution process intensity for reaction rate is connected with interaction between mass, heat, electrical discharge transportation and hydrodynamic parameters. In electrochemical machining process supported by laser radiation it is very important to evaluate the interaction between electrode reaction kinetics and heat transportation.

One of characteristic properties of electrochemical reaction carried out with high current density and high overpotential is the surface heat generation. The difference between surface and volume temperature can be calculated from heat balance equation. The relationship between constant of reaction rate K and activation energy and temperature is given by Arrhenius equation. It was presented in previous paper [Zybura, Ruszaj 2000].

2   EXPERIMENTAL INVESTIGATIONS OF THE PROCESS

The primary investigations carried out at the Institute of Metal Cutting have proved that the electrolyte temperature is decisive for the run of dissolution process [Zybura, Ruszaj 2000]. The temperature of machined area has significant influence on the activation energy and main electrochemical reaction. However other technological parameters in electrochemical machining, such as: interelectrode voltage, electrode feed rate, are also responsible for the run of dissolution process. They decide, in which state the dissolution process is running. It is expected that during electrochemical machining with proper assistance of laser beam it should be possible to increase the dissolution rate and dissolution process localization. This estimation will be taken into account during the investigations, in order to prove that the laser beam activates the electrochemical dissolution process.

Taking into account the rules of the electrochemical machining, we can conclude that in order to obtain the satisfactory dissolution process localization it is necessary to choose the process parameters such as:

- velocity of dissolution process in the area of laser beam influence should be as high as possible and constant;
- velocity of dissolution process outside of the area of laser beam influence should be as small as possible (its optimal value is zero);
- velocity of dissolution process on the border between above mentioned areas should change its value in jump way.

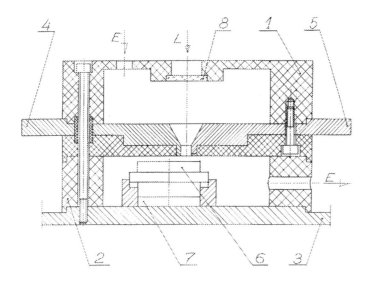

Fig. 1 Scheme of chamber for carrying out electrochemical machining assisted with laser beam, where: L - Nd-YAG laser beam, E - electrolyte inlet and outlet, 1 - upper housing, 2 - lower housing, 3 - base plate, 4, 5 - electrode with insulation (version I and II), 6 - workpiece, 7 - mechanism for setting of the gap thickness, 8 - special optical glass for transmission laser beam

At the Institute of Metal Cutting, the special test stand with industrial laser Nd – YAG was built. This test stand consists of the chamber for ECLMM process realization, special electrolyte circulation system for electrolyte supplying to the interelectrode gap and two kinds of electrode tools.

The scheme of test stand for ECLMM process realisation is presented in Fig. 1. In presented construction the laser beam can be displaced along x, y axis on the area 20x20 mm. It is also possible to apply stable laser beam and displaced along x, y axis working table.

In our test stand very important are possibilities of: setting of the gap thickness (special mechanism for setting of the thickness); changing the electrolyte pressure inside the machining gap (pump gives the required electrolyte pressure on the inlet of the electrolyte to the special chamber); setting the different electrode tools. There are electrode tools with insulated (electrode B) and non-insulated electrode face (electrode A).

The aim of investigations was to find the relation between the results of electrochemical machining (ECM) and electrochemical machining supported by laser beam (ECLM) for drilling process. Experimental tests have been carried out for two ways of machining and also for two types of electrode tools (electrode A and electrode B). The first one was the classical electrochemical machining process. The other one was the electrochemical machining process supported by laser beam. Experiments have been carried out for the following range of process parameters: electrolyte concentration $C_e = 2.5 - 22.5\%$; interelectrode voltage $U = 6 - 10$ V for electrode A and $U = 10 - 30$ V for electrode B; aperture for laser system $A = 1.6 - 2.4$; laser current intensity $I = 10 - 22$ A; electrode diameter $D = 5 - 9$ mm (outer diameter) for electrode A and $D = 1 - 5$ mm for electrode B. As machined material the stainless steel type NC6 has been applied, electrode has been made of copper. During the experiments the industrial Nd-YAG laser with maximal power of 60 W has been applied. Electrolyte was water solution of $NaNO_3$, initial electrolyte temperature $T = 22^0C$, initial interelectrode gap was 2 mm for the stable electrode tool – workpiece system and the same time of machining in all cases.

3 EXPERIMENTAL TESTS RESULTS

The results of primary experiments have shown that together with process parameters current density J and electrode potential E_a change significantly [Zybura, Ruszaj 2000]. By changing temperature T and electrolyte concentration C_e coefficient ηk_v is also changed and the same current density J. The electrode potential E_a also depends on T and C_e. So, changing for instance temperature T it is possible to change E_a and the same kind and velocity of main reaction. The estimated activation energy W for NC6 steel was in range $24 - 45$ kcal/mol.

The primary investigations have proved that the electrolyte temperature is decisive for the run of dissolution process. Other technological parameters, such as: interelectrode voltage, electrode feed rate, are also responsible for the run of dissolution process. They decided, in which state the dissolution process is running. It is expected that during electrochemical machining with proper assistance of laser beam it should be possible to increase the dissolution rate and dissolution process localization.

This assumption has been proved as the results of investigation with activation of electrochemical dissolution process by the laser beam.

The results of experiments were presented in the form of relations with neural nets application, MATLAB approximation and classical regression estimation by the regression equations. The examples of achieved results for electrode A are presented in Figs 2 and 3; and for electrode B in Figs 4 and 5. The main tested parameters were the current density j and shape deviation to the applied electrode diameter ΔD. The additional parameter was surface roughness R_a.

A003/027/2003 © IMechE 2003

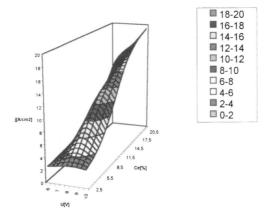

Fig. 2 Relation $j = f(U, C_e)$ for electrochemical machining in case when electrode type A with diameter $D = 7$ mm was applied

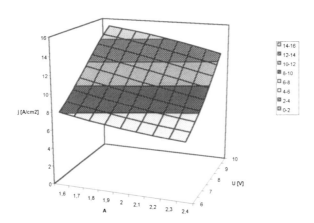

Fig. 3 Relation $j = f(A, U)$ for electrochemical machining supported by laser beam where: $I = 16$ A, $C_e = 12.5$ % in case when electrode type A with diameter $D = 7$ mm was applied

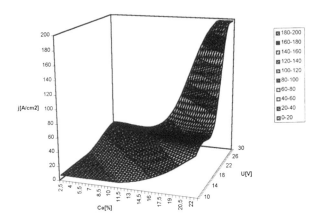

Fig. 4 Relation $j = f(U, C_e)$ for electrochemical machining in case when electrode type B with diameter $D = 3$ mm was applied

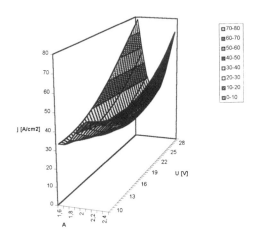

Fig. 5 Relation $j = f(A, U)$ for electrochemical machining supported by laser beam, where: $I = 16$ A, $C_e = 12.5$ % in case when electrode type B with diameter $D = 3$ mm was applied

A003/027/2003 © IMechE 2003

4 ANALYSIS OF EXPERIMENTAL RESULTS

Taking into account the test results, it was possible to estimate the relation, for example between R_a and j to interelectrode voltage. There are the following relations for estimation the current density value for electrode A with $D = 7$ mm for ECM machining and for electrolyte concentration $C_e = 12.5\ \%$:

$$j = -54.2050 + 12.0042U' - 0.10834U'^2 - 0.0408U'^3 \tag{3}$$

$$R_a = -9.415 - 0.024U' + 0.508U'^2 - 0.039U'^3 \tag{4}$$

These relations have been estimated using the MATLAB approximation. They are presented in Figs 6 and 7.

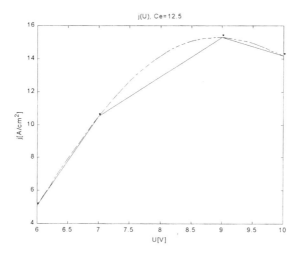

Fig. 6 Approximation of current density j for ECM machining with electrode A for $D = 7$ mm and $C_e = 12.5\ \%$

Using the approximation method, it is also possible to estimate the relations for electrochemical machining supported by laser beam.

It is seen (see Figs 6 and 7) that using the above presented approximation it is possible to estimate the machining conditions for obtaining the machined surface with required surface quality. It could be very useful in solving the special tasks.

It is worth to underline that the tests have been realized according to the statistical plan of experiments. The obtained results have given us information that all received relations are adequately, so we can also apply for comparison the regression equations. We received the good agreement for the above presented methods for estimation the test results.

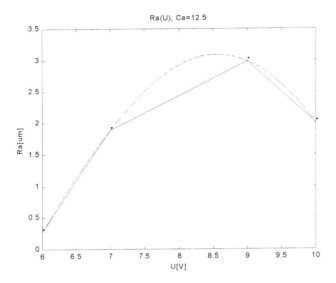

Fig. 7 Approximation of surface roughness R_a for ECM machining with electrode A for $D = 7$ mm and $C_e = 12.5$ %

Analysing the test result we can say, that the focusing laser beam has gone through the electrolyte layer and heated the machined area on the workpiece surface. The surface temperature increase is influencing on the increase of dissolution rate. For estimation the result of laser beam influence, the comparison of the current density in ECLM to ECM for these same electrochemical machining parameters have been carried out. The coefficient of the electrochemical dissolution rate increase k_r has been defined:

$$k_r = \frac{j_{ECLM}}{j_{ECM}} \qquad (5)$$

where: j_{ECLM} – current density during electrochemical machining supported by laser beam,
j_{ECM} - current density during electrochemical machining.

In investigated case when the coefficient k_r is higher than 1 it means that the rate for the electrochemical machining supported by laser beam is greater than for electrochemical machining. For the same machining parameters the higher value for current density gives information that dissolution rate is higher. For the electrode type A the coefficient k_r is smaller than 1. It means that this type of electrode is not good for electrochemical machining supported by laser beam. For electrode B the coefficient k_r is greater than 1 or close to 1. The maximum value for $k_r = 6$ has been received for the electrolyte concentration $C_e = 7.5$ %, $U = 15$ V and for laser power density $W = 4.14$ W/mm^2. The best results have been obtained for law electrolyte concentration $C_e \leq 12.5$ % and $U \leq 15$ V and higher laser power density W. In this case we also obtained the decreasing of surface roughness R_a from 2.15 μm (for ECM machining) to about 0.7 μm (for electrochemical machining supported by laser beam).

A003/027/2003 © IMechE 2003

From the analysis of the test results we have concluded that the laser beam can give the increase of the velocity of dissolution process.

In case when dissolution process is supported by laser radiation the main influence on electrochemical reaction run has first of all the heat generating in machined area.

In case of high current density transportation processes in direction to electrode limit the electrochemical reaction rate. This rate depends on coefficient of diffusion of ions, which are created the electrode potential. When the concentrations of reagents near the electrode are constant and not dependent on current density and time, the concentration polarisation occurs. It means that the rate of transportation and additional reactions is significantly higher than rate on main electrochemical reaction.

One of characteristic properties of electrochemical reaction carried out with high current density and high overpotential is the surface heat generation. The difference between surface and volume temperature can be calculated from heat balance equation. The relationship between constant of reaction rate K and activation energy and temperature is given by Arrhenius equation. It was analysed in the previous paper [Zybura, Ruszaj 2000].

5 CONCLUSIONS

The rate of electrochemical dissolution process depends mainly on the kinetics of electrode reactions and the activation energy of the electrode reaction. According to the Arrhenius law, it is possible to increase the rate of electrochemical dissolution by increasing the temperature in the interelectrode gap. The increase of temperature in the interelectrode gap is possible thanks to support the electrochemical machining by laser beam. The laser beam heats the electrolyte in the interelectrode gap and also the machined surface. The localized increase of temperature in the interelectrode gap gives the increase of current density and machining accuracy. One of the efficient way for increasing the temperature in the interelectrode gap is to apply the heating of machined surface by the laser beam. The laser beam helps to increase the temperature on the machined surface and localisation of the ECM process.

The better conditions for providing electrochemical machining supported by laser beam is in case, when we use electrode with isolated face. In the further investigation the best conditions for providing the machining process and best resolution for practical applications could be selected.

6 ACKNOWLEDGEMENTS

The support of the Polish State Committee for Scientific Research (KBN – grant No 7 T07D 002 18) is gratefully acknowledged. Authors wish to thank Directors of the Institute for creating good atmosphere for scientific research. Authors also wish to thank Colleagues from The Institute of Metal Cutting from the Department of Electrochemical Machining for help in carrying out experiments.

REFERENCES

1. Datta M., Romankiw L.T., Vigliotti D.R., Gutfeld R.J. (1989): Jet and laser - jet electrochemical micromachining of nickel and steel. *J. Electrochem. Soc.*, Vol. 136, No 8, p.2251-2256.

2. Davydov A.D. (1994): Lazerno-elektrochimiczeskaja obrabotka metallov. *Elektrochimija*, Vol. 30, No 8, p.965 - 976.

3. Kozak J. (1998): Mikrokształtowanie elektrochemiczne z laserową aktywacją procesów elektrodowych. Wyd. *Pol. Warsz.*, *Program Priorytetowy, Nowe Technologie, Prace naukowe*, Zeszyt 1, p.25 -33.

4. Kozak J., Rozenek M., Dąbrowski L., Zawora J. (1999): Badania wpływu promieniowania laserowego na proces kształtowania elektrochemicznego. Wyd. *Pol. Warsz.*, *Program Priorytetowy, Nowe Technologie, Prace naukowe*, Zeszyt 2, p.233-242.

5. Rajurkar K.P., Zhu D., McGeough J.A., Kozak J., De Silva A. (1999): New developments in electrochemical machining. *Annals of the CIRP*, Vol. 48/2, p.567-580.

6. McGeough, Tang Y., De Silva A., Kozak J., Ruszaj A., Altena H. (2000): Hybrid laser – electrochemical micromachining. *Report for INCO – COPERNICUS Programme*, *Contract No IC15-CT98-0801*, pp.57.

7. Zybura-Skrabalak M., Ruszaj A. (2000): Investigations aiming tio increase the rate of electrochemical dissolution process. *Proceed. 16th Intern. Conference on Computer Aided Production Engineering CAPE 2000*, Edinburgh, UK, p.163-172.

A003/027/2003 © IMechE 2003

Primary investigations on USECM–CNC process

S SKOCZYPIEC, A RUSZAJ, J CZEKAJ, and **M ZYBURA-SKRABALAK**
Department of Electrochemical Machining, The Institute of Metal Cutting, Krakow, Poland

ABSTRACT

The primary investigations for the electrochemical machining when the universal electrode vibrating with ultrasonic frequency are presented. The influence of the amplitude of ultrasonic vibrations, interelectrode voltage, electrode feed rate, interelectrode gap thickness on final metal removal rate and surface roughness have been taken into account. The tests indicated that introduction of electrode ultrasonic vibrations can change conditions of dissolution process and cause the increase of allowance thickness a and material removed rate V_W without surface quality decreasing.

1 INTRODUCTION

Electrochemical machining (ECM) - an important technology in machining difficult for cutting materials has traditional fields of application in space, aircraft and domestic industries. It results from the fact that after electrochemical machining it is possible to receive high quality of surface layer (1), (2), (9).

One of the kinematics variants of electrochemical process is machining with using universal electrode tool (ECM - CNC). In ECM – CNC material is removed by electrode with a three dimensional movement as shown in Fig. 1. The advantages of this method are simple shape of electrode and the increase in machining accuracy and workpiece surface quality in comparison to classical sinking. This is achieved by the decrease of working area what significantly reduces the influence of heat and gas generation on the electrolyte properties in the interelectrode gap. The main disadvantage of ECM – CNC is relatively small metal removal rate in comparison to classical sinking. It is a reason that ECM – CNC should be used in finishing machining of sculptured surfaces initially machined by other methods.

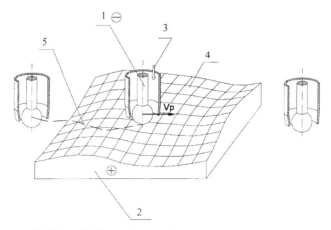

Fig. 1 Scheme of ECM – CNC machining with spherical universal electrode tool: 1 – electrode tool, 2 – workpiece, 3 – electrolyte inlet, 4 – machining surface, 5 – electrode tool path, v_p - *velocity* of electrode displacement (feed rate) (5)

2 PROBLEM FORMULATION

Taking into account results of investigation presented in (4), (6), (7), (8), (10), (11), (12) it is right to state that ultrasonic vibrations and ultrasonic field, which is creating during tool vibrations, have a significant influence on the conditions of electrode processes. The main effects of ultrasound derives from acoustic cavitation, the formation, growth and implosive collapse of bubbles in liquids irradiated with ultrasound (13). Kozak (4) suggested that ultrasonic waves gives possibility for creating the cavitation micro bubbles near the workpiece surface. Process of micro-bubbles collapsing in area adjacent to electrode gives the possibility of increasing the intensification of mass and electric charge transportation and increases the dissolution rate. Also results of investigations carried out in The Institute of Metal Cutting (10), (11), (12) showed that thanks to ultrasonic vibrations the intensification of electrochemical process by increasing the diffusion of metal ions take place. As a result the intensification of electrochemical processes can take place.

The main disadvantages of ECM – CNC machining is small material removal rate and problems with creating appropriate hydrodynamic conditions in order to remove machining products from interelectrode gap. Taking into account facts presented above it can be assumed that one of the ways for solving this problem can be introduction of electrode ultrasonic vibrations into interelectrode gap. Below the results of primary investigations on ECM – CNC process assisted by universal electrode ultrasonic vibrations (USECM - CNC) are presented.

3 MATHEMATICAL MODELLING

Mathematical modelling of electrochemical machining with universal ball-ended electrode was presented in (5). This model was developed for given conditions of dissolution, initial shape of workpiece and tool trajectory. Equations describing surface shaping process was solved for displacement of machined surface according to the normal to machined surface. In this case of modeling it is possible to determine the successive distances Δa_n with which machined surface is displaced in time Δt (Fig. 2):

$$\Delta a_{n_A} = \eta k_{vA} i_A \Delta t = \eta k_{vA} \kappa \frac{U - E}{D} \Delta t = V_n \Delta t \tag{1}$$

where: ηk_{vA} – coefficient of electrochemical machinability in point A, κ - electrolyte conductivity, V_n – velocity of metal removal rate in direction perpendicular to machined surface, i_A – current density in point A, D – distance between electrode and workpiece, U – interelectrode voltage, E – potential drops in the electrolyte films adjacent to the electrode and workpiece.

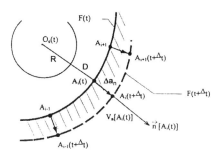

Fig. 2 Scheme used for describing machined shape changes in time Δt; Δa_n – anode surface displacement in Δt time (5)

Based on this model the special software was worked out at The Institute of Metal Cutting. This software was used for predicting the changes of allowance thickness a with the voltage drop in the electrolyte films adjacent to the electrode and workpiece during one electrode path. Results of these simulations were presented in Fig. 3a from which it is clear that a decrease with E increase.

a)   b)

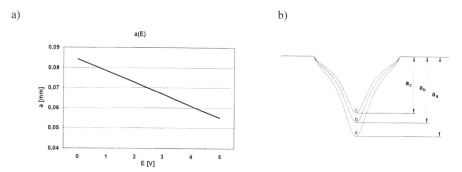

Fig. 3 Results of ECM – CNC computer simulation: a) relation between maximal thickness of removed allowance a and voltage drop E for: $U=14$ V, $v_p=30$ mm/min, $S_o=0.5$ mm, $R=5$ mm, $\kappa=0.0136$ $1/\Omega mm$, $k_v=0.0213$ mm^3/As; b) machining surface crossections in direction perpendicular to electrode displacement for $E=0, 3, 5$ V

This relation is evident and well known, however can be very helpful for analysis results from USECM – CNC investigations. Also changes of the machined surface intersection shape with potential drop E (Fig. 3b) should be taking into account in analysis of ultrasonic vibrations influence. In case of electrochemical machining with universal ball-ended electrode assisted by ultrasonic vibrations the vibrations impact on workpiece surface depends on angle α (Fig 4a) in following way:

$$\frac{A_n}{A} = \cos \alpha \Rightarrow A_n = A \cos \alpha \tag{2}$$

where: α range is $(0, \alpha_{max})$ and α_{max} results from condition of minimal current density, which is necessary for dissolution process to proceed. In similar way can change potential drops E. Therefore increasing of removed allowance $\Delta a(x)$ causing by ultrasonic vibrations influence is not constant for all points but decreasing with α increase (Fig. 4b). This fact and above presented results of ECM – CNC simulation were useful for preliminary explanation of the differences in machined surfaces shaped after ECM – CNC and USECM – CNC machining.

a)   b)

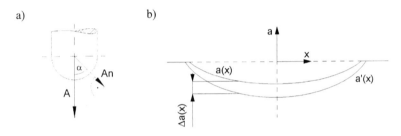

Fig 4 Ultrasound impact on workpiece surface: a) relation $A_n(\alpha)$; b) distribution $a(x)$ and $a'(x)$; $a(x)$ – thickness of machined allowance in ECM – CNC, $a'(x)$ – thickness of machined allowance in USECM - CNC

4 DESCRIPTION OF EXPERIMENTS

The ultrasonic head and equipment for workpiece clamping have been mounted in the working chamber of the electrochemical machine type EOCA 40 produced by The Institute of Metal Cutting. Ultrasonic head and generator have been also worked out at The Institute of Metal Cutting. The technical specification of head is presented in Table 1.

Table 1 Technical specification of ultrasonic generator

Working frequency	$f=22\pm1.5\ kHz$
Maximal power of transducer vibration	$P=160\ W$
Maximal amplitude of electrode vibration	$A=16\ \mu m$
Maximal current	$100\ A$
upply	$220\ V,\ 50\ Hz$

The tests have been carried out for two ways of machining for ball ended electrode tool ($R=5$ mm). The first one was for the case with electrode ultrasonic vibration (USECM - CNC). The second was the classical ECM – CNC machining. The following parameters have been taken into account: interelectrode voltage $U=8 \div 20$ V, electrode feed rate: $v_p=1 \div 59$ mm/min, starting electrode distance from workpiece $S_0=0.1 \div 0.9$ mm, electrolyte – water solution of NaNO$_3$, $C_e=10 \div 25\%$, power of ultrasonic vibrations $P=30 \div 150$ W (it corresponds to amplitude range: $2.73 \div 9.75$ μm). The resulting factors were: thickness of removed allowance a, material removal rate V_w and surface roughness parameters R_a, R_z. As the machined material NC6 tool steel has been applied, electrode has been made of copper. The machined material is a martensitic steel and consists of Fe 1.4 % C, 0.6 % of Mn and 1.4 % of Cr. The scheme of test stand is presented in Fig 5a. In this primary part of experiments only one electrode path over machined surface has been carried out (Fig 5b).

a)   b)

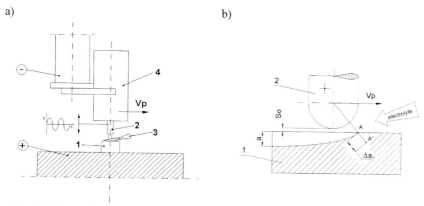

Fig. 5 Scheme of test stand for experiments (a) and machining in experiment (b); 1 – workpiece, 2 – electrode tool, 3 – electrolyte inlet, 4 – ultrasonic head, v_p –direction of machining, a - removed allowance thickness

It has been assumed that function of investigated object can be the equation in form of polynomial. As function of investigated object also neural nets have been taken (3). The results of analysis show that polynomials were adequate for all investigated factors, but errors of approximation when using neural nets were lesser than errors obtained from classical polynomials approximation. Below the results and analysis of investigations are presented.

5 ANALYSIS OF EXPERIMENTAL RESULTS

From Fig. 6 it results that it is possible to choose such USECM – CNC process parameters that increase of material allowance thickness removed take place. When electrode ultrasonic vibrations are introduced the allowance thickness for values of S_0 between $0.1 \div 0.7$ mm considerable increases.

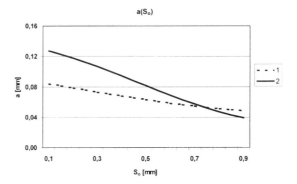

Fig. 6 Relationship *a(S₀)* when *U=14 V*; curve 1 – ECM – CNC machining, curve 2 –
USECM – CNC machining, amplitude of electrode vibrations *A=5.70 μm*

From equation (1) results that differences in values of a in case of machining with and without electrode ultrasonic vibrations can be explained by differences in dissolution process conditions. For the same parameters (v_p, U) of both processes these differences can result from the fact that ultrasonic vibrations change the course of electrode reactions and change the values of E, ηk_v, κ and V_n. During the tests the increase of current density for USECM – CNC machining was observed what can indicate that ultrasonic cause decreasing of E. The differences in shape of electrode path crossections for the investigated ways of machining (Fig. 7) are similar to differences between shapes obtained from simulation for various E (Fig 3b). It confirms, that the assumption about decreasing E was correct.

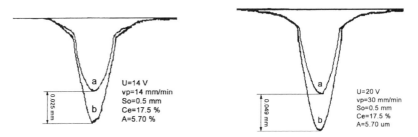

Fig. 7 Comparison of crossections from experiments for ECM – CNC (a) and USECM –
CNC (b)

The decreasing of E can be explained by cavitation bubble collapse in liquid, which results in an enormous concentration of energy from the conversion of the kinetic of liquid motion into heating of the bubble. This collapse in multi-bubble cavitation field produces hot spots with effective temperatures of $\approx$5000 K, pressures of $\approx$1000 atmospheres, and heating and cooling rates above 10^{10} K/s (13). Thus, cavitation can create extraordinary physical and chemical conditions in liquids. Cavity collapse near an extended solid surface becomes non-spherical, drives high-speed jets of liquid into the surface, and creates shockwave damage to the surface. Thus, micro-jets and shock wave impact on surface have a substantial effects on the chemical composition and physical morphology of solid that can enhance chemical reactivity (13). Above presented phenomena cause changing value of potential drops in the layers adjacent to electrodes (electrode polarization E).

In general, properties of specific energy source determine the course of chemical reactions. Ultrasonic energy differs from traditional energy sources (such as heat, light or ionizing radiation) in duration, pressure, and energy per molecule (13). A huge local temperatures and pressures together with extraordinary heating and cooling rates generated by cavitation bubble collapse mean that ultrasound provides a unique conditions in electrolyte. Therefore, as a result of ultrasonic vibrations in machining area, the sonolisys of aqueous solutions may occur, according to which H atoms and OH radicals are formed in the course of ionization and excitation events in solvent:

$$H_2O \rightarrow H+OH$$

These radicals either combine to form H_2 or H_2O_2 or attack soluted molecules, which in this way are reduced or oxidized. These particles mainly take part in the redox reactions of dissolved metal ions. Above mentioned phenomena can also cause changes of values of dissolution process indicators (E, ηk_V)

The comparison of relations $V_W(S_o)$ showed in Fig. 8 also confirm that introduction of ultrasonic vibrations change the course of dissolution process and, in the consequence of it, improve material removal rate. The relation $V_W(S_o)$ for USECM - machining is compatible with relation $V_W(S_o)$ from simulation for ECM – CNC with $E=0$ V what prove that ultrasonic vibrations decrease voltage drops on anode and cathode.

$$V_w(S_o)$$

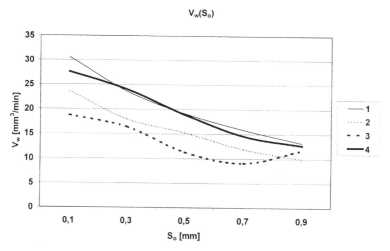

Fig. 8 Relations $V_w(S_o)$ for $U=20$ V, $Ce=17.5$ %, $v_p=30$ mm/min; 1 – curve from simulation for $E=0$ V, 2 – curve from simulation for $E=5$ V, 3 – curve from experiment ECM – CNC, 4 - curve from experiment USECM - CNC

One of the investigated factors was surface roughness parameters. Analysis of it showed that despite increasing of current density there is no differences in R_a and R_z for ECM – CNC and USECM – CNC machining. In ECM the mechanism of smoothing process results from differences in dissolution process for tops and bottoms of roughness During the ECM supported by electrode ultrasonic vibrations the micro-jets, which are created during collapse of cavitation bubbles, can create uniform electrolyte properties for tops and bottoms of roughness. So, in this case, it causes the equalization of dissolution rate.

6 CONCLUSIONS

Above presented results of primary investigations of USECM – CNC process indicate that introduction of electrode ultrasonic vibration in case of machining with universal electrode can change conditions of dissolution process and cause the increase of allowance thickness a and material removed rate V_w without surface quality decreasing. Preliminary analysis of phenomena in electrolyte irradiated with ultrasound show that probably acoustic cavitation is responsible for dissolution intensification. However, in order to explain this problem in more detailed way, the further investigations are necessary. They should also include recognizing of relation between Δa and angle α because this relationship can be useful not only for machining with universal electrode but also for electrochemical sinking assisted by ultrasonic electrode vibrations. In this case of machining relation $\Delta a(\alpha, A)$ can be useful on stage of designing electrode tool shape and dimensions.

Presented in this paper way of improvement of electrochemical machining with universal electrode can cause that ECM – CNC will wide its range of practical applications. Despite of increase of material removal rate in ECM – CNC process, this way of improvement doesn't cause that electrochemical machining with universal electrode starts competition with classical electrochemical sinking of sculptured surfaces. USECM – CNC should be used in finishing operations of surfaces initially machined by other methods.

7 ACKNOWLEDGEMENTS

The support of the Polish Science Research Committee (KBN – grant No 7T07D 002 18) is gratefully acknowledged. The authors wish to thank the Directors of The Institute of Metal Cutting for creating good conditions for scientific research and the author's colleagues from Department of Electrochemical Machining, who helped to build the experiments.

REFERENCES

1   **Davydov, A. D., Kozak, J** (1990) High Rate Electrochemical Shaping. Nauka, Moscow. (in Russian).
2   **McGeough, J.A.** (1976) Principles of Electrochemical Machining. Chapman and Hall, London.
3   **Gawlik, J., Karbowski, K., Ruszaj, A.** (1999) The neural – genetic optimisation of the electrochemical machining. In *Computer Integrated Manufacturing*, Vol. 1, WNT Warszawa, pp. 166 – 173.

4   **Kozak, J.** (1997) Analiza procesu obróbki elektrochemicznej elektrodą drgającą. In *Obróbka erozyjna – Materiały Konferencyjne*. Bydgoszcz – Golub Dobrzyń, pp. 204 – 212 (in Polish).

5   **Kozak, J., Chuchro, M., Ruszaj, A., Karbowski, K.** (1999) The computer aided simulation of electrochemical process with universal spherical electrodes when machining sculptured surfaces. In *Proceedings of the 15th International CAPE Conference*, pp. 425 – 430.

6   **Kozak, J., Rajurkar. K. P., Malicki S.** (2000): Study of electrochemical machining utilizing a vibrating tool electrode. In *Proceedings of the 16th International Conference on Computer – Aided Production Engineering*. pp. 173 – 181.

7   **Perusich, S. A., Alkire, R. C.** (1991) Ultrasonically induced cavitations studies of electrochemical passivity and transport mechanism. I. Theoretical. In *J. Electrochemical Soc*. Vol. 138/3, pp 700 – 707.

8   **Perusich, S. A., Alkire, R. C.** (1991) Ultrasonically induced cavitations studies of electrochemical passivity and transport mechanism. II. Experimental. In *J. Electrochemical Soc*. Vol. 138/3, pp. 708 – 713.

9   **Rajurkar, K. P., Zhu, D., McGeough, J. A., Kozak, J., De Silva A.** (2000) New Developments in Electro-Chemical Machining. In *Annals of the CIRP*. Vol. 48 (2), pp. 567 – 579.

10   **Ruszaj, A., Zybura – Skrabalak, M., Skoczypiec, S., Żurek R.** (2001) Electrochemical machining supported by electrode ultrasonic vibrations. In *Proceedings of the 13th International Symposium of Electromachining*. pp. 953 - 964.

11   **Ruszaj, A., Zybura, M., Żurek, R., Skrabalak, G.** (2001) Some aspects of electrochemical machining process supported by electrode ultrasonic vibrations optimisation. *2nd International Conference on Advances in Production Engineering*. Warsaw, Poland, vol. II, pp. 281 - 290.

12   **Ruszaj A., Zybura – Skrabalak M., Żurek R., Skoczypiec S., Skrabalak G.** (2001) Electrochemical machining supported by electrode ultrasonic vibrations. In *Proceedings of The Symposium on Research on Clean Hybrid Micromachining (HMM) Processes*. Krakow, Poland, paper number B6.

13   **Suslick, K.S, Didenko, Y., Fang, M.M., Hyeon, T., Kolbeck, K.J., Mcnamara, W.B., Mdleleni, M.M., Wong M.** (1999) Acoustic cavitation and its chemical consequences. *Phil. Trans. Roy. Soc*. A, vol. 357, pp. 335 – 353.

Design of ultrasonic block horns by finite element analysis

A CARDONI and **M LUCAS**
Department of Mechanical Engineering, University of Glasgow, UK

ABSTRACT

For tuned block horns that act as intermediate components between an ultrasonic transducer and a working tool, using standard block horn slotting and profiling techniques can be detrimental to performance. Slots in block horns, while allowing the output face vibration to be controlled, increase the number of vibration modes, increasing the risk of modal coupling at the operating frequency and resulting in nontuned combination modes being excited. These block horns must therefore be designed based on the output requirements of the attached tools. Finite element models are used to characterise the vibration behaviour of block horns and propose improved designs.

1 INTRODUCTION

High power ultrasonic devices are used in many industrial applications such as cutting, forming and welding. Ultrasonic devices are constructed using tuned components, usually driven by a piezoelectric transducer. The tuned components consist of the tool, used at the interface with the workpiece, and intermediate horns, used to introduce amplitude gain into the device or allow several tools to be operated from a single transducer. For most industrial processes using high power ultrasonics, the resonant components are tuned to a longitudinal, radial or torsional mode of vibration, and operate at a selected frequency between 20 and 100 kHz.

Ultrasonic tools consisting of several tuned components driven by a single transducer are gaining popularity, especially for cutting operations, as they offer an economic advantage in continuous multi-blade cutting processes. However, such systems are complex and are prone to nonlinear modal interactions that can result in high audible noise levels, component failures and poor cutting performance.

A typical cutting device will have several tuned cutting blades connected to the transducer via a block horn. Conventional block horn design relies on the use of standard slotted geometries and profiles. Block profiles are largely based on the required amplitude gain for the output face, while slots are used to control the vibration amplitude uniformity on the output face and control the separation of non-tuned modal frequencies from the tuned frequency. For systems where the block horn is an intermediate component, the same requirement for amplitude uniformity does not apply, and the increased number of modes due to slotting increases the risk of modal coupling and can result in nonlinear modal interactions. For such block horns, the design strategy must be based on the required vibration behaviour of the attached tools and on eliminating the adverse effects of nonlinear responses.

Initially, an experimental investigation is presented to characterise the modal interactions between the tuned mode and non-tuned modes of a three-bladed cutting system resonating in a longitudinal mode. Subsequently, finite element models are developed to predict the vibration behaviour and influence of the block horn and to propose design solutions to eliminate the adverse effects of modal interactions.

2 MODAL INTERACTIONS IN AN ULTRASONIC CUTTING SYSTEM

Modal interactions in parametrically excited systems can occur if special relationships (combination resonances or internal resonances) between one or more modal frequencies and the excitation frequency exist. For systems characterised by many modes, such as a multi-blade ultrasonic cutting device, external excitation of the tuned mode may excite one or two other modes through internal resonance. Each of these modes may excite further modes through internal resonance and the result is a frequency cascading effect [1-3]. Evidence suggests that systems tend to leak energy into lower frequency modes, with the consequence for ultrasonic systems being that energy is leaked into audible modal frequencies and that the response of a tuned, longitudinal mode, cutting device includes coupled flexural and torsional mode responses.

In Figure 1, a typical multi-blade cutting head is presented, where a half-wavelength double-slotted block horn is attached to three half-wavelength tuned blades. An experimental characterisation of the modal interactions in the cutting head assembly was carried out. A swept-sine excitation was used, with the vibration response being monitored by a 3D laser vibrometer, allowing an out-of-plane and two in-plane responses to be measured. A multi-channel data acquisition analyser connected to a portable computer enabled calculation of frequency response functions (FRFs) and identification of system responses via signal processing software (Figure 2).

A frequency response function measured at the tip of an outer blade, obtained through a fast sweep of the excitation frequency over a 0–50 kHz range, is shown in Figure 3(a). The tuned longitudinal mode frequency is measured at 35.1 kHz, with numerous other modal frequencies excited. A complete experimental modal analysis of the cutting head enabled classification all the modes in the range. Modes due to the blades, the block horn and the whole assembly were identified. The large number of modes is a result of the complexity of the system, but also the presence of mode families. These are excited because of the use of multiple, identical blades and also by the creation of multiple columns in the block horn by the use of slots. Mode families are multiple frequencies characterised by a common mode shape of the blade (or column) but different spatial phase variations between the blades (or

A003/034/2003 © IMechE 2003

columns). A mode family usually appears as a cluster of modes in a narrow frequency band and the excitation of these families significantly enriches the frequency response spectrum.

Modal interactions were identified from a slow frequency sweep over a narrow frequency band around the tuned longitudinal mode frequency at 35.1 kHz. The excitation was swept over a range of 300 Hz and the swept-sine measurement was repeated for 5 V increments of the excitation voltage. In Figure 3(b) it is observed that at 15 V excitation, when the sweep-down frequency reaches 35.2 kHz, two responses are excited in two lower frequency modes. These modes, corresponding to a bending mode of the assembly and a torsional mode of the blades, occur at 11.6 kHz (f_b) and 23.6 kHz (f_t), respectively and satisfy the combination resonance, $f_e \approx f_b + f_t$. When the excitation frequency was further decreased to 35.15 kHz the response became enriched with sidebands around all three modal frequencies, indicating a modulated response as shown in Figure 3(c). The modulation frequency was present in the lower end of the spectrum at 750 Hz, at the first system natural frequency. This type of response is characteristic of multi-component ultrasonic tools and different mode combinations, multiple mode combinations and modulations have been measured on a range of ultrasonic cutting heads.

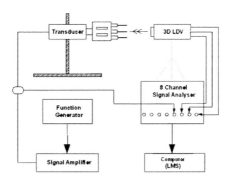

Figure 1. Wavelength three-blade cutting head

Figure 2. Experimental set-up for vibration response measurements

3 DESIGN STRATEGIES FOR IMPROVED VIBRATION BEHAVIOUR

A simple approach to redesigning a system exciting a combination resonance is to uncouple the tuned mode from the other modes involved in the combination resonance by making small geometry modifications. For instance, altering the positioning and/or length of the threaded studs connecting the block and the blades can provide the required mode frequency shifts to remove the internal resonance. However, this detuning strategy, which can be useful to eliminate modal interactions in systems characterised by a few modes of vibration, is inadequate for multi-component systems characterised by many modes of vibrations. For such systems, a slight detuning of a resonance condition alters the modal behaviour of the system, thus favouring the excitation of one or more other modal interactions. Hence, attention focused on finding a design strategy to reduce the number of modes so that, in the case of exciting an internal resonance, the removal of the response contributions of non-tuned modes became straightforward.

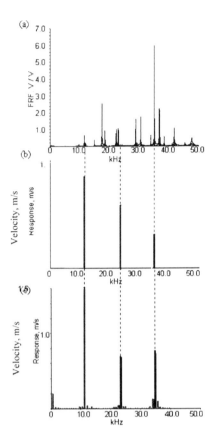

Figure 3. (a) Frequency response function from wavelength cutting head; (b) combination resonance; (c) combination resonance with modulation

A characterisation of the modes of vibration led to the following findings. Firstly, the mode families associated with spatial phase variations of the three identical blades, each consist of a cluster of very close modal frequencies. Therefore, the possibilities for exciting internal resonances is not significantly increased by increasing the number of identical blades mounted on the block horn. The number of modes in the family increases but the their modal frequencies tend to couple, with the response dominated by one or two of the modes. Secondly, the number of modes is directly affected by the length of the whole assembly. It is usual to construct ultrasonic systems from a series of tuned half-wavelength components. Where a block horn acts as an intermediate component, the system is one wavelength. An opportunity therefore exists to reduce the number of modes by designing a block and blade assembly within a half-wavelength. Finally, block horns are responsible for the presence of a large number of modes and the incorporation of slots results in a highly enriched response spectrum. Reducing the number of slots in block horns for wavelength systems also provides a strategy for eliminating modal interactions.

A003/034/2003 © IMechE 2003

3.1 Effect of number of slots on block horn modes

For multi-blade cutting assemblies, participation of blade flexural responses in the system longitudinal mode often results in poor operating performance and increased stress in the blades [4,5]. In fact, the response of blades in the system longitudinal mode of a multi-blade cutting head, is often characterised by flexural and longitudinal motions. The flexural responses are not due to coupling of the longitudinal mode with a flexural mode, but are connected to the shape of the longitudinal mode of the block horn. For a block horn with no slots, the longitudinal mode exhibits curvature on the faces of the block due to Poisson's effect. For the three-blade cutting head this leads to the outer blades exhibiting longitudinal and flexural responses while the middle blade response is purely longitudinal. For slotted block horns, the output face vibration amplitude in the longitudinal mode is very uniform but, consequently, the outer columns of the block horn exhibit flexural responses which, again, result in flexural and longitudinal motion of the outer blades and purely longitudinal motion of the middle blade. The solution, for design aimed at reducing the number of modes, is to use castellations to reduce the effects of curvature in the response at the output face of the block horn and reduce the number of slots. Three block horns are shown in Figure 4. The double-slotted block horn is designed based on standard slotting configurations, whereas the other two blocks are a single-slotted and solid block geometry incorporating castellated faces to improve uniformity of the output face vibration amplitude and constrain the flexural responses in the outer blades. The single slotted block horn in Figure 4 exhibits 20% fewer modes than the double-slotted block horn and the solid block exhibits 40% fewer modes.

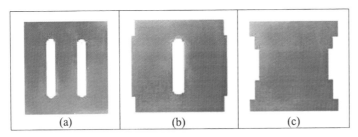

Figure 4. (a) Double-slotted block horn; (b) single-slotted block horn; (c) solid block horn, for wavelength cutting head

3.2 Short-block design for a half-wavelength system

A wavelength block horn and blades assembly is tuned to operate in the second longitudinal mode. In order to reduce the number of modes, a novel half-wavelength three-blade cutting system was designed by finite element modelling, tuned to the first longitudinal mode. The redesign was achieved without altering the blade cutting length and the system is shown in Figure 5. The finite element model predicted the existence of about half of the number of modes of the wavelength system in the 0-35 kHz frequency range.

**Figure 5. Half-wavelength
three-blade cutting head**

3.3 Incorporating slits to eliminate flexural responses

For the half-wavelength cutting system, it is possible to accommodate slits into the design, which constrains flexural responses of the outer blades and results in a pure longitudinal mode response of the system with parallel, in-phase responses in the blades. The predicted longitudinal mode response for a half-wavelength cutting head, without and with slits, is shown in Figure 6. The half-wavelength cutting head provides straightforward opportunities to remove the adverse effects of combination modes, where they exist, by making small geometry modifications to the design.

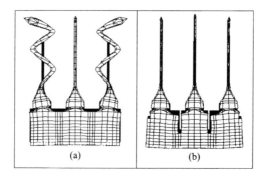

**Figure 6. Predicted longitudinal mode response of half-wavelength
cutting head with (a) solid block horn, (b) double-slitted block horn**

4 DESIGN STRATEGIES FOR REDUCING STRESS

An important consideration in the design of ultrasonic components is the stress condition. In particular, amplitude gain is built into the cutting system by using steep section reductions in the blade profile and by altering the thickness and profile of the block. The result is that weak points exist that become failure locations in operation, and the most common failure location in the cutting systems is at the steepest section reduction on the blades. The removal of flexural responses in the cutting blades, as discussed previously, significantly reduces the number of blade failures but other geometry modifications can also improve the reliability of the cutting assemblies. The highest stress in the half-wavelength cutting system excited in the longitudinal mode occurs at the blade step due to the section reduction. The effect of variation

 A003/034/2003 © IMechE 2003

of the block thickness is investigated. Normalised displacement and stress distributions along the axis of the system FE model for different horn thicknesses are presented in Figure 7. The highest stressed section of the blade coincides with the maximum slope of the modal displacement curve. The stress can therefore be reduced by relocating the node of the longitudinal mode further back into the block. One way this can be achieved is by increasing the thickness of the block. Figure 7 illustrates the reduction in stress at the failure location due to variation of block thickness, where increasing the thickness moves the node backwards, thus reducing the slope of the modal displacement curve. Altering other geometry parameters can also be used to improve the stress condition by judicious repositioning of the longitudinal mode node.

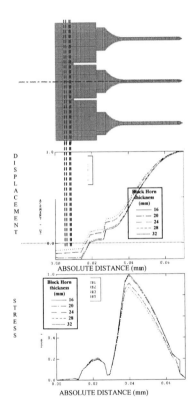

Figure 7. Predicted normalised axial displacement and stress for half-wavelength cutting head

5 CONCLUSIONS

An experimental characterisation of the vibration behaviour of a typical multi-blade ultrasonic cutting head demonstrated that modal interactions are particularly difficult to control in systems exhibiting a very large number of modal frequencies below the tuned frequency. A design strategy is proposed which focuses on reducing the number of modes. Block horns incorporating a reduced number of slots or a solid geometry are modelled by finite element analysis and block horns are designed with castellated faces to reduce the number of modes while maintaining purely longitudinal cutting blade responses. A half-wavelength system tuned to the first longitudinal mode is designed which halves the number of modes and the design incorporates slits to maintain purely longitudinal responses in the blades. Finally, a strategy for controlling the stress in multi-component assemblies is proposed, showing that stress reduction at a failure location can be achieved by geometry modifications which relocate the longitudinal mode node away from the failure location.

REFERENCES

1   **Cardoni, A., Lim, F.C.N., Lucas, M.** and **Cartmell, M.P.** (2002) Characterising modal interactions in an ultrasonic cutting system. Forum Acusticum paper ULT-02-003-IP.

2   **Barr, A.D.S.** (1980) Some developments in parametric stability and nonlinear vibration. International Conference on Recent Advances in Structural Dynamics, pp. 545-567.

3   **Anderson, T.J., Balachandran, B.** and **Nayfeh, A.H.** (1992) Observation of nonlinear interactions in a flexible cantilever beam. AIAA, pp. 1678-1685.

4   **Cardoni, A.** and **Lucas, M.** (2002) Enhanced vibration performance of ultrasonic block horns. *Ultrasonics*, Vol. 40, pp. 365-369.

5   **Cardoni, A.** and **Lucas, M.** (2002) Strategies for reducing stress in ultrasonic cutting systems. BSSM International Conference on Advances in Experimental Mechanics, pp. 101-104.

A simulated investigation on the surface functionalities in precision machining processes

K CHENG and **X C LUO**
School of Engineering, Leeds Metropolitan University, UK
X K LUO
North China University of Technology, Beijing, People's Republic of China

ABSTRACT

The surface functionalities resulted from manufacturing processes have long been recognized as having significant impacts on the product performance. It is of great significance to get a comprehensive understanding on how the machining process directly affects the functional performance of the component/product manufactured.

In this paper, the characteristic parameters for assessing surface functionalities of precision gear teeth surfaces are proposed in respect to the surface friction, wear resistance, fatigue and lubrication capacity. The generation of the gear surfaces and associated surface functionalities formation in the precision machining process are modelled and simulated by an integrated model which includes the dynamic cutting force model, the machine structural response model (transfer function), and the Molecular Dynamics (MD) model. The research also investigates the effects of nonlinear factors from machine tool, cutting tool and workpiece material on the surface functionalites formation. Simulation results show that the approach proposed is very promising in modeling and simulation of the surface generation and its functionality in relation with the machining process.

1 INTRODUCTION

There is a great demand for precision products, such as precision gears, bearings, precision lead screws, etc. The performance of these products much depends on their surface quality. But this might not suggest that 'the smoother the better' would apply because other factors, such as lubrication capacity and wear resistance, must be considered according to the function of the product. The surface functionalities resulted from manufacturing processes are more and more recognized as having significant impacts on the product performance. Therefore, surface functionalities should be taken into account by controlling the operation conditions in precision machining processes.

The aim of precision machining is to produce component/product with high dimensional accuracy and high quality surfaces and so as to enable it to perform well as desired. There is no doubt that a better understanding the surface generation and associated functionality formation would be of great benefit to the achievement of high quality surfaces more suitable to their function.

In this paper, the surface generation model is introduced as an intermediate step in the manufacture loop to predict the surface function. It can also be further developed to control and optimize the machining process. It can overcome the disadvantages in conventional manufacturing loop, since on-line measurement in the complex machining environment is difficult and the production will be delayed by off-line measurement. In section 2, surface functionality characteristic parameters for precision gear teeth surfaces are classified in the light of their functional performance. The modeling and simulation of surface generation and surface functionality formation are carried out in section 3. The effects of nonlinear factors including environmental vibrations, spindle run-out and built-up-edge are all investigated in section 4.

2 SURFACE FUNCTIONALITIES CHARACTERIZATION

The complexity of surface functionalities characterization is obvious since the deflection of the machine tool or workpiece, vibration, chatter, the flexibility of the machine, the error in the slideway, etc. will all leave their marks on the machined surfaces and will affect the surface functionality formation. Indeed, the surface conveys a vast amount of information, which can be used to control the machining process. The characterization of surface functionality should clarify the information and the effects on the functional performance of the product.

On the other hand, surface functionalities are different from that of the operational performance surfaces. For instance, the tribological functionality will become dominant when translational surfaces are used; The joint stiffness, contact and adhesion function will behave significantly for static contact surfaces; finishing, reflective or hygiene function are important for some non-contact surfaces [1]. Therefore, it is not feasible to get a group of versatile surface functionalities characteristic paremeters for precision product.

Precision gears are very popularly used in industry, so they are chosen as the target for characterizing surface functionality. They would subject to wear, friction, fatigue if they are not properly manufactured or lubricated. In the light of their functional performance, surface functionality characterization parameters of precision gear teeth surfaces are classified as follows:

Table 1. Surface functionality characterization parameters of gear teeth surfaces.

Application	Functionalities			
	Wear	Friction	Lubrication	Fatigue
Gear teeth surfaces	S_{qp}, S_{ku} S_{bi} S_{fd}, S_H	S_{qp}, S_{ku} S_{bi} S_{mr}, S_{mo}	S_{qv} S_{ci}, S_{vi}	S_{rs} S_H

Where, S_{qp} - Root-mean-square deviation of peaks;
        S_{qv} - Root-mean-square deviation of valleys;
        S_{ku} - Kurtosis of surface height distribution;
        S_{mr} – Area mean spacing of 3D motifs;
        S_{mo} – 3D motifs orientation distribution;
        S_{bi} – Surface bearing index;
        S_{ci} – Core fluid retention index;
        S_{vi} – Valley fluid retention index;
        S_{fd} – Surface fractal dimension;
        S_{rs} - Surface residual stress;
        S_{H} - Surface microhardness.

The reason to choose S_{qp} and S_{qv} is because they can reflect the peaks and valleys characteristic of the surface roughness respectively, and they have some different functional importance. S_{qp} relates to the wear resistance ability and friction characteristic of the gear teeth surfaces, and S_{qv} affects the lubrication capacity of the gear teeth surfaces. S_{ku} is a measure of the peakedness or sharpness of the surface height. It will affect the interaction of the working surfaces, and relate with the wear resistance and friction characteristic of the gear teeth surfaces.

Surface bearing index S_{bi} reflects the bearing property of the gear teeth surfaces, and it also relates to the surface wear resistance ability. Core fluid retention index S_{ci} and valley fluid retention index S_{vi} indicate the ability of lubricant retention of the gear teeth surfaces. The reference [1] has provided the definitions for these parameters.

Area mean spacing of 3D motifs S_{mr} and 3D motifs orientation distribution S_{mo} are proposed to assess the connectability of surface summits. They relate to the friction characteristic of the gear teeth surfaces. Although we can predict the number of summits or valleys and their mean size, as a function of height from the mean line, from existing theories, the "connectedness" of summits is still not known. Knowledge of the connectedness of summits would be interesting from the point of view of contact mechanics [2]. Kweon and Kanade introduce the concept of a topographic change tree based on motif method to describe the connectability of a surface [3]. To convert the qualitative information contained in a change tree to quantitative information, S_{mo} and S_m are proposed to indicate the direction of connection and spatial distribution of the connected summits. The direction of motif is colinear within the limits of the vectorial product of a pit with its two neighbour summits. S_{mo} is the polar histogram of the motif direction. S_{mr} is defined as the average Cartesian distance of the direction of the motif which coincides with the general direction of a principle manufacturing scratch. It is expressed as:

$$S_{mr} = \frac{1}{2n}\sum_{i}^{n}\sqrt{(x_i - x_{in})^2 + (y_i - y_{in})^2} \tag{1}$$

where n is the 3D motifs number. x_i, y_i, x_{in} and y_{in} stand for the Cartesian co-ordinates of the i-th motif and its closest motif neighbour respectively.

The fractal dimension is an indication of the complexity or intricacy of the machined surface texture. It is found that the tool wear process reduces the fractal dimension of the tool surface and the machined surfaces at the same time [4]. Therefore surface fractal dimension can be used to characterize the wear resistance ability of the gear teeth surfaces. Here the simplest approach based on the definition of the box dimension is adopted. The fractal dimension is expressed as:

$$S_{fd} = \frac{\log N(\varepsilon)}{\log(\varepsilon^{-1})} \tag{2}$$

where ε is the small side length of the little cubs and $N(\varepsilon)$ denotes the total number of cubes of the ε-tessellation containing sampled surface points.

The desired surface residual stress S_{rs} and surface microhardness S_H can decrease the fatigue failure of the gear, so they are chosen as the characteristic parameters for assessing the fatigue resistance ability of the gear teeth surfaces. The surface microhardness will also affect the wear resistance ability of the gear.

3 SIMULATION ON SURFACE FUNCTIONALITY FORMATION

The precision gear teeth are usually manufactured by more than one manufacturing process which results in a surface texture and its associate functionality that are based on the combined processes. The common finishing process is shaving, used after hobbing or shaping to improve finish and accuracy. In this study, a comprehensive modeling approach is used to model the multi-process manufacturing of a precision spur gear tooth, including hobbing and shaving. The surface generation and its associate functionality formation are then simulated.

3.1 Modeling the multi-process manufacturing of gear

3.1.1 Inputs to the multi-process manufacturing model
The achievement of high precision surfaces requires careful attention to all aspects of the

Table 2. The inputs of the machining process model.

Source	Linear Inputs	Nonlinear Inputs
Machine tool	Stiffness, damping rate, and nature frequency of machining system Straightness error of the slideway	Spindle rotational run-out Spindle axial run-out
Cutting tool	Initial rake angle Tool nose radius Teeth number Cutter radius	Built–up–edge
Operation condition	Axial/radial depth of cut Spindle speed Feed rate per tooth	Coolant Environmental vibrations
Material property	Initial shear stress at workshop temperature Initial shear angle	Hard grain Variation of shear stress Variation of shear angle

machining process, such as machine tools, operation conditions, cutting tools and workpiece materials. Each aspect includes a number of factors which directly affect the surface machined and produced. Obviously, these factors are the inputs to the model of multi-process manufacturing of gear. They can be divided into linear and non-linear inputs based on their behavior and properties. Table 2 has listed major linear and nonlinear factors associated with each of four primary aspects.

3.1.2 Modeling the overall multi-process manufacturing of gear

As shown in Figure 1, the cutting action of hobbing is that of continuous milling. The gear blank rotating while the hob, also rotating, is moved across the face of the gear. It is worth mentioning that the regenerative vibration will take place in the hobbing process. It is excited by the cutting forces, as a wavy surface finish left at the previous revolution, is removed during

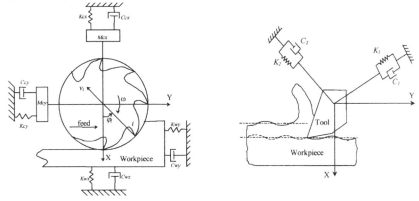

Figure 1. Hobbing process model.	Figure 2. Shaving process model.

the successive revolution which also leaves a wavy surface owing to machine structural vibration. And it will result in the variation of the chip thickness and width and to excite the variations of the cutting forces vice versa. Therefore, the regenerative vibration is generated in the cycles.

The preshave hobs will remove stock below the active profile of the teeth to provide clearance for the tips of the shaving cutting's teeth. Shaving cutters are helical gears with serrated teeth to provide the cutting edges. Figure 2 illustrates the cutting action of shaving. Here the regenerative vibration is still inevitable since the shaving cutting will cut the surfaces left by

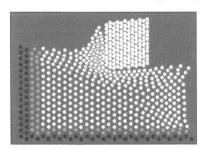

Figure 3. Molecular Dynamics model for shaving process.

previous bobbing and it will generate wavy surface due to machine structural vibration as well. In addition to, an Molecular Dynamics (MD) model (shown in figure 3) is used to simulate the interaction between the tool tip and the workpiece and calculation of the residual stress of the machined surfaces. The Morse potential function is applied to calculate the interatomic cutting forces.

As shown in Figure 4, the complicated machining processes of gear can be expressed by block diagram. The inputs of hobbing model include the linear and nonlinear inputs from operation

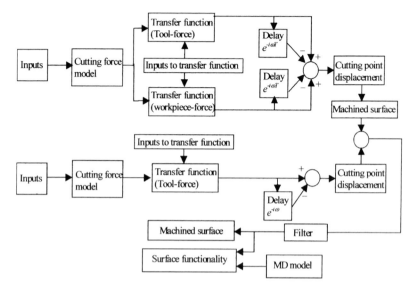

Figure 4. The whole modeling process of machining processes.

factors for hobbing process, the geometry of the hob, the workpiece material properties and structural properties of the hobbing machine. The dynamic cutting forces can be acquired through the cutting force model. The machine structural response model is used to get the response of the cutting point to the dynamics cutting force, i.e. the displacement of the workpiece and the hob at the cutting point by their respective transfer functions. The transfer functions can be determined by machine structural dynamics test. Based on the displacement of the cutting point and tool edge profile of hob, the intersections points of the sequence tool path on the workpiece surface are calculated by the following equation:

$$
\begin{cases}
r_{ti,ti} = \dfrac{2R(y_{ti,tj} - y_{ti+1,tj}) - (x_{ti+1,tj}^{\,2} - x_{ti,tj}^{\,2})}{2(x_{ti,tj} - x_{ti+1,tj})} \\[2mm]
H_{ti,tj} = y_{ti+1,tj} + \dfrac{(r_{ti,tj} - x_{ti+1,tj})^{2}}{2R}
\end{cases}
\tag{3}
$$

Where r and H are the projection on plane XZ and height vector of the intersections points of the sequence tool path, respectively. t_i is the index for the tool tip position in the tangential force direction. t_j is the index for the tool tip position along the feed direction. R is the tool nose

A003/043/2003 © IMechE 2003

radius. Trimming the line above the intersection points, the machined surfaces will be generated. It will be another input to the shaving process model for generating the machined surface.

The shaving process model is very similar to that of the hobbing process except the machine structural response model. In shaving process, the machining structure has multiple degree of freedom. So oriented transfer function of the system is used. After getting the displacement of the cutting point, the shaved surface can also be generated using the similar method expressed in equation (3). But there is vector overlap for r and H from the input of precious hobbing process. Linear least square filter has been used to transfer the curved surface to a nominally flat surface in order to make the characterization easier. The machined surfaces and the surface functionality parameters will be the outputs of the model.

In hobbing and shaving process, the variable transport delay functions are used to simulate the variation of the chip thickness and width due to the regenerative vibration. Their frequencies are all 240 Hz. The following nonlinear factors are also included:
• The spindle rotational run-out, which imposes an additional force applying on the machining system, is simulated by a sinusoidal function with frequency about 26.8Hz and 28Hz.
• The environmental vibrations and the spindle axial run-out will result in the relative vibrations between the tool and workpiece. They are emulated by two sinusoidal functions with frequency about 28.8 Hz and 28.7 Hz respectively.
• The step function is adopted to simulate the variation of friction angle due to the using of coolant.
• The intervally repeated ramp function denoted for the variation of effective rake angle due to the generation of build-up-edge and its removal at the tool's rake surface. Its frequency is assumed about 5.25 Hz.
• The variation of shear stress with temperature rise in the cutting zone is emulated by a curve function.
• A series of impulse values randomly generated at regular intervals is applied to emulate the workpiece hard spots and their effects on the shear stress of workpiece material.
• The chirp signal function is used to simulate the variation of shear angle due to the microstructure, such as crystal defects and different crystallographic orientation.

The whole model has been easily completed by MATLAB SIMULINK. The non-linear factors can be switched on or off very easily and their effects on the surface generation can thus be interactively visualized and quantitatively investigated on a basis of individual isolated event.

3.2 Simulation on machined surface and surface functionality

3.2.1 Simulation on machined surface
At first, the hobbing process simulation is carried out using a hob with helix angle = 30°, radial rake angle = 5°, and diameter = 25.4 mm. The operation condition are as follows: spindle speed = 2400 rpm, feed rate = 0.204 mm/tooth, axial depth of cut = 5.08 mm, and radial depth of cut = 9.525 mm. The geometry of the cutting edge of the shaving cutters are as follows: tool nose radius = 1 mm, rake angle = 5°. The spindle speed =1500 rev/min; depth of cut = 1.0 mm; federate = 0.1mm/tooth. It is assumed the friction angle = 30°, the shear angle = 25°. During the simulation process, all of the nonlinar inputs are switched off. After the linear least square filter, the simulated machined surface is shown in Figure 5.

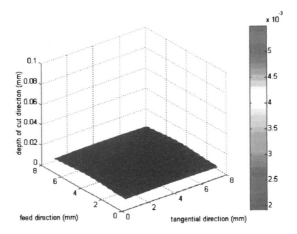

Figure 5. Simulated machined gear tooth surface (without nonlinear inputs).

3.2.2 Simulation on surface functionality

By statistic calculation and equation (1) and (2), it is very easy to get most of the surface functionality parameters. After the MD simulation, the residual stress can be calculated using the following equation:

$$S_{rs} = \frac{F_{sur}}{A_s} \qquad (4)$$

where F_{sur} is the force applied on atoms in the upper layer of surface after machining, A_s is area of the interface between the tool and workpiece.

The surface microhardness can be calculated after the MD simulation of the indentation to the machined surface, it can be expressed as [5]:

$$S_H = \frac{P_{max}}{25.4h_c} \qquad (5)$$

where P_{max} is the max load during the indentation process, h_c is the contact depth. The surface functionalities of the simulated machined surfaces are listed in table 3.

Table 3. Surface functionalities (without nonlinear inputs).

Wear	Friction	Lubrication	Fatigue
$S_{qp} = 0.25$ μm	$S_{qp} = 0.25$ μm	$S_{qv} = 1.85$ μm.	$S_H = 5.6$ Gpa
$S_{ku} = 2.73$	$S_{ku} = 2.73$	$S_{ci} = 1.87$	$Srs = 11.3$ MPa
$S_{bi} = 0.8026$	$S_{bi} = 0.8026$	$S_{vi} = 0.85$	
$S_{fd} = 2.3$	$S_{mr} = 0.323$ mm		
$S_H = 5.6$ Gpa	$S_{mo} = 90°$		

 A003/043/2003 © IMechE 2003

From these parameters, we can draw the conclusion that the machined surface has good friction characteristic and wear resistance ability since the small values of RMS of the peaks and the Kurtosis of surface height distribution show that the surface has relative flat top. The high values of surface bearing index, microhardness and fractal dimension also prove this. The value of 3D motif orientation distribution shows the dominant motif direction nearly parallels to the feed direction, and it means the friction resistance ability of the surface will be good in this direction. The values of core fluid retention index and valley fluid index indicate the surface has a good fluid retention property in the core zone but not good enough in the valley zone. The high value of RMS of the valleys also shows that. The high value of microhardness and positive residual stress indicates that the surface has good fatigue resistance ability.

4 DISCUSSION

To investigate the effects of nonlinear factors on the machined surface quality, the simulation is run with nonlinear inputs switched on, and the other operation conditions keep the same as those in Section 3. Figure 6 is the simulated machined surfaces with nonlinear inputs. It can be

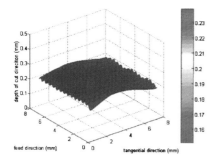

Figure 6. Simulated machined surface (with nonlinear inputs).

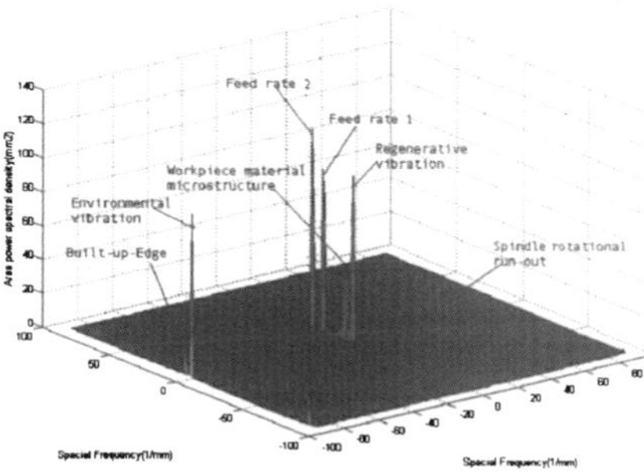

Figure 7. APSD diagram of the simulated machined surface.

seen that the surface integrity obviously has been demolished. The value of RMS of the peaks, valleys and the Kurtosis of surface height distribution are increased to 5.8μm, 8.2μm and 20.45 respectively, which show that the surface has sharp top and valley. It means the wear resistance abilityof the surface decreases. The decreasing values of surface bearing index, fractal dimension and microhardness are also consistent with the conclusion. The value of 3D motif orientation distribution is still 90°, which indicates that the nonlinear factors don't change the good friction resistance direction. On the other hand the increase of value of area mean spacing of 3D motifs shows the friction resistance ability of the surface is decreased. The values of core fluid retention index and valley fluid retention index indicate the surface has a bad fluid retention property in the core zone but merit fluid retention property at valley zone. The residual stress has changed to -3.4Mpa. The presence of a tensile residual stress is detrimental and make the gear subject to fatigue failure.

Figure 7 shows the diagram of area power spectral density of the machined surfaces with nonlinear inputs. Several significant low spatial frequency components can be found in it. The high value of power spectral density of the feed rate of shaving process ($1/0.1=10mm^{-1}$) is the most dominant component in the machined surfaces. The feed rate of hobbing process $(1/0.2=5\text{mm}^{-1})$ is also noticed. The other significant components are the regenerative vibration, of which spatial frequency in feed rate direction is about 20 mm^{-1}, and the environmental vibrations, of which spatial frequency in the feed direction is about 7.6 mm^{-1}. The area power spectral density of the variation of the shear stress and the shear angle due to the microstructure of workpiece material is lower than the three components mentioned above. Although, the variation of rake angle due to built-up-edge and the vibration due to the spindle rotational run-out have lower magnitude of the power spectral density compared with others, they still leave their marks on the machined surfaces. It shows that some nonlinear factors, such as the regenerative vibration and the environmental vibrations have the most significant effects on the generation of the machined surfaces. The effects of other nonlinear factors, such as the built-up-edge, the workpiece material microstructure and the spindle rotational run-out are less significant, but still cannot be ignored.

CONCLUSIONS

The surface generation and surface functionalities formation in precision machining process of gear teeth are modelled and simulated by a novel approach proposed. It is based on combining numerical computing methods, macro/micro cutting mechanics, block diagrams and non-linear functions to simulate the complexity of the machining system as a whole. Comprehensive parameters are classified to characterize the gear teeth surfaces functionality in the light of their function. The simulation results show that feed rate plays an important role in surface functionality formation. Some nonlinear factors, such as the environmental vibrations and the regenerative vibration have significant effects on the surface functionality formation. The impacts of other nonlinear factors, such as the microstructure of the workpiece material, the spindle run-out and the built-up–edge cannot be ignored as well.

REFERENCES

1   **Stout, K. J.** and **Blunt, L.** (2000) Three Dimensional Surface Topography, Penton Press, London.

2   **Thomas, T. R.** (1998) Trends in surface roughness, International Journal of Machine Tools and Manufacture, Vol. 38, No. 5-6, pp. 405-411.

3   **Kweon, I. S.** and **Kanade, T.** (1994) Extracting topographic terrain features from elevation maps, CVGIP: image understanding, Vol. 59, No. 2, pp. 171-182.

4   **Russ, J. C.** (1998) Fractal dimension measurement of engineering surfaces, International Journal of Machine Tools and Manufacture, Vol. 38, No. 5-6, pp. 567-571.

5   **Liu, X.** and **Gao, F.** (2001) Multi-function evaluation of surfaces at micro/nano scales by a new tribological probe microscope. Proceeding of 2nd euspen International Conference, Turin, Italy, May 27th-31st 2001, pp. 508-509.

A003/043/2003 © IMechE 2003

Manufacturing of automotive parts using active hydromechanical sheet metal forming

J BUCHERT and **H BAUER**
Fachhochschule Aalen, Germany
D K HARRISON and **A K M DE SILVA**
School of Engineering, Science, and Design, Glasgow Caledonian University, UK
R KOLLECK
Schuler SMG, Waghäusel, Germany

ABSTRACT

Conventional deep drawing is pushed to its limits by new development trends in the automobile industry for the economic forming of parts with more individuality in very small lot sizes. In order to meet these demands Active Hydromechanical Forming (AHMF), a sheet metal forming process which uses fluid as the working medium is developed. At present, the start of series production using the new AHMF technology is carried out successfully. Research done in feasibility studies show that there is tremendous potential for AHMF technology in novel application areas of deep drawing.

1 INTRODUCTION

More and more automobile manufacturers are embracing the luxury car market to meet the increasing demand for vehicles in the premium class and for special clientele specified vehicles. Current trends in the automobile industry therefore demand much more variants with different design and with higher material specifications, in parts produced by deep drawing. The number of units of cars made from these parts is rather small (1000 – 5000 units per year) compared to the high volume cars, such as VW Golf 3 and BMW 3 series.

Nowadays, conventional press lines are not economical for production volumes of less than 30000 units per year. In addition, new high strength and stiffness materials, which are difficult to form, are increasingly being used for automobile parts. The sheet metal forming industry has responded to these demands by developing new processes for metal forming. In the area of the deep drawing the working media based sheet-metal forming was rediscovered and developed for new tasks [1]. The Active Hydro Mechanical Deep Drawing (AHMF) process tries to satisfy the needs of the customers and gives an alternative to the conventional deep drawing.

2 SHEET METAL FORMING USING WORKING MEDIA

The AHMF process was developed by Schuler SMG in the early 70´s, in order to overcome the limitations of the conventional deep drawing process. With the new technique, the drawing ratio is increased up to 2.7 in comparison to the conventional technology which reached 2.0 in the single course [2]. In normal production it is possible to reduce drawing steps and annealing operations with the AHMF technology, so the manufacturing process becomes very economical. AHMF process is widely used for producing cups and containers for nearly all applications. Another advantage of the AHFM method is that it can be used for forming plane blanks as well as already pre-formed parts.

2.1 The principle of AHMF

The structure of the tool is similar to the conventional deep drawing process. It is defined by the punch, the water chamber, filled with the working media, which adopts the attribute of a die. In the first step, the blank holder locks the blank and a pressure between 5-20 bar - it depends on the material - prebulges the blank. The following step is the punch movement. In the prebulging step the blank gets the first contact with the punch, directly after the first mm of displacement the pressure grows up and forms the sheet metal against the shape of the punch. Within this process chain it is possible to control the pressure of the working media to improve the results. During the complete manufacturing the outer surface of the part is in contact with the fluid. This causes the class A surface without rabbles and damages and implements that also painted sheet metals, which already have their finish coating, can be manufactured.

An increase of the mechanical properties is the result of the prebulging step. At the beginning of this step the punch position is over the blank. In this free space the sheet metal has the possibility to increase plastic strain in the middle of the part. To control the plastic strain, it is possible to regulate the blank holder force, the pressure or the distance between punch and blank [3]. The aim of this process is a better use of the material properties. Figure 1 shows the individual steps of the process.

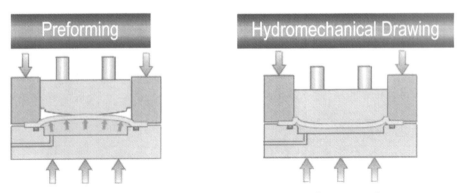

Fig. 1 Principle of AHMF for the example roof outer panel

  A003/069/2003 © With Authors 2003

3 ADVANTAGES OF AHMF

The increasing requirements of strength and the complexity of construction units lead to hydromechanical deep drawing. In the automobile industry there is a huge area with large outer surface parts, such as roofs, doors and hoods. These structures have a very small dynamic and static denting resistance after the conventional deep drawing, which is caused by the low deformation degree in the middle of the part [4]. This low component stability has a negative effect on crash resistance of vehicles and resistance to impact from hailstones. By using the AHMF technology, a consistent plastic strain distribution can be brought into the part and therefore the stability rises. The aim is to use the complete material properties to develop new stiffer parts with less sheet thickness.

Another point of view is the deep drawing process of complex structures, which is positively affected by the one-side tool contact and the pre-bulging step. This causes a better material flow and a smaller total friction. A result of all these factors is a higher deformation degree reached in only one drawing step. The AHMF method can be used for a wide range of parts, including all large panels such as hoods, roofs doors and fenders.

In the future there will be more and more complex parts in small lot sizes and also materials which are extremely difficult to handle. The AHMF process is able to reduce the existing problems with aluminium and high strength steel.

4 FORMING OF SMALL RADII

Several tests of forming small radii show that there is a lot of forming pressure needed. In an experimental tool the correlation between pressure and radii was tested. The following pictures show the upper and the lower tool. The geometry is in the upper tool that has to be formed with the pressure of the working media. In the lower tool there is a seal to define the area in which the working media can react. The results of the tests are shown in a diagram for the material DP 600 and a thickness of 0.6 and 0.7 mm.

 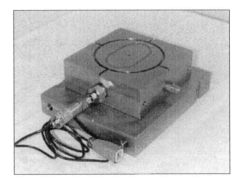

Fig. 2 Experimental tool for correlation tests between pressure and formed radii

State of the art is radii between 4 and 10 mm. To form this outlines in the part a pressure up to 1000 bar is necessary [5]. For example, a door outer panel has a punch area of one square meter. The required closing force for the press must be 100000 KN, but such high closing forces are very uneconomical for the process. This problem can be overcome by the use of elastic cushions in the ground of the water chamber. These cushions produce a force when they get compressed. In the process of AHMF the punch presses the blank against the elastic cushion and the reagent forms the radii. This technique needs less closing force, and a deep drawing process can be done with max. 200 bar. The absolute value of the force produced by the cushion depends on the elastic deformation.

The forming of small radii can be made with the help of elastic cushions without defects in the outer surface. This is an important point for the production of parts with class-A surface.

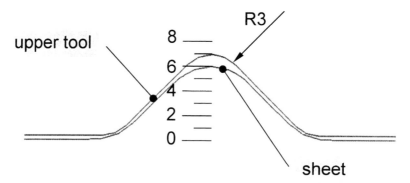

Fig. 3 Section through the experimental tool

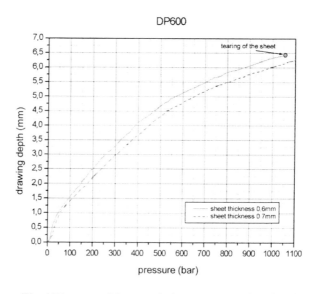

Fig. 4 Diagram of the correlation pressure – drawing depth

   A003/069/2003 © With Authors 2003

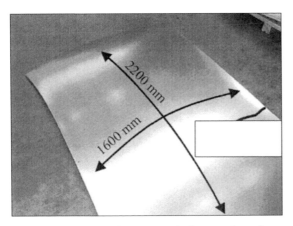

Fig. 5 Aluminium roof outer panel of a premium class car

5 PART DIMENSION AND GEOMETRY

At present there is a wide range of parts which are manufactured as feasibility studies, but also a roof as a series product. The parts start at the size of a door outer panel with 1300mm x 900mm and end with a roof of the size 2400mm x 1600 mm. The used materials are aluminium and high strength steal e.g. DP 600 or Trip 700. Many simulations are made with various parts and most of them can be manufactured successfully. Figures 5 and 6 illustrate examples where the AHFM process has been proved successful.

Fig. 6 Aluminium hood of a sports car, punch and part after deep drawing

6 CONCLUDING REMARKS

The feasibility of AHMF sheet metal forming process for the manufacture of automotive outer panels has been demonstrated successfully. In some cases, series production has already started. In the future more and more complex parts will be designed by the automotive industry and in this case the new technology can demonstrate its advantages.

REFERENCES

1. Schuler, Metal Forming Handbook, ISBN 3-540-61185-1.
2. Umformtechnik, Handbuch für Industrie und Wissenschaft, K. Lange
   ISBN 0-387-50039-1
3. Finite-Element-Simulation wirkmedienbasierter Blechumformverfahren als Teil der virtuellen Fertigung, Ralf Kolleck, Shaker Verlag, D290 (University Dortmund)
4. FEM-Leitfaden, Peter Fröhlich, Springer, ISBN 3-540-58643-1.
5. Hydroforming of Tubes,Extrusions and Sheet Metals, Edited by Klaus Siegert
   ISBN 3-88355-301-8.

A003/069/2003 © With Authors 2003

Study of self-dressing and sequence of operations in abrasive electro discharge grinding (AEDG)

J KOZAK, K P RAJURKAR, and P S SANTHANAM
Center for Nontraditional Manufacturing Research, University of Nebraska-Lincoln, Nebraska, USA

ABSTRACT

In Abrasive Electrical Discharge Grinding (AEDG) process the mechanical abrasion of grinding is combined with the electro-erosion of electrical discharge machining, leading to higher material removal rate. The major benefit of this process is that the electrical discharges simultaneously erode the workpiece and dress the wheel. This paper presents a thermal model and sequence of operations to analyze this self-dressing of the wheel. The model indicates that the embedded or lodged chips melt well before the end of the pulse on-time and the experimental results confirm the presence and effect of self-dressing in AEDG. Performance characteristics of Grinding-AEDG-Grinding sequence and the process optimization have been discussed.

1 INTRODUCTION

Abrasive Electrical Discharge Grinding (AEDG) is a hybrid machining process where conductive material is removed by a combination of rapid, repetitive spark discharges between workpiece and rotating tool, and by a mechanical action of irregularly shaped abrasive particles on the rotating tool [1-4]. This process uses an electrically conductive metal bonded wheel impregnated with electrically non-conductive abrasive grains. The erosion process in AEDG decreases the cutting forces on the wheel and helps to effectively dress the metal bonded grinding wheel (self-dressing) [5,6]. Advanced materials like PCD, super alloys and engineering ceramics have been successfully machined on a coborn EM1 EDG machine by replacing the graphite wheel with a metal bonded diamond wheel. Increase in performance measures of the AEDG process become evident when machining these materials The machining parameters such as peak current, wheel speed and pulse on-time were found to significantly influence the process performance [5,7 and 8].

This paper reports the analysis of self-dressing phenomenon and the effect of sequence of grinding and AEDG on the process performance. The thermal model development is presented in section 2. The experimental details and results are presented in section 3. This section also describes the effect of Grinding-AEDG-Grinding sequence. The last section summarizes this paper.

2 ANALYSIS OF SELF-DRESSING PHENOMENON IN AEDG PROCESS

One of the most important benefits of AEDG process is self-dressing, where the electrical discharges simultaneously erode workpiece and tool (wheel). Therefore, there is no need to interrupt the machining for dressing the wheel, resulting in considerable saving of productive machining time. Figure 1 shows the schematic diagram of workpiece and wheel contact region with abrasive grits, metallic bond and lodged workpiece chips on the wheel. These lodged chips melt due to the thermal energy generated by series of electrical discharges and occasional short circuits.

An experimental analysis of debris after machining indicated the occurrence of melting of chips. In addition, it was observed that the material removed during AEDG was in powdered form for certain values of pulse on-time and pulse current. The experimental observation shows the occurrence of melting of embedded chips, within a time that is less than the applied pulse on-time. To verify this observation and to estimate the time required for chip temperature to reach melting point, a thermal model of a chip (length L at voltage U) has been developed (Figure 2).

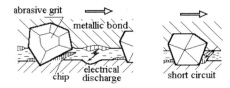

Figure 1. Workpiece and wheel contact region

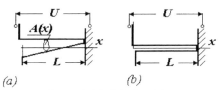

Figure 2. Model of a chip

Following assumptions were made in developing the thermal model:
1. The cross-sectional area A of the chip in a model shown in Figure 2(a) changes along the length of the chip and in a model shown in Figure 2(b) the value of A is constant throughout the length of the chip.
2. The changes in temperature, current density and electrical properties in the cross sectional area of the chip are negligible.
3. The changes in the specific capacity C and density ρ of the chip with temperature are negligible.
4. The change in the electrical resistance r of the chip material is given by $r = r_0(1 + \alpha\theta)$, where r_0 and α are the electrical resistance and the temperature coefficient of resistance at temperature T_0, respectively, and $\theta = T - T_0$ is the increase in temperature of the chip.

A003/049/2003 © IMechE 2003

5. The current is determined by ohmic drop voltage i.e. electromagnetic effects are neglected.

Considering the joule heat, the increase in temperature of the chip can be described as

$$\frac{\partial \theta}{\partial t} = \frac{a}{A} \frac{\partial}{\partial x}\left(A \frac{\partial \theta}{\partial x}\right) + r \frac{i^2}{\rho \cdot C}$$

(1)

Where $i = \dfrac{I}{A(x)}$ is current density, and a is the thermal diffusivity. The Current I, determined from ohm's law is,

$$I\left(r_0 \int_0^L \frac{1+\alpha \cdot \theta}{A(x)} dx + R_c\right) = U$$

(2)

Where $A(x)$ is the cross-sectional area of the chip at x and R_c is contact resistance. At increasing temperature of the chip and the assumption that copper bond of wheel plays the role of a perfectly good conductor; the contact resistance is described by the expression [9]:

$$R_c = r_0(1 + \frac{2}{3}\alpha \cdot \theta_c)/d$$

(3)

where θ_c is temperature increase at the contact spot and d is the characteristic dimension of contact spot. For circular contact, d is the diameter of the spot. Combining equations (2) and equation (3), the current density can be estimated by,

$$i = \frac{U}{r_0\left[\dfrac{1+0.7\alpha \cdot \theta_c}{d} + \int_0^L \dfrac{1+\alpha \cdot \theta}{A(x)} dx\right]}$$

(4)

For simplification, it is necessary to determine the magnitude of terms on the right side of equation (1) for typical condition of AEDG. For example, during machining of cobalt matrix composite (such as PCD and tungsten carbides) the following data is needed for estimation: density $\rho = 8900$ kg/m^3, specific heat $C = 0.42$ kJ/Kg K, specific electrical resistance $r_0 = 5.81 \times 10^{-8}$ Ω m, melting point $T_m = 1768$ K, thermal coefficient of electrical resistance $\alpha = 0.004$ 1/K, $\theta = T_m$ -To. For short circuits U=2.5V, grit size 240 (grain size 63μm) and the chip with L=50μm and d = 20 μm, the order of mentioned magnitudes in K/s are

$$a \frac{\partial^2 \theta}{\partial x^2} \sim a \frac{T_m - T_0}{L^2} = 2.36 \times 10^7$$

and

$$r \frac{i^2}{\rho \cdot C} = 4.28 \times 10^{16}$$

Therefore, the first term, which is connected with heat conduction along axis x, can be neglected, and therefore equation (1) after transformation, becomes:

$$\frac{\partial \theta}{\partial t} = \frac{r_0 \cdot U^2}{\rho \cdot C \cdot A^2(x)\left[\frac{1+0.7\alpha\cdot\theta_c}{d} + \int_0^L \frac{1+\alpha\cdot\theta}{A(x)}dx\right]^2} \qquad (5)$$

with initial condition $\theta(t=0) = 0$.

For AEDG process when voltage during short circuit is approximately stable, the integration of equation (5) leads to the time for melting as

$$t = \frac{r_0 \cdot \rho \cdot C \cdot L^2 \cdot \theta}{U^2}\left[1 + \frac{1}{2}\alpha\cdot\theta + \left(1 + \frac{1}{3}\alpha\cdot\theta\right)\frac{d}{L}\right] \qquad (6)$$

Figure 3 shows time required for melting for reaching different temperature values, for chip dimensions of L=16μm and 20μm, d = 10μm and 15μm. For θ = 1495 K (melting point), the time for melting of the chip is 0.078 μsec for chip of L=16μm and d = 10μm, but for chip of L= 20μm and d = 10μm the time for melting is 0.2 μsec.

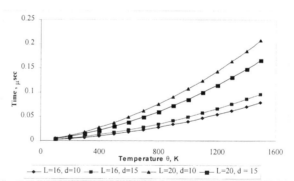

**Figure 3. Time vs Temperature for PCD
(Various chip length and diameter)**

For the case of AEDG when pulse generator with constant amplitude of pulse current, I, was used, the heating can be approximately described by

$$\frac{\partial \theta}{\partial t} = \frac{r_0(1+\alpha\theta)i^2}{\rho \cdot C} \qquad (7)$$

$\theta(t=0) = 0$ where $i = \dfrac{I}{n \cdot A(x)}$, I is setting current and n is number of chips contacted with the wheel bond. Solution of equation (7) for this case is

A003/049/2003 © IMechE 2003

$$t = \frac{\rho \cdot C \cdot n^2 A^2}{\alpha \cdot r_0 \cdot I^2} \ln(1 + \alpha\theta) \tag{8}$$

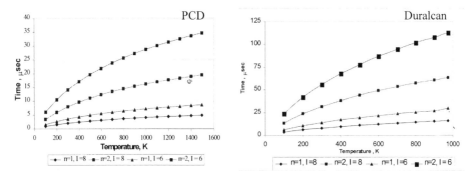

Figure 4. Time vs Temperature
(Chip crosssectional area 100 μm^2)

The Figure 4 shows the required time for certain temperature value for the chip of cross sectional area 100 μm^2 for PCD and Duralcan . The figure also illustrates the change in melting time for different currents and for different value of n, i.e. number of chips in contact with the wheel bond. For a current of 8A, the estimated time for melting a single chip is 5 μsec but for n =2 it is 20 μsec. Similar estimates can be made for chip of 225μm^2 area.

Similar estimates for different workpiece materials using Coborn EM1 machine (pulse on time of 15 μs to 120 μs and pulse current of 5 Amp to 25 Amp) confirmed that the embedded or lodged chips melt well before the end of the pulse on time. Visual observation and analysis of AEDG debris also confirmed this conclusion. An absence of chips was noticed at the end of machining as opposed to presence of chips after the grinding operation.

3 EXPERIMENTAL DETAILS

Experiments have been performed on a modified Coborn EM1 EDG machine tool by incorporating a dielectric system for de-ionized water developed at University of Nebraska. Figure 5 shows the experimental AEDG system [7,8]. The present research was conducted using de-ionized water as the dielectric fluid. To incorporate the water based dielectric system the existing system, which carried the oil from dielectric tank to the machine and vice versa were disconnected. A new system is designed where a separate tank with a pump is used along with de-ionizing chambers and a filter.

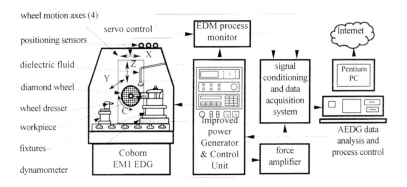

Figure 5. Experimental AEDG system [8]

3.1 Experimental analysis of self-dressing phenomenon

The experiments were also performed, by switching off the current to correspond to pure grinding condition. However a current of approximately 0.2A flows through the gap for the purpose of gap sensing. Experiments were performed on polycrystalline diamond (PCD) and Duralcan (F3S.20S). Initial experimental analysis was conducted for duration of 70 minutes of pure grinding and 70 mins of AEDG process. Figure 6 shows that the material removal rate during grinding (zero current) of polycrystalline diamond (PCD) is high in the beginning phase of the experiments. However, material removal drops to almost zero as the experiment progresses beyond the initial 10 minutes of machining. The hard PCD workpiece rounds the cutting edges of the tool faster preventing any grinding action to occur. As there is no current, self-dressing doesn't occur and the wheel losses its machinability substantially.

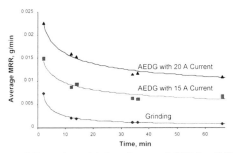

Figure 6. Effect of machining time on MRR for PCD [8]
(speed=2000 rpm, pulse on-time=60 µs, duty factor=0.5, grit size=320)

When the AEDG process is performed at 15 Amp current, the material removal rate slightly reduces with machining time. This can be attributed to the self-dressing effect, where the electrical sparks sharpens the blunt edges of grains on the periphery of the wheel. However, it seems that electrical sparks were not sufficient to dress the wheel to the required level. This leads to a slight drop in the material removal rate over the machining time. At 20A current, MRR was almost double the MRR achieved at 15A current and 3 times higher than MRR of grinding at beginning of machining. The Minimum MRR at 20A current (i.e. after 70 min of

A003/049/2003 © IMechE 2003

machining) is almost 2 times higher than maximum MRR of grinding at the beginning. Thus the self-dressing effect assists the material removal mechanism even without external dressing.

3.2 Experimental analysis of Grinding-AEDG-Grinding sequence

Additional experiments were performed with Duralcan in the sequence of: Grinding–AEDG–Grinding–AEDG–Grinding operations in order to analyze the self-dressing characteristics of the process. The machining conditions during the investigation is set as given in the Table 1

Table 1. Machining conditions for Grinding-AEDG-Grinding sequence

Machining conditions for Duralcan and PCD		
Parameters	Grinding	AEDG
Current , A	0.8	10 , 20
Pulse on–time , μsec	15	15
Thresh %	50	50
Gain	6	6
Wheel speed , rpm	1000, 3000	1000,3000
Abrasive grit size	320	320
Work piece dimension Duralcan, mm	25.4 x 25.4 x 4.2	25.4 x 25.4x 4.2
Work piece dimension PCD, mm	17.0 x 17.0 x 3.2	17.0 x 17.0x 3.2
Pulse off- time , μsec	15	9999

The experiments were conducted for a total duration of 52 minutes for a wheel speed of 1000 rpm and 3000 rpm and the current employed while doing AEDG was 20A and 10A.The abrasive wheel was dressed mechanically using Norton's silicon carbide dressing stick at the start of each experiment and the wheel should not be dressed mechanically thereafter until the completion of the sequence of operations. The material removal rate was calculated from the change in the mass of the workpiece after each machining cycle per unit time with respect to the density of the material. Table 2 lists the sequence and duration of conducted experiments [10].

Table 2. Sequence of operations for current 20A

Steps	Sequence of Operation	Machining time(min)	Number of machining cycle	Total Duration (min)
I	Stage I Grinding	4	4	16
II	Stage I AEDG	2	4	8
III	Stage II Grinding	4	4	16
IV	Stage II AEDG	2	2	4
V	Stage III Grinding	4	2	8

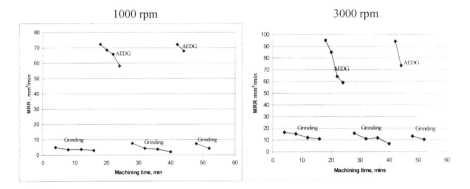

Figure 7. Self-dressing phenomenon – duralcan, effect of Grinding-AEDG-Grinding sequence on MRR at Various speeds
(gain =6, on-time=15μs, grit size=320, thresh = 50%, current = 20 A)

Figure 7 shows the effect of operation sequence on MRR at 1000 rpm and 3000 rpm for grinding and AEDG processes with increasing machining time. Initially the diamond wheel was dressed mechanically and then first stage of grinding process was started with approximately zero current. The initial MRR was observed to be 5 mm³/min and at the end of 16 minutes of experiments, the material removal rate was 35% less than the initial MRR. This was because the wheel got loaded on with the workpiece material, blunting the diamond grains. The second process was the AEDG process performed at 20A with same wheel without any mechanical dressing. This process was performed for four machining cycles at the rate of 2 minutes per machining cycle for a total duration of 8 minutes. The initial MRR during the AEDG process increased to about 72.32 mm³/min, which is about 15 times more than the initial MRR at the start of the grinding after mechanical dressing. This might be due to electrical discharges of AEDG process, which sharpened the already blunt wheel. However, the electrical discharges that were developed where not sufficient to make the self-dressing effect predominant in the process. In the mean time, MRR dropped by 20% compared to the MRR at the start of the AEDG process.

The third process in the sequence was grinding, the initial MRR here was 7.5 mm³ /min which was 50 % more compared to the MRR at the start of the stage I grinding process after mechanical dressing. This might be due to the presence of some uniform grain protrusion because of self-dressing effect of the AEDG process. From this it can be inferred that the self-dressing was more effective than mechanical dressing in terms of grain protrusion and also the amount of material removed as a result. As grinding process proceeded for 8 minutes, material removal rate dropped 3 times as compared to the MRR at the start of this grinding process and final MRR at this time was same as the final MRR after stage I grinding (after mechanical dressing). When the sequence of operations (grinding-AEDG-grinding-AEDG-grinding) was followed, the similar trends of results were observed.

A003/049/2003 © IMechE 2003

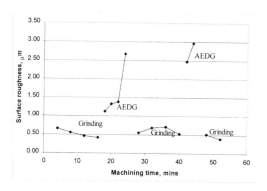

Figure 8 Self-dressing phenomenon – duralcan, Effect on Ra at 3000 rpm
(Gain =6, on-time=15μs, Grit size=320, Current = 20A, Thresh = 50%)

Figure 8 shows the effect of operation sequence on surface roughness at 3000 rpm The surface roughness observed was better after initial mechanical dressing, but when the operational sequence was continued for the total machining time of 52 minutes, observed the surface deterioration increased. Therefore, it can be concluded that self-dressing phenomenon did not assist in producing a better surface finish; it only assisted in higher material removal rate. From the investigation of sequence of operations it is clear that the AEDG process was very effective in rough machining the workpiece with a high MRR, and the grinding process without any electrical discharge is more desirable for finishing and improving the quality of surface.

SUMMARY

This paper reports a theoretical analysis and experimental investigation of self-dressing mechanism in Abrasive Electro Discharge Grinding (AEDG). A visual inspection of debris and quantitative comparison of material removal variation with grinding process support the theoretical analysis. The investigation of sequence of operations Grinding-AEDG-Grinding shows that the application of AEDG is very effective for first stage of machining i.e. mainly for rough machining the workpiece with high MRR, but for finishing and improving the quality of surface, the process of grinding without any electrical discharges is recommended. The observation of lack of chips during AEDG indicates the presence of self-dressing process.

ACKNOWLEDGEMENTS

The National Science Foundation (Research Grant #DMI-9908219) and the State of Nebraska (Nebraska Research Initiative Fund) are gratefully acknowledged for their support of this study.

REFERENCES

1. **Kozak, J., and Rajurkar, K. P.** Selected Problems of Hybrid Machining Processes Advances in Manufacturing Science and Technology, Vol 24, 2000, pp. 25-50.

2. **Wei, B., and Rajurkar, K. P.** "Abrasive Electro Discharge Grinding of Superalloys and Ceramics", 1st International machining and Grinding Conference, Dearborn, Michigan, 1995, pp. 493-500.

3. **Rajurkar, K. P., Shrivastava, S., and Zhao, W. S.** "Rotary Electro Discharge Machining of Polycrystalline Diamond", Transactions of the North American manufacturing Research Institution of SME, Vol. XXVII, 1999, pp. 41-146.

4. **Nooka, S. R.** "Abrasive Electro Discharge Grinding of Advanced Materials", M. S. Thesis, University of Nebraska-Lincoln, 1994.

5. **Rajurkar, K. P., Wei, B. Kozak, J., and Nooka, S. R.** "Abrasive Electro Discharge Grinding of Advanced Materials", International Symposium for Electromachining, ISEM-X1, Lausanne, Switzerland, 1995, pp. 863-870.

6. **Muruganandham, R.** "Abrasive Electro Discharge Grinding of Advanced Materials using De-ionized Water as Dielectric Fluid". M. S. Thesis, University of Nebraska Lincoln, 2001.

7. **Kozak, J., Rajurkar, K. P., and Muruganandham, R.** "Neural Network Prediction of Abrasive Electrodischarge Grinding (AEDG) Process Performance", International Symposium for Electromachining, ISEM-XIII, Vol.1, Bilbao, Spain, 2001, pp. 405-420.

8. **Kozak J., Rajurkar K.P. and Khilnani H.G.** "Performance Characteristics in Rotary Abrasive Electrodischarge Machining", Transaction of the North American Manufacturing Research Institution of SME (NAMRI/SME), vol. 30, 2002, pp.145-152.

9. **Holm R.** Electric Contacts, Spring-Verlag, New York Inc., New York, 1967.

10. **Sathish P. S.** "Self-dressing Characteristics and Abrasive Electro Discharge Grinding of Advanced Materials". M. S. Thesis, University of Nebraska Lincoln, 2002.

A003/049/2003 © IMechE 2003

Rapid Prototyping

Surface roughness simulation for FDM processed parts

P M PANDEY, N V REDDY, and **S G DHANDE**
Department of Mechanical Engineering, Indian Institute of Technology Kanpur, India

ABSTRACT

Stochastic surface roughness model based on the measured surface profiles of Fused Deposition Modeled part is used as a key to adaptively slice a tessellated (STL) CAD geometry. It is ensured while slicing that at any location of the part the surface roughness does not exceed the specified value of surface roughness. Surface roughness value over the part surface is graphically shown using different colors for different Ra values. The developed system also predicts the average part surface quality in terms of standard centerline average surface roughness (Ra) value.

1. INTRODUCTION

In Layered Manufacturing (LM) part is produced by layer-by-layer material deposition or addition. This layer-by-layer deposition leads to *staircase* (figure 1) on part surface and detracts part surface finish. Staircase is a geometrical constraint for LM parts and constraints their functionality. Therefore, there is a need to enhance surface finish or control the surface roughness of LM parts. Fused Deposition Modelling (FDM) [1] is one of the important LM process in which a movable nozzle deposits a thread of molten polymeric material onto a substrate as shown in figure 2. The build material (commercially available polymeric spool based material, ABS) is heated slightly above its melting point so that it solidifies within reasonable time after extrusion and cold-welds with the previously deposited layers. Factors to be taken into consideration are the necessity for a steady nozzle speed and material extrusion rate, the addition of support structure for overhanging parts, and the speed of the head, which affects the overall layer thickness. An advantage of a FDM system is that it may be viewed as a desktop prototyping facility, cheap, non-toxic, not smelly   and environmentally safe. The disadvantage is the surface finish is inferior as compared to Stereo-

---

* Ph.D. Scholar, Corresponding author, Email: pmpandey@iitk.ac.in
∞ Assistant Professor
⁺ Director, Indian Institute of Technology Kanpur (INDIA)

lithography (SL) parts. This is because of resolution of the process is lower as this is dictated by the filament diameter [2].

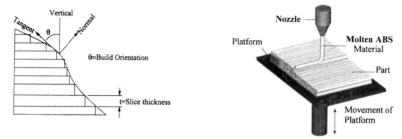

Figure 1. Staircase Effect in LM Parts   **Figure 2. FDM Process [2]**

Issues related to poor surface finish of LM parts have been addressed by a number of researchers [3-10]. Kruth et al. [3] reported that accuracy and surface finish are major handicaps of LM parts than their strength. Bharath et al. [4] reported that the best possible surface finish on a FDM part could be obtained by choosing the optimal FDM process parameters. They carried out fractional factorial experiments considering layer thickness, road width, air gap, build orientation and polymer melt temperature as process parameters. Using analysis of variance (ANOVA), they concluded that layer thickness, build orientation and interaction of these two parameters are factors with significance index 23.8828, 41.5151 and 18.4001% respectively, which affect the surface finish. Gautham et al. [5] studied the surface roughness of FDM parts considering slice thickness, build orientation, edge profile (assumed elliptical without measuring), layer composition and sub-perimeter composition as the parameters. They concluded that the rate of variation of surface roughness in the range 40 to 90 degree build orientation is less as compared to 0 to 40 degree region. It is reported that surface roughness is higher for downward facing surfaces as compared to upward facing surfaces. This may be due the gravity effects or presence of support structure that is removed after the prototype is deposited completely. There have been some attempts to study the effect of process parameters on part quality produced by SL using statistical techniques. Diane et al. [6,7] measured Ra values of SL parts and concluded that layer thickness and part orientation are important parameters. Perez et al. [8,9] developed surface roughness model for SL parts by assuming the edge profiles as filleted for values of φ between $30°$ and $85°$ and rounded for for values of φ between $85°$ and $90°$. For $\varphi = 0°$, a constant value of surface roughness (Ra) is proposed based on experimental results. Here, φ is angle between horizontal and surface tangent. They concluded that theoretical models proposed for surface roughness agree well with experimental values. They also explored the possibilities of manufacturing SL parts with constant slice thickness or with surface roughness confined within given values of surface roughness with variable layer thickness. Campbell et al. [10] presented a computer graphics based visualization system for surface roughness of LM parts (SL 250, Actua 2100, FDM 1650, LOM1015 and Z 402). A part named "Truncheon" is fabricated using different LM processes with constant slice thickness. Surface roughness value is measured using *contact Talysurf* system. Experimentally obtained surface roughness values were used to generate the graphical output.

          A003/068/2003 © IMechE 2003

It is clear from the previous discussion that poor surface finish is a major drawback of FDM parts and surface roughness varies significantly across its total surface area. It is often desirable to have less surface roughness, particularly in the aesthetically and/or functionally important areas. Even though the surface roughness can be reduced, by reducing the slice thickness but the build time of the prototype increases drastically. This problem can be handled by adaptive slicing of CAD models [11-25] in which variable slice thickness is used (slice thickness is determined based on geometry of part and LM machine specifications) instead of constant slice thickness. It is also possible to improve surface finish by post finishing operations however; it is time consuming and at the same time leads to geometrical inaccuracies on prototype, hence such operations should be avoided. In addition, this problem can also be solved to some extent through careful orientation of the model during deposition. Therefore, in the present work the surface roughness simulation for FDM processed parts is implemented with adaptive slicing to provide an overall idea of variation of surface roughness on the part surface before fabricating the actual prototype. Designer can have an idea of surface roughness at any location of the part well in advance for a given orientation of part deposition and it is possible to decide about suitable orientation for part deposition or acceptance of a product design. Effect of support structure and effect of upward facing and downward facing surfaces on surface roughness have not been considered in this work.

2. SURFACE ROUGHNESS MODEL

Pandey et al. [25] has considered layer thickness and build orientation, as the two most significant process variables that effect surface finish, based on the previous literature. To study the surface roughness of FDM part surface, a part with different build orientation (angle between surface tangent and vertical direction as shown in figure 1) was fabricated on FDM-1650 machine with 0.254 mm slice thickness, $270°$ C polymer melt temperature of part material i.e. ABS, $265°$ C polymer melt temperature of support material, $70°$ C envelope temperature and zero air gap. Surface roughness (Ra) value of the different faces have been measured by them using *Surf-Analyzer5000*. Surface profiles obtained over a 2 mm sample length are shown in Figure 3. It can be seen from figure 3(a) (between $0° \leq \theta \leq 70°$) that there is no gap between deposited roads while there is gap between deposited roads (between $70° \leq \theta < 90°$) in figure 3(b), where θ is build orientation. The surface profiles presented in Figure 3(a) clearly indicate the geometry of build layer edge profiles, which can be approximated as a parabola. Stochastic model is developed in the range $0° \leq \theta \leq 70°$ by approximating the layer edge profile by a parabola with base length ($t / \cos \theta$) and height as η percentage of base length, where t is slice thickness [25]. Constant η is established to follow normal distribution and hence for a 99% of confidence level the expression for centerline average surface roughness (Ra) is obtained as

$$Ra(\mu m) = (69.28 - 72.36)\frac{t\ (mm)}{\cos \theta} \qquad \text{for } 0° \leq \theta \leq 70° \qquad (1)$$

Surface profile for $\theta = 90°$ (horizontal surface or surface parallel to FDM platform) is idealized as semicircle (instead of parabola) with base length t and height $0.5 \times t$. Surface roughness value for $\theta = 90°$ is thus obtained as

$$Ra(\mu m) = 112.6 \times t(mm) \qquad \text{for } \theta = 90° \qquad (2)$$

Surface roughness for $70^{\circ} \leq \theta < 90^{\circ}$ is calculated by assuming linear variation of surface roughness between $Ra_{70}{}^{o}$ and $Ra_{90}{}^{o}$ i.e.

$$Ra(\mu m) = \frac{1}{20}\left[90Ra_{70^{\circ}} - 70Ra_{90^{\circ}} + \theta(Ra_{90^{\circ}} - Ra_{70^{\circ}})\right] \quad \text{for } 70^{\circ} \leq \theta < 90^{\circ} \quad (3)$$

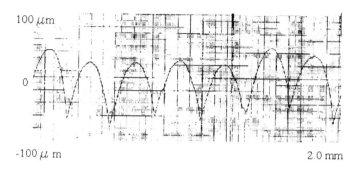

(a) Build Orientation = 45°

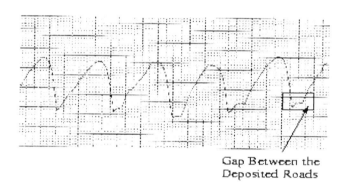

Gap Between the
Deposited Roads

(b) Build Orientation = 70°

Figure 3. Surface Profiles of FDM Processed Part, Slice thickness = 0.254 mm

3. ADAPTIVE SLICING

Limited cusp height [11] and limited area deviation [18] are the two criteria used by researchers [11-24] to slice a CAD model in most of the works. These two criteria have limitations, as a standard manufacturing measure of surface quality is not considered in these works. Pandey et al. [25] developed stochastic model of surface roughness (expression 1) and used it as a key to slice a tessellated CAD model. In the present work slice thickness is also calculated by the following expression as used by Pandey et al. [25]

A003/068/2003 © IMechE 2003

$$t(mm) = \frac{Ra(\mu m) \times \cos\theta}{70.82} \qquad (4)$$

where Ra is maximum possible value of surface roughness (bound kept on Ra) permitted on part. The average part surface roughness (surface quality) has been calculated using following expression

$$Ra_{av} = \frac{\sum Ra_i A_i}{\sum A_i} \qquad (5)$$

where Ra_{av} is average surface roughness of the part, Ra_i is the roughness and A_i is the area of the ith triangular facet of STL file. Slicing of the tessellated CAD model is performed by the procedure proposed by Pandey et al. [25] and the slice thickness and corresponding height of slicing plane (which are parallel to FDM platform, xy palnes) are stored.

4. SURFACE ROUGHNESS SIMULATION

Simulation of part surface roughness is implemented to have an idea of variation of part surface roughness over entire part. Adaptive slicing of tessellated CAD model gives variable slice thickness depends on the local geometry of tessellated CAD model, and a triangular facet may be common to more than one slice as shown in figure 4. Therefore, average slice thickness for a triangular facet is considered for calculation of Ra over a triangular facet. Build orientation of each triangle is calculated by the information of surface normals form STL file of the part. Surface roughness corresponding to each triangle is calculated using expressions (1) to (3), and average part surface roughness is calculated using expression (5). Surface roughness values over the part are graphically shown by rendering triangular facets with different colors according to their Ra values. The system is implemented using C, OpenGL and GLUI 2.1 [26].

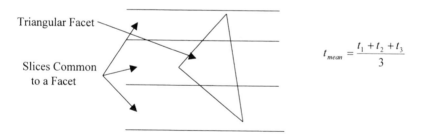

Figure 4: A Triangular Facet Common to Many Slices

CASE STUDY

The surface roughness variation of a typical part for slice thickness 0.254 mm (adaptive slicing is not used here) is presented in figure 5. Maximum, minimum and average surface roughness are predicted as 52.27, 18.03 and 24.45 μm respectively. Same part is sliced with

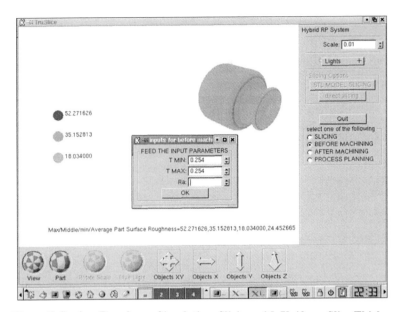

Figure 5. Surface Roughness Simulation, Slicing with Uniform Slice Thickness

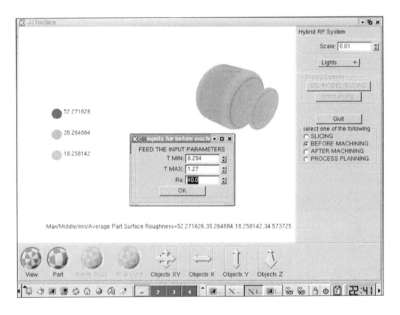

Figure 6. Surface Roughness Simulation with Adaptive Slicing, Bound on *Ra* = 40 μm

A003/068/2003 © IMechE 2003

minimum and maximum slice thickness 0.254 and 1.27 mm respectively for bound kept on *Ra* as 40, 50 and 60 μm respectively. It is observed that for 40 and 50 μm bound kept on *Ra* the maximum surface roughness is found to be 52.27 μm (refer figure 6 for 40 μm bound kept on *Ra*). It indicates that at certain critical locations, minimum slice thickness i.e. 0.254 mm is used to deposit the slice while the surface roughness could not be controlled below the bound kept. For these two cases (40 and 50 μm bound kept on *Ra*), the average part surface roughness is found to be 34.57 and 40.57 μm respectively. It indicates that at other locations, which are not critical, comparatively thicker slices are used, and hence average part surface roughness increased. It is also observed that for 60 μm bound kept on *Ra* surface roughness can be controlled below the bound kept everywhere on the part. At critical locations, the maximum surface roughness is found to be 59.87 μm. In this case, the average part surface roughness increased to 46.72 μm because at non-critical locations thicker slices are used for deposition.

5. CONCLUSIONS

Surface roughness simulation for FDM processed parts is successfully implemented with adaptive slicing. For a particular orientation of a CAD model, the areas of higher surface roughness can be identified and if the surface roughness is unacceptable, the part can be checked with different orientation or product design modification can be suggested. A significant decrease in manufacturing lead time and cost of the prototype may be achieved by using the developed system in the integrated design and manufacturing. The work is in progress to add the effects of support structures and upward and downward faces on surface roughness of a FDM processed part.

REFERENCES

1. **Kai, C.C., Fai, L.K.** (2000) Rapid Prototyping: Principles and Applications in Manufacturing, World Scientific.

2. **Pham, D.T., Dimov, S.S.** (2001) Rapid Manufacturing, Springer-Verlag London Limited.

3. **Kruth, J. P., Leu, M. C., Nakagawa, T.** (1998) Progress in Additive Manufacturing and Rapid Prototyping, Annals of the CIRP, Vol. 47, no.2, pp. 525-540.

4. **Bharath, V., Dharam, P. N., Henderson, M.** (2000) Sensitivity of RP Surface Finish to Process Parameters Variation, Solid Free-form Fabrication Proceedings, pp 251-258.

5. **Gautham, K., Novi, M. I., Henderson, M.** A Design Tool to Control Surface Roughness in Rapid Fabrication, http://prism.asu.edu/publication/Manufacturing /abs_design.html.

6. **Diane, A. S., Kou-Rey Chu, Montgomery, D. C.** (1997) Optimizing Stereolithography Throughout, Journal of Manufacturing Systems, Vol. 16, no. 4, pp. 290-303.

7. **Diane, A. S., Mongomery, D. C.** (1997) Using Experimental Design to Optimize the Stereo-lithography Process, Quality Engineering, Vol. 9, no. 4, pp. 575-585.

8. **Perez, C. J. L., Vivancos, J., Sebastian, M. A.** (2001) Surface Roughness Analysis in Layered Forming Processes, Precision Engineering, Vol. 25, pp. 1-12 .

9. **Perez, C. J. L., Calvet, J. V., Perez, M. A. S.** (2001) Geometric Roughness Analysis in Solid Free-Form Manufacturing Processes, Journal of Material Processing Technology}, Vol. 119, pp. 52-57.

10. **Campbell, R. I., Martorelli, M., Lee, H. S.** (2002) Surface Roughness Visualisation for Rapid Prototyping Models, Computer Aided Design, Vol. 34, pp. 717-725.

11. **Dolenc, A., Makela, I.** (1994) Slicing Procedure for Layered Manufacturing Techniques, Computer Aided Design, (1994), Vol. 1, no. 2, pp. 4 –12.

12. **Sabourin, E., Houser, S.A., Bohn, J.H.** (1996) Adaptive Slicing Using Stepwise Uniform Refinement, Rapid Prototyping Journal, Vol. 2, no. 4, pp. 20-26.

13. **Tyberg, J., Bohn, J.H.** (1998) Local Adaptive Slicing, Rapid Prototyping Journal, Vol. 4, no. 3, pp.118-127.

14. **Tyberg, J., Bohn, J.H** (1999) FDM systems and Local Adaptive Slicing, Materials and Design, Vol. 20, pp. 77-82.

15. **Sabourin, E., Houser, S.A.** (1997) Bohn, J.H., Accurate Exterior Fast Interior Layered Manufacturing, Rapid Prototyping Journal, Vol. 3, no. 2, pp. 44-52.

16. **Tata, K., Fadel, G., Bagchi, A., Aziz, N.** (1998) Efficient Slicing for Layered Manufacturing, Rapid Prototyping Journal, Vol. 4, no. 4, pp. 151-167.

17. **Cormier, D., Unnanon, K., Sanni, E.** (2000) Specifying Non-Uniform Cusp Heights as a Potential for Adaptive Slicing, Rapid prototyping Journal, Vol. 6, no. 3, pp. 204-211.

18. **Jamieson, R., Hacker, H.** (1995) Direct Slicing of CAD models for Rapid Prototyping, Rapid Prototyping Journal, Vol. 3, no. 1, pp. 12-19.

19. **Kulkarni, P., Dutta, D.** (1996) An Accurate Slicing Procedure for Layered Manufacturing, Computer Aided Design, Vol. 28, no. 9, pp. 683-697.

20. **Hope, R.L., Jacobs, P.A., Roth, R.N.** (1997) Rapid Prototyping With Sloping Surfaces, Rapid Prototyping Journal, Vol. 3, no. 1, pp. 12-19.

21. **Hope, R.L., Roth, R.N., Jacobs, P.A.** (1997) Adaptive Slicing with Sloping Layer Surfaces, Rapid Prototyping Journal, Vol. 3, no.3, pp. 89-98.

22. **Mani, K., Kulkarni, P., Dutta, D.** (1999) Region-Based Adaptive Slicing, Computer Aided Design, Vol. 31, no. 5, pp. 317-333.

23. **Weiyin, M., Peiren, H.** (1999) An Adaptive Slicing and Selective Hatching Strategy for Layered Manufacturing, Journal of Material Processing Technology, Vol. 89-90, pp.191-197.

A003/068/2003 © IMechE 2003

24. **Lee, K.H., Choi, K.** (2000) Generating Optimal Sliced Data for Layered Manufacturing, International Journal of Advanced Manufacturing Technology, Vol. 16, pp. 277-84.

25. **Pandey, P. M., Reddy, N. V., Dhande, S. G.** (2003) A Real Time Adaptive Slicing for Fused Deposition Modelling, International Journal of Machine tools and Manufacture, Vo.43, no. 1, pp. 61-71.

26. **GLUI 2.1**, http://gd.tuwien.ac.at/hci/glui/

RP building blocks – let's stack them up!

T LIM, J R CORNEY, J M RITCHIE, and **J B C DAVIES**
School of Engineering and Physical Sciences, Heriot-Watt University, Edinburgh, UK

ABSTRACT

Rapid Prototyping and Manufacturing (RP&M) technologies have become increasingly advanced through the years but the basic principle underlying them all remains. Parts are built gradually by adding materials in layers. Building by layers present a simple way of creating components with complex geometry that pose difficulties for machining. Unfortunately, for a majority of RP systems, the process of building parts is often slow and does not support the use of dissimilar materials. Ideally, RP systems should be able to create prototypes within hours, rather than days, while catering for a variety of build materials (e.g. MDF, plastics and metals). This paper describes a feasibility study of a system (RPBloX) that may be able to deliver this *ideal* performance. RPBloX has the potential to supplement current RP techniques and/or functions as a standalone method that reduces the design-manufacture cycle time.

1 INTRODUCTION

Rapid prototyping systems are known by the generic names of *solid freeform fabrication* or *layered manufacturing*. Compared to classical subtractive fabrication methods such as milling or turning, rapid prototyping related technologies fabricate physical objects directly by adding and bonding materials together in a series of layers. The ability to construct geometrically complex or intricate objects without the need for elaborate machine setup or final assembly has seen its progressive use in manufacturing, life sciences and many other disciplines that need to quickly produce physical models from 3D digital data [1][2]. There are several research issues pertaining to current RP systems – material properties, fabrication speed and bond strength, computational geometry and surface quality to list a few. Obtaining geometric accuracy of layered parts becomes increasingly important for *Production* prototypes i.e., where the prototype is the final part [3][4].

Throughout the RP industry, the data file format .STL, has dominated and has become the de facto standard [5]. Some loss of information is inevitable in STL and this greatly influences surface quality, form error, build-time and contributes significantly to post processing time [6][7]. The STL data is sectioned into a series of thin horizontal slices. The dimension of these slices depends on the technology applied and the resolution required. The geometric complexity of the part has significantly less impact on the fabrication process - a simple cube and a sculptured solid are equally easy to manufacture [8]. Although the quality of the faceting varies the local accuracy with which surfaces are reproduced there is no option to generate a grossly approximate model in lesser time than a precise one.

A survey carried out by Wholers [3] found that around 23.4% of RP parts were used as *Visual* prototypes (i.e. physical models used to reveal design defects and gain marketing clearance). Approximately 27.5% accounted for *Material* and *Production* prototypes (e.g. master patterns for secondary manufacturing processes and direct tooling). In industry, 15.6% of *Functional* prototypes are mainly used to detect problems in the "form and fit" of assemblies, 16.1% for functional tests and the reminder for proposals, quotes, ergonomic evaluation, etc.

We propose here, a novel idea and describe a feasibility study for the prototyping of large components. The implementation is focused on thick-walled components and also has the potential to support and complement existing RP technologies.

2 LET'S RPBloX IT!

Prototyping time can be minimised by using new prototyping technology and improving traditional methods [9]. This paper reports our preliminary research work towards using conventional machine tools such as CNC machines and robots to manufacture prototype models. The research is motivated by several observations on existing RP systems:

- ➢ 2D profiles (in the form of layers) are stacked to produce 3D shapes.
- ➢ Build rates are typically slow (around 12mm an hour along the build axis [10]).
- ➢ High cost of raw materials and lack of support for dissimilar material build.
- ➢ RP systems generally work at a limited number of resolutions (i.e. there is no option to generate a 'rough' model in much less time than a precise one).

RPBLoX employs a cellular approach to build up a part. Rather than slicing up the 3D CAD model into numerous thin sheets, RPBloX segments the model into 3D cells (or Bloxs) of various sizes. In contrast to current RP technology, conventional machine tools such as CNC machines and robots are used to manufacture and assemble these Bloxs.

Rapidity and flexibility means that new products and/or variations within the product family can be developed and produced in shorter times [11]. Flexibility and rapidity can be provided through the use of high-speed CNC machines and robots along with associated software to enable ease of programming and modification. Mass-produced bar stock material of both organic and inorganic can be used thus reducing material costs. It is envisaged that this concept will significantly reduce production costs without forfeiting accuracy and speed.

3 RPBloX FUNCTIONAL OVERVIEW

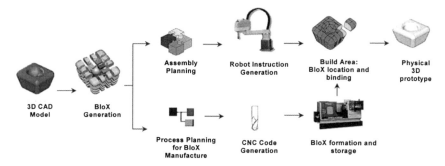

Fig. 1 Functional overview of the RPBloX methodology.

The RPBloX technique proposed applies the concept of creating 3D blocked shaped objects (Blox) in preference to working with 2D profiles (Fig. 1). 3D solid models are more suitable for RP systems since they are unambiguous and provide a complete mathematical representation of the shape. Furthermore, it provides a natural interface to well-established technologies of CNC part programming, machining and robotics [1][12].

In order to generate the Blox set, RPBloX interrogates the 3D model and applies an adaptive slicing algorithm to the model. CNC machines can interpret this data to automatically machine the individual Blox. The shape of each individual Blox must allow the automatic generation of a part-program, i.e. CNC code, for manufacture from bar stock.

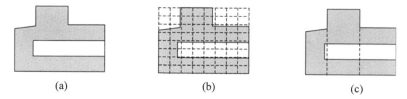

(a)   (b)   (c)

Fig. 2 Subdivision strategies. (a) Example component. (b) Uniform subdivision. (c) Adaptive subdivision for assembly and machinability.

Fig. 2 illustrates an example of subdivision strategies. A naïve subdivision is adopted in Fig. 2b where an equidistant slicing grid is used for compartition. This method produces numerous blocks of which some pose both difficulty in machining and assembly. In Fig. 2c, geometric reasoning and machining knowledge is used to improve the result to create three units that can be more easily machined and assembled.

As speed and accuracy is essential, each Blox generated by the system should contain simple features (e.g. planar faces, axis aligned holes) that can be machined in one pass on the CNC machine. To help eliminate the "staircase" effects on conventional RP systems, C-axis machines can be used to generate the required free-formed surfaces on a Blox. The output of the adaptive subdivision algorithm is represented using a combination of cellular topology and graph theory described in the next section.

3. 1 Subdivision representation

The RPBloX data structure uses the Cellular Topology (CT) component of ACIS® [14] to represent the subdivided solid CAD model. CT allows the modelling of sub-regions in a solid and cells can be either 2D sheets or 3D solids. Unique information can be associated with each cell in the form of 'attributes' supported by the kernel modeller. Examples of the types of information stored include the centre of gravity, robot pick up location, mass properties and spatial adjacency, geometry and topology. The subdivision algorithm generates all information automatically.

In order to easily manipulate, modify and perform search routines, the RPBloX data structure uses graph theory to define connectivity relationships. The graph structure employed by the RPBloX system is similar in many ways to the FEG and is referred to as the Blox Adjacency Graph (BAG). Vertex nodes represent Bloxes while edges represent the connectivity of adjacent faces of each Blox. A length attribute is associated with each edge in the graph to record the spatial proximity between Blox centres. Fig. 3 illustrates the RPBloX graphical structure. Blox B1 has been chosen at random simply for illustration purposes and has a degree of connectivity of 3 (i.e. three cells are immediately adjacent to it).

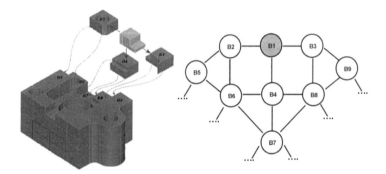

Fig. 3 Blox adjacency graph (BAG) representation. Arrows indicate Blox adjacencies.

The BAG supports both assembly sequence generation and subdivision optimisation. For example, Bloxes with geometry conforming to cuboids are merged into larger Blox thereby reducing the overall number of Blox to be machined and assembled. Longer term it is planned that the BAG will support algorithms that modify the geometry of Blox to incorporate mechanical locks (e.g. those found on jigsaw pieces or a dowel and pin location system).

4.0 IMPLEMENTATION AND EXAMPLE

To test possible subdivision and assembly sequence strategies a GUI for the BAG has been developed using Visual C++ and ACIS® geometric modelling kernel from Spatial Technology [14]. Fig. 4 shows the interface referred to as 'The HWUTestBed', which allows interactive exploration of the BAG via a tree-control. Several subdivision methods are available to the user via the menu system and vertical toolbar to manually, semi- or automatically, subdivide the component model (Fig. 4). The user has a choice of generating lamina slices similar to current RP&M techniques or creating a customised slicing lattice that will subdivide the 3D model into block shaped units.

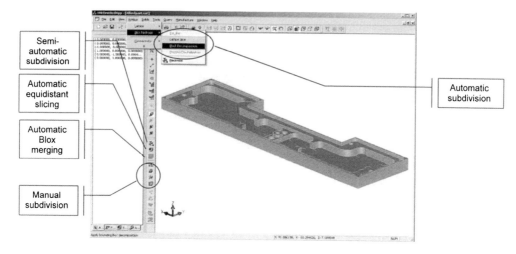

Fig. 4 The HWUTestBed.

A test component downloaded from the National Design Repository [16] is used as example to illustrate both the BAG structure and its exploration via a tree-control. Fig. 5a shows the resulting graphical representation of the BAG after the model is automatically subdivided. The algorithm used to segment the model into cubical units is detailed in [15].

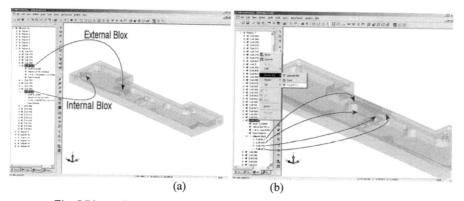

(a)   (b)

Fig. 5 Blox Adjacency Graph (BAG) representation and Interactive menu.

All Blox are ordered according to the *degree of connectivity between the cells* as seen in the partially expanded tree in Fig. 5. Each Blox is uniquely identified and records information such as its *level* and *type*. Level represents the positive distance of the Blox from the datum (in this case the base of the model) while *type* indicates whether the Blox is *external* or *internal*. An internal Blox will typically have the highest *degree* value indicating that it has many other adjacent Bloxes in spatial proximity. In this example, a total of 386 Blox were generated and the degree of the Blox set range between 2 to 24. Fig. 5b shows how the user can investigate a Blox's immediate neighbours by invoking the associated menu. Our example shows that Blox (Cell 250) has a four adjacent Blox that are of type *external*.

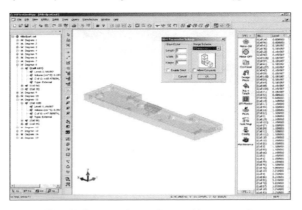

Fig. 6 Blox merging and assembly sequencing

To reduce the overall number of Blox, several merging algorithms have been developed to support both manual and automatic merging. One of the automatic merging algorithms is based on looking for cut vertices and shortest cycles within the BAG. Fig. 6 shows a dialog interface for defining merging parameters and the result of a single activation of the automated merging algorithm. Comparison of the selected Blox in Fig. 5b to the result after merging in Fig. 6 shows that it has incorporated some of its neighbours. Overall the total number of Blox is reduced from 386 to 108. Fig. 6 also shows the assembly sequence for the Blox set. From this interface, the user can generate the robot assembly instructions. A further reduction can be achieved by performing a series of merging operations (Fig. 7).

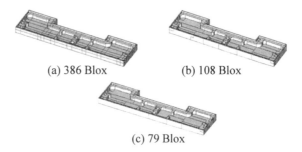

(a) 386 Blox   (b) 108 Blox

(c) 79 Blox

Fig. 7 Blox reduction through consecutive merging.

A003/016/2003 © IMechE 2003

5 DISCUSSIONS

Early in a products' life cycle, the focus is on design and manufacturing cost. There are many new technologies that can be exploited and RP techniques are useful in bringing all the operational units in the enterprise together early in the design cycle. In this context the benefits offered by the RPBloX system can be assessed against the test component illustrated in Fig. 6.

Machining time: Each Blox could be manufactured and positioned in a matter of seconds but overall build time depends on adhesive cure rate. A rough estimate of possible performance can be made by calculating the time required to machine and assemble the Blox shown in Fig. 7a (approx. 140x50x10mm) that has been subdivided into 386 units. The production time calculated assumes machining will take place on a high-speed C-axis machine tool with automatic stock feed and tool change, thus set-up time can be considered negligible. Assume an average CNC machining (including loading/unloading) time of 8 seconds/block, implies a total machining time of approximately 51.4 minutes.

Post-processing of the finished prototype would not be required if the Blox are manufactured with a high tolerance and accuracy. Therefore, the characteristic "staircase" effect inherent in RP systems can largely be eliminated.

Assembly time: Assuming a working volume $0.5m^3$, an average robot traversal speed of 1.26m/s, an average return distance travelled to locate a block of 0.75m, an average cure time of 5 seconds/block (for ABS plastic) and a time for block pick-up and adhesive application of 1.5 seconds. Implies that, once manufactured, 386 blocks could be positioned in around 100 minutes. The approximate total production time would be 41.8 minutes, potentially much faster than traditional RP methods on components of a similar size (~3.02 hours).

Other less easily quantifiable advantages include:

Variable Resolution and Material: Assembly based systems would be able to trade accuracy against speed. For example, if an approximate model is needed quickly, the system could use larger cuboids.

Non-homogenous materials are increasingly being used in consumer products, mechanical components, structures, etc. Unfortunately current RP&M methods rarely have this support. There are also issues with different cure rates, risk of contamination, and compatible resins to ensure robust bonding. Fortuitously, Blox can be made from different materials and then bonded together using, for example, cyanoacrylate adhesives.

Strength: Assembly based systems can exploit the trade-offs between curing time and adhesive strength. For example a slow build time may be acceptable if the resulting solid has superior strength and controlled anisotropy.

5.1 Implementation challenge

The feasibility of the individual elements of the process described is not in doubt, since the technologies already exist as commercial products. For example:

BloX Creation: The RPBloX system has shown that adaptive subdivision and intelligent merging are feasible. Currently, the test component (Fig. 6) takes approximately 10 minutes for initial segmentation and under 2 minutes for merging running on a P4 1.5Ghz machine. This time can most certainly be reduced by optimising the code or by parallel processing.
A plethora of intelligent slicing algorithms for CAD models exist for RP but these are mainly for layering [17][18]. Devising efficient and robust algorithms for adaptive subdivision and merging of a model continues to be an interesting challenge.

Blox Storage and Indexing: Storage and feeding of individual blocks between the machine tool and assembly cell is at present manual. In the longer term an automated storage and indexing facility will be incorporated.

Blox Assembly: The assembly process uses an EPSON SCARA robot [18] with a high accuracy and a C++ interface for direct communication with the CAD system. The robot has a linear resolution of 0.02mm. Unlike conventional RP&M, which requires a fixed, build axis, once the Bloxes are produced they can be simply assembled together in any axis to form the prototype. The robot arm is fitted with an automatic adhesive dispenser and a manipulator for picking up Blox (see Fig. 8). Depending on the material from which the Blox are made, various other methods of bonding can be applied such as brazing or welding.

The challenges presented by generation of CNC code and robot instructions for the manufacture and assembly of the blocks are essentially similar to many problems found in Computer Aided Production (CAP) research. Although the systems' accuracy should be superior to other RP methods this hypothesis can only be verified experimentally.

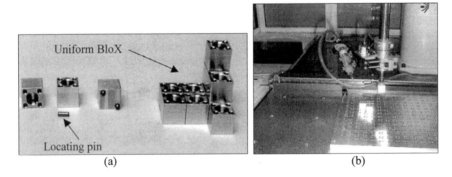

Fig. 8 Prototype RPBloX assembly system for Blox stacking.

Fig. 8 shows some initial work to establish performance parameters (e.g. assembly, gripping and positioning). The system uses uniformed Blox (Fig. 8a) to build arbitrary structures on a jig. Blox are fed into the assembly area via a slide feeder (Fig. 8b) and locked together using locating pins.

    A003/016/2003 © IMechE 2003

Prototype fabrication: There is a very large range of industrial adhesives available, which vary in the strength of bonds they form and speed with which they cure. The application in this case would generally require speed rather than strength. Consequently some form of cyanoacrylate adhesive represents one feasible bonding technology. Requiring no special surface preparation these adhesives have fixture times between 5 (plastics) and 15 (metals) seconds. There is no curing time needed if the Bloxes were machined with mechanical locks.

6 CONCLUSIONS

RPBloX demonstrates how commonplace workshop technology (e.g. CNC machine tools and robots) and CAD/CAM packages can be used to quickly and cheaply fabricate thick-walled prototypes. Depending on the required resolution the Blox set could be machined such that when assembled it can either represent a rough approximation or the exact dimensions of the final product. Bloxes of dissimilar materials could also be bonded together.

As reported by Yan and Gu [1] the two fundamental issues in RP&M are the slicing strategy and mechanism of support. Slicing influences the quality, efficiency and cost of the process while the support mechanism is essential to produce shapes of any complexity. However, the majority of RP processes still rely on uniform slicing with 0th order edge approximation [17]. Further research is therefore needed into adaptive slicing and merging techniques. Progress is being made in traversing the RPBloX graph structure so that robot assembly instructions and CNC code generation can be done automatically.

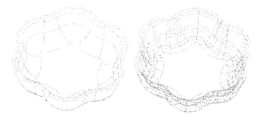

Fig. 9 Iso-parametric lines on a complex component.

There are still many hurdles but the progress so far is very encouraging. The current algorithm is mainly applicable to thick-walled double-sided 2.5D components. However, future work aims to create Blox decomposition based on iso-parametric lines (Fig. 9).

ACKNOWLEDGMENTS

This work is supported by the EPSRC grant GR/R35285/01. The authors also gratefully acknowledge the support of the following industrial partners: BAE Systems, Bridgeport Machines, C.A Models, Pathtrace and Renishaw.

REFERENCES

[1]   **Yan, X. and Gu, P.** (1996) A review of rapid prototyping technologies and systems. CAD, Vol. 28, No. 4, pp.307-318.

[2] **Ashley, S.** (1995) Rapid prototyping is coming of age. Mechanical Engineering Conference on Rapid Prototyping, pp. 63-68, Dayton, USA.

[3] **Wholers, T.** (1999) Rapid Prototyping State of The Industry – 1999 World Wide Progress Report, RPA-SME.

[4] **Electrolux Rapid Development Information Leaflet.** (1996) The shorter countdown.

[5] **Koc, B., Lee, Y.S., Ma, Y.** (2001) Max-Fit Biarc to STL Models for Rapid Prototyping Processes. In Proc. Of the 6th ACM Symp. On Solid Modeling and Applications 2001, pp.225-233, Ann Arbor, Michigan, USA.

[6] **Alexander, P., Allen, S. and Dutta, D.** (1998) Part orientation and build cost determination in layered manufacturing. In CAD, Vol. 30, No. 5, pp.343-356.

[7] **Tata, K., Fadel, G., Bagchi, A. and Aziz, N.** (1998) In Rapid Prototyping Journal, Vol. 4, No. 4, pp.151-167.

[8] **Kulkarni, P., Dutta, D.** (1996) An accurate slicing procedure for layered manufacturing. In CAD, Vol. 28, No.2, pp.683-697.

[9] **Tse, W.C. and Chen, Y.H.** (1997) A robotic system for rapid prototyping. In Proc. Of the 1997 IEEE Int'l Conf. On Robotics and Automation, pp.1815-1820, Albuquerque, New Mexico.

[10] CA Models Ltd., http://www.camodels.co.uk.

[11] **Glardon, R. and Stagno, A.** (1999) Facing the challenge of rapidity and flexibility in manufacturing organization. In Proc. of the Conf. CAD/CAM, pp.308-314, Neuchatel, Switzerland.

[12] **Ji, Q. and Marefat, M.M.** (1997) Machine Interpretation of CAD Data for Manufacturing Applications. In *ACM Computing Surveys*, No. 3, Vol. 24, pp264-311.

[13] **Lim, T., Corney J. and Clark, D.E.R.** (2001) Laminae-Based Feature Recognition. In IEEE Transactions on Pattern Analysis and Machine Intelligence, Vol. 23, No. 9, pp.1043-1048.

[14] **Spatial Technology**, 2425 55th Street Building A, Boulder, CO 80301-5740. ACIS® Geometric Modeler, V7.0.

[15] **Lim, T., Corney, J., Ritchie, J.M. and Davies, B.J.** (2002) RPBloX rapid prototyping - More than just layers. In Proc. of DETC'02 ASME 2002 Design Engineering Technical Conferences and Computers and Information in Engineering Conference Montreal, Canada, September 29-October 2.

[16] **National   Design   Repository**,   Drexel   University: http://repos.mcs.drexel.edu/frameset.html.

[17] **Maekelae, D.A.** (1994) Simple procedures for Layered Manufacturing techniques. CAD, Vol. 26, No.5, pp.119-126.

[18] **de Jager, P.J., Broek, J.J. and Vegeest, J.S.M**. (1997) A comparison between zero and first order approximation algorithms for layered manufacturing. Assembly Automation, Vol. 17, No. 3, pp.233-238.

[19] **EPSON Robots**, http://www.epson.co.jp]

Fundamental experimental study on free fabrication of nanocrystalline copper bulk by selective electro deposition with electrolyte jet

J ZHAO, Y HUANG, J ZHANG, and C YU
Rapid Prototyping Centre, Nanjing University of Aeronautics and Astronautics, China

ABSTRACT

This paper describes the fundamental experimental study on free, direct fabrication of nanocrystalline copper (Cu_{nc}) bulk with selective electrodeposition with electrolyte jet. The experiment results shows that the Cu_{nc} bulk with free shape can be fabricated, the shape of it is better accordant with that of designing and the average grain size of it is distributed to the range of 40-60 nm. With the mechanical performance of high break strength and high microhardness, the other performances of the Cu_{nc} bulk such as extending rate and plastic modulus are decreased sharply because of the co-deposition with hydrogen.

1 INTRODUCTION

Many studies revealed that compared to their coarse grained counterparts, the mechanical property of nanostructured (nano)-metal enhances remarkably. Besides the mechanical performance of nano-metal, the preparation technique is one of the main content of nano-metal. At present, there are mainly two kinds of preparation of nano-metal, one is inert gas condensation technique which can synthesize large size specimen, but the shape of the specimen is limited with the special mould and its fatal problem is that the specimen can be contaminated easily during its synthesizing. The other is electrodeposition, which is used

A003/019/2003 © With Authors 2003

widely in today's nano-metal materials studies. Many important conclusions on nano-metal materials, such as the low temperature creep and the superplastic extensibility, are acquired based on the full density and no-contamination specimen fabricated by electrodeposition (1). Compared with the inert gas condensation technique, the weakness of electrodeposition technique is that the shape of the specimen can't be controlled.

Introducing the principle of Rapid Prototyping Manufacturing (RPM) to electrodeposition, its machining capability would be extend to free fabrication of free shape, which means that the metal part with discretional shape can be fabricated by electrodeposition. In 1995, Bocking first brought forward the concept of Rapid Prototyping by electrodeposition. In the research, the electrolyte jet within a metal nozzle was used as the anode and copper bulks were gotten. Then, in 1998, Masanory and Kunieda built the first system of rapid prototyping by electrodeposition and got some metal specimens with simple shape (2). Base on the research above, this project aims developing a system of Selective Electrodeposition with Electrolyte Jet (SEEJ), which can be used for fabrication of Cu_{nc} specimen with free shape. With this system, the fundamental experimental researches are carried out.

2 SELECTIVE ELECTRODEPOSITION WITH ELECTROLYTE JET

2.1 Principle
As the first step of SEEJ, the CAD model of a part is divided into a series of slices with a certain thickness in the height direction. Then, scanning traces of each slice is calculated and translated into NC code according to the scanning mode, machining parameters and geometric information of the slice. The electrolyte jet discharged from a nozzle moves on the cathode surface with its scanning trace to electrodeposit metal ions which are dissolved in the electrolyte (Fig.1). The metal ions are consequently electrodeposited on the cathode surface exclusively in the certain zone. With the moving of nozzle, a metal bar can be deposited. Continuous metal bars form a layer. Its shape is consistent with the slice of CAD model and its thickness is consistent with the design dimension. When the deposition of a layer completed, the cathode is promoted with the distance of a slice thickness and scan the next layer. Layer by layer, a metal part consistent with the CAD model can be obtained.

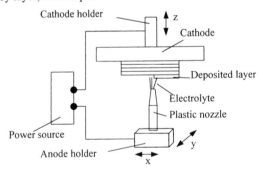

Fig. 1 Principle of selective electrodeposition of electrolyte jet

A003/019/2003 © With Authors 2003

2.2 Experimental system

Fig. 2 shows the experimental system used in this research. It is consist of machine body, control system and software system. All three moving axes are controlled by the control system. The control of the X-Y axes enables 2-D positioning of the nozzle over the cathode. The Z-axis is controlled to adjust the gap between the nozzle and deposit automatically with the feedback unit. A coulometer is integrated in the feedback unit as a feedback signal. Since a constant current power source is used in the system, the thickness of the deposit over the cathode surface can be kept constantly together with the coulometer. Different from the Masanori's research, cathode in this system is designed to be inversion and the electrolyte is jetted upward to the cathode surface. The electric current is supplied to the jet through the nozzle holder, which is made of stainless steel. The anode is divided into two parts: a main anode and an assistant anode. Nozzle is the main anode, which is made of plastic. Some pieces of copper within a basket made of Ti, which is put into the electrolyte bath; act as the assistant anode to supply enough metal ions in electrolyte.

Fig. 2 Experimental system

3 EXPERIMENTAL PRODUCE

The specimens were designed with the shape of square and right-angled triangle. The side of square was 20mm and the long side of triangle was 25mm with the acute angle of 30 degree. The electrolyte was consisted of $CuSO_4$ (250 g/L) and H_2SO_4 (50g/L). Additive was not used in this electrolyte. The temperature of electrolyte bath was kept at 30 ± , the flux was 80 L/h and the voltage of bath was altered from 25 to 5oV. The inter size of nozzle was 1.5mm and the scanning speed varied from 0.1 to 1 mm/s and the scanning space is 0.3mm. The average grain size and the microstructure of the specimens were characterized by means of XRD analysis and SEM. Quantitative XRD measurements of the Cu_{nc} specimen was carried out in a Rigaku D/MAX-Rc x-ray diffractometer with Cu Ka radiation. Ttensile tests were performed on a tensile device attached in a type WDW-100 and the hardness was measured with a sclerometer.

4 RESULTS AND DISCUSSION

4.1 Forming capability

Fig.3 shows the specimens on the stainless steel cathode fabricated with SEEJ. The dimension deviation of them is about ±0.2 mm/10mm, and the thickness is 1.2mm. The consistency of the current density distribution in electrolyte jet and the deposit shape on the cathode surface determine the dimension accuracy in X-Y plane. Kunieda's research indicated that the electrodeposition takes place in a round district, which is about twice of the nozzle dimension. With the inverse cathode in this system, the electrolyte is spayed upward to the cathode and dispersed in all directions immediately under the jet pressure and the weight of the electrolyte. It means that the deposition of copper ions could only take place in the dimension district of the nozzle on the cathode; outside the district, there was hardly any ions deposition.

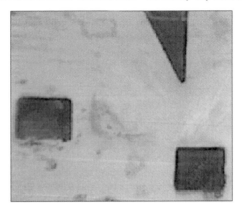

Fig.3 Specimens on the stainless steel substrate

The distribution of current density in electrolyte jet leads to the unequal deposit thickness in the deposit layer (3). If a metal nozzle used, the concentration of electric field around the nozzle on cathode will further strengthen the unequal trend. When the thickness of whole deposit is up to a certain value, the deposition process can't continue. Masanori Kunieda used the electrochemical process to level the surface of deposit, but the depositing time is increased greatly. In this system, the deposit thickness is controlled real-timely through a coulometer. Without any additives and evening process but a plastic nozzle, the specimens with thickness of 1.2mm were fabricated. The results show that this system performs the better forming capability.

4.2 Grain size

The data of average grain size determined by a Fourier analysis x-ray diffraction is listed in table 1. Changing of scanning speed and bath voltage, the average grain size of deposit is within the range of 40-60nm. The grain size decrease with the increasing scanning speed or the decreasing bath voltage. According the trend, the grain size would further decrease with the higher scanning speed and the lower bath voltage. It is unfortunately that the movement capability of step motor used in this system limited the farther experiment research. Bath voltage can not be decreased at discretion because the enough high voltage is needed to keep the overpotential of anode, which is necessary to the fine grain formed.

A003/019/2003

Table 1 Average grain size of Cu$_{nc}$ metal specimen under the varied parameters

Experiment one	U=25 V	Scanning speed: v/(mm/s)	0.45	0.65	1
		Grain size: D/nm	57	50	42
Experiment two	v=0.1 mm/s	Bath voltage: U/V	25	30	45
		Grain size: D/nm	44	48	60

4.3 Mechanical performance

The mechanical performances of the Cu$_{nc}$ specimen fabricated with SEEJ are listed in table 2. The elastic modulus, fracture strength, extending ratio and microhardness of the specimen are separately 0.313, 1.9, 0.015 and 3 times as that of the coarse grained copper. With these characteristics, such as fracture strength and microhardness increasing sharply and extending ratio and elastic modulus decreasing greatly, the Cu$_{nc}$ specimen fabricated in this experiment just perform those of Cu$_{nc}$ sample synthesized by the inert gas condensation (4). In tension testing, no creep and no bending happened before fracture, which indicates that the sample isn't full density and 100%pure. The SEM micrograph shows that the micropores are distributed uniformly in the whole depositing region and the size of them is about 1um or less (Fig.4). These micropores are the main reason to cause the low extending ratio and elastic modulus.

Table 2 Mechanical performances of Cu$_{nc}$ metal specimen

Fracture strength σ_F/ MPa	Elastic modulus E/MPa	Extending ratio δ/%	Microhardness HV /Gpa
243.55	40.45	0.602	2.1

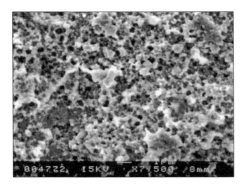

Fig. 4 SEM micrographs from the cross-section of specimen

Because of the higher surface energy of the nano crystal, the hydrogen on the cathode produced with the high overpotential was combined with metallic crystal to form the impurity attached in the deposit layer. The impurity in deposit layer occupies the location on which the ions would be deposited, and caused the further aberration of electric field, thus the growing procedure of metal crystal was affected greatly. With the increasing of the deposition layer thickness, the absence of ions around the impurity caused the micropores in the deposit.

CONCLUSIONS

The Cu_{nc} specimen with desired shape was fabricated by the SEEJ technique. The average grain size of it is under 100nm and the performance of this sample is similar to the Cu_{nc} sample synthesized by the inert gas condensation. Low extending ratio and elastic modulus is because of micropores in deposit layer with the co-deposition of hydrogen.

REFERENCES

1 **Lu Ke and Lu Lei**, Progress in mechanical properties of nanocrystalline materials, ACTA METALLURGICA SINICA, 2000, Vol.36, No.8, pp.785-789,

2 **Masanori Kunieda, Ritsu Katoh and Yasushi Mori.** (1998) Rapid prototyping by selective electrodeposition using electrolyte jet   Annual of the CIRP, Vol. 47, No. 1, pp. 161-164

3 **Karakus C, Chin D T.** Metal distribution in jet plating. Electrochem. Soc., 1994, Vol. 141, No. 3, pp.691-697

4 **X.J. Wu, L.G. Du, H.F. Zhang, J.F. Liu, Y.S. Zhou, Z.Q. Li, L.Y. Xiong and Y.L. Bai** Synthesis and tensile property of nanocrystalline metal copper, Nanostructured Materials, 1999, Vol. 12, pp.221-224,

A003/019/2003 © With Authors 2003

Robotics in Medicine

Safety in application of robots to surgery

E GENTILI, A ATTANASIO, E CERETTI, and C GIARDINI
Department of Mechanical Engineering, University of Brescia, Italy

ABSTRACT

We would like to consider the recent developments regarding the safety of the use of robots in surgery and to outline the research project we have been working on in this field. We have pursued the task of making safe and so certifiable a medical device which is able to perform automatically, by the use of a robot, a liver biopsy.

Our work began with the study of the actual regulation in force regarding safety and product certification of medical devices, and went on to study the actual procedure in a liver biopsy normally executed in hospital, in order to understand the risks of this operation, the context in which it is performed and the possible ways of improvement. This analysis led us to define three different robotised bioptic processes and among these, following technical and economical criteria, we chose the one which was supposed to be the best. Having clarified specifications and methodologies, we made the risk analysis, which is not only a document to be presented for the certification, but also a guide to find solutions which follow principles of safety, in relation to the state of technology.

Finally, we demonstrated that, as requested by the norms, the benefit is greater than the risk to the patient and the operating team.

1 INTRODUCTION

In recent years there has been a fast development of new applications of robotical systems, widening the already vast field of use of such machinery to the service of man [1]. The area which has attracted, maybe more than the others, the interest of research workers is that of medical and surgical applications. In fact, such systems are the object of many studies, because they offer enormous potentialities of development. Starting from the actual use of prevalently "passive" robots as an aid for the surgeon during the execution of surgical operations, they will become in a few years completely automated and flexible systems, which will perform, under the supervision of a surgeon, an ample range of operations. The latent advantages of such systems in terms of precision, repeatability, rapidity of execution and elimination of latent causes of contamination is easy to imagine. However there is to consider the possible advantages that could be obtained, thanks to an elevated productiveness, also in terms of costs and of efficiency of the health system. Additionally, this solution would allow the execution of teleoperations [2], a precious help in distinctive situations where there is a shortage of surgical personnel, for example in case of war or in undeveloped countries. The main problems to face in order to realise such systems are the development of technical solutions that will provide performances suitable for the applications, the reduction of the elevated costs that are often involved and above all, dealing with systems that operate in narrow contact with man, the safety study that must guarantee the absence of risks with reasonable certainty. In fact, the safety study has been a critical problem in every research that has been analysed and it always goes hand in hand with the degree of automation of the systems. Safety is the real problem to face, apart from the technical and economic ones, for the progress of these applications from the phase of scientific study to that of commercial production and it that must be involved in all the phases in the development of the device. The following paragraphs provide the description of the suggested flexible robotical system. In the methodology followed by the authors for the safety analysis of this device is explained, focusing mainly on the critical fundamental problems faced and the principal solutions found to guarantee its safety and allow the certification according to the recent European Directives, necessary for a future commercialisation of the commodity in the European Union.

2 SYSTEM DESCRIPTION

The system [3] has been developed to perform the operations in an automated and mini-invasive way, acting under an ecographic guide. In the procedure followed, similar for the four operations, the physician characterises the target using the probe of the ecograph, and then a second operator starts the system confirming the target position on the PC monitor. The computer calculates the position of the point in the space and the ideal trajectory to follow, and then the robotical arm proceeds automatically to move and to insert the needle in the patient to take the desired tissue sample. The device consists of four main components: a robot with its control system, an ecograph, a tracking device and a supervising master PC. For this new device we have used an industrial robot with six degrees of freedom, which presented technical characteristics already substantially fit for the purpose. It has a sophisticated control system, its own programming language and it can be connected with normal PCs. The diagnostic tool chosen for the identification of the operation target is an ecograph that, as well as guaranteeing a sufficient precision and the capability to carry out the operation in real time, avoids the problems connected with the TAC. This choice has involved the necessity for an identification system for the probe position in the space, as it is moved and oriented manually by the physician, and so it is doesn't give the absolute co-ordinates of the points identified, but only those in its own relative reference system. The device developed is an optic one, which consists of two stereoscopic cameras that, framing the area of work, give the position in the space of four leds mounted on the diagnostic tool and then to calculate the position and the orientation of the probe. The supervising and control system adopted is a common PC that, as well as guaranteeing all the required characteristics, develops a command and control software in Microsoft Windows® environment, which is very intuitive and easy to use. The use of UNIX operating system can increase the reliability of all the system. The surgical system consists of many other components, first of all the sample device, that is mounted on the robot's end-effector, and on which it is possible to mount different needles according to the specific operation. The described application already has technical specifications that fit the surgical operations, for example in terms of accuracy, repeatability, load capability and functionality of the software. Anyway for the passage of a similar system from the phase of experimental study to that of a future commercialisation and use in hospitals, an accurate study of the problems connected with the existing European normative is necessary, and then, as we will see, of all the aspects related to the safety of the device. In figure 1 a general scheme of the system is represented.

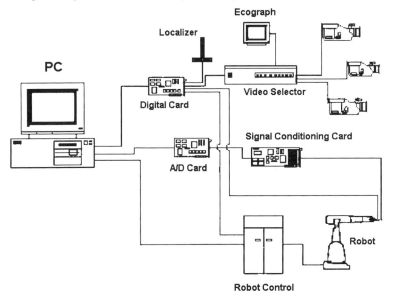

Figure 1. General scheme of the system

3 ROBOT GUIDED BIOPSIES

The introduction of robotisation in biopsies [4] can give a number of advantages, among which are, a greater precision (our device is able to reach a target inside the patient with a maximum error of 2 mm), greater reliability and repeatability due to the process standardisation, lower execution time and probability of catching infections and, finally, the possibility of "tele-operations".

There are three procedures to follow before the intended medical device can be put to use. These procedures correspond to the degree of automation of the robot and the different ways of pinpointing the target-point co-ordinates.

The three procedures are:
  1. echo-guided biopsy - teach-in

2. echo-guided biopsy - stereoscopic localisation
3. TAC-guided biopsy

The first distinction we made, echo or TAC guided, refers to the possible visualisation system adopted. In the echo-guided biopsy, the most important problem is the ecograph probe location [5]. The probe supplies the internal images of the patient and the co-ordinates of every image point in its reference system, but the ecograph cannot locate the probe spatial position. With this in mind, we thought to exploit the teach-in possibilities of the robot, which after being conducted to the probe position, is able to memorise its co-ordinates and reach the biopsy point optimising the learnt path. In the stereoscopic localisation case, the probe is located by a system of at least two cameras.

With the TAC-guided biopsy, the maximum degree of automation is reached: the tomograph is provided by an autonomous reference system which can directly communicate to the robot the target point co-ordinates. There is a greater precision, but there are also problems to be taken into account such as the radiation, the contrast-liquid injected and the higher cost, as well as the difficulty for the robot of inserting in the tomograph.

The best method in our opinion is the echo-guided system with stereoscopic localisation, because, not only does it maintain the ecograph use advantages (biological risk absence, low cost, easy use etc.), but it also allows a faster operation even in tele-communication.

4 SAFETY AND THE EUROPEAN NORMATIVE

Safety [6], which is the condition that avoids that an unfavourable event might lead to a danger, which in time could cause a damage to people or things, is the real foundation of most recent European Directives. The European Union has in fact established in 1985 with the strategy of the "new approach" that it must be warranted the free circulation in its territory only of the commodities answering to the essential safety features described in the directives. These are submitted then to the competent European Organisms (CEN and CENELEC [4]) which issue the harmonised standards that interpret these principles, considering the technological progress at the moment and translating them in agreed technical guide-lines. These rules are voluntary, but their application activates the "presumption of conformity" principle for the technical solutions adopted for the device (that must be respected by the Member States until proved otherwise, that notably simplifies the demonstration of the directive essential features respect. In particular, the Product Directives (there are also Social Directives and General Directives) are binding for the Member States as they must be acknowledged without any modification and they specify in a detailed way the modes to follow to obtain the attestation of conformity and the CE mark, obligatory for all the ruled commodities.

As regards the surgical system, the reference directive is the 93/42/CEE, concerning all medical devices (except the implantable active ones). The directive proposes a classification of devices in four danger classes, and it contains specific certification modes which become more complex with the growing level of risk for man. The surgical system considered belongs to the IIB class, for which an examination by an Advised Certification Organisation is obligatory not only for a prototype of the commodity and all its documentation, but also for the procedures followed by the firm in all the phases of the product realisation, from planning to production and to final inspections. Considering also the substantial absence of technical C norms for the device, mainly due to its great innovation compared to the traditional ones, it is even more important to perform an accurate risk analysis which examines each aspect and component starting from planning. In fact, the impossibility to follow agreed guidelines for the final system blocks the possibility of activating the "presumption of conformity" principle for the solutions adopted, and involves the necessity of demonstrating the conformity to the safety features in a more systematic and detailed way.

5 THE METHODOLOGY FOR THE SYSTEM ANALYSIS

In order to elaborate a correct analysis of the risks of the robotical system it is very important to refer, each time it is possible, to harmonised technical standards. In our case, even if we could not count on specific norms, it is possible to refer to general standards, A and B ones. In particular, the ones which guided the elaboration of an analysis methodology are: EN 292-1/2, which contains fundamental safety concepts and general design principles, EN 1050 which deals with risk evaluation and, above all, recent EN 1441, which is a revision of EN 1050 specific for medical devices. For what concerns B standards, that are useful in the resolution of problems regarding specific aspects or safety devices, many ones can help, but the more suitable for the presented case is EN 60601-1, dealing with electromedical devices.

The methodology elaborated for the analysis device follows substantially the one presented in EN 1441 standard, and it articulates in the following macro-phases:

1. Identification of qualitative and quantitative characteristics of the device: we have listed all the characteristics that could affect safety, on the base of a complex series of evaluations (use destination, type of operators, type of contact with the patients, characteristics of the materials used, forms of involved energy, etc.);
2. Identification of all possible dangers: starting from the lists presented in the I enclosure of directive 93/42 and in the C enclosure of the norm EN 1441, we have realised a list of dangers potentially present in the device, both in normal conditions and in first breakdown ones. To clarify the analysis, dangers have been divided in 6 classes, which are further divided, to involve all the possible dangers in a schematic way. Here are the 6 macro-classes we have established :
   A) Dangers due to forms of energy in the device
   B) Environmental dangers
   C) Biological dangers
   D) Dangers connected to the use of the device
   E) Dangers connected to the operation of the device
   F) Dangers connected to the human element.
3. Risks assessment for each danger: risk has been evaluated in relation to each danger, using data coming both from the normative, both from actual scientific acquaintances and from specific analysis. In order to obtain a reliable assessment for risks, we referred to a formulation with strong international agreement, the semi-quantitative method presented also in EN 1050, which estimates risks considering two principal factors, the entity of the damage and its probability of occurrence. The correspondent risk assessment is then obtained from a chart of the type in table 1 (the number of probability classes has been increased in order to improve the precision of the method).

The norms don't explain how to obtain these two values, and so we have found three elements which are simpler to evaluate and which, appropriately combined, furnish an evaluation of the probability of occurrence of the damage, while the entity of the damage has been simply divided in 4 classes in relation to the gravity of the injury and to the duration of its effects. These elements, which have been then sub-divided again in order to obtain more elementary entities, are: the probability of occurrence of a dangerous event (connected to the expected number of events in the useful life of the device), the amount of time spent by the subject in the dangerous zone compared to the total time of use of the device, and finally an index of prevention, which involves factors like the capability to avoid the damage when the danger occurs, the preparation of the operators and the existence of protections. We have then prepared some charts, as in figure 2, to obtain the value of the desired index. In this chart, in particular, we obtain the probability of occurrence of the damage, with one of the 7 indicated values, starting from the values of the 3 constitutive elements, expressed also in discrete values: for example, the probability of occurrence of the negative event go from W1 (minimum) to W3 (maximum).

Table 1. Risk assessment table

	Entity of the Damage			
Probability of Occurrence	Catastrophic	Critical	Secondary	Negligible
Total	8	6	4	2
Frequent	7	5	3	1
Probable	6	4	2	0
Occasional	5	3	1	0
Rare	4	2	0	0
Improbable	3	1	0	0
Incredible	2	0	0	0

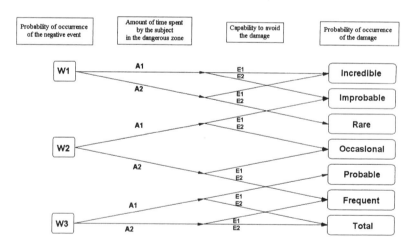

Figure 2. Index of prevention chart

4. Acceptability of the risk: this evaluation is one of the critical phases of the safety study, as it involves aspects which are different from usual technical considerations,

A003/021/2003 © IMechE 2003

like the definition of the "state of technological art" of the moment, the comparison between risks and benefits introduced by the device and those consequent to the clinical mode, and also evaluations of economic feasibility. In the cases where it was possible to refer to solutions presented in technical norms this problem was quite easy to solve, while in the other situations we had to consider all the above-mentioned aspects and establish a level of acceptability varying from case to case.

5. Risk's reduction: in order to reduce risks to an acceptable level we can use: direct safety (that is to say a different design of the device), indirect safety (obtained through means of safeguard, like protections) and finally, when we still have consistent "residual risks", descriptive safety (and so prescriptions that give instructions or restrictions on the use of the device). The "safety integration principle" that's one of the foundations of the directive 93/42/CEE, requires that risks must be reduced first in the design phase, and only when that is not possible, considering technical considerations and economic feasibility, with the adoption of protections or, as last resource, with prescriptions of use. It's important to remark that the discretion of the device builder in these evaluations has been strongly reduced from the modes of certification that require, for the most dangerous commodities, the examination of the device and of all the documentation by a Certification Organisation Advised to the European Union.

6. Generation of other dangers: we have to evaluate if new solutions adopted could lead to the introduction of new dangers or not.

7. Evaluation of all dangers: the process is repeated, returning to the third point, till when all identified dangers have been evaluated.

8. Final relation of the analysis: we have to prepare the final document with the results of the risk analysis, in order to establish whether the risks associated with the device use are acceptable or not.

9. Revision of the analysis: each time new information from technological development or from normative are available, or the device or its destination of use are altered, a revision of the analysis is necessary.

6 RESULTS OF THE ANALYSIS

The document entitled "Risk assessment and management" contains the result of the analysis, which consists of the risk assessment for patient, operators and other people, and also of the risk management process, which establishes the acceptable level of risk in each specific case and contains the list of solutions found in order to reduce this value. The document is organised according to the 6 classes of danger indicated, which are then sub-divided up to the elementary dangers found in the system. In order to give more detail, for each danger, we have valued the latent risk for patient, surgeon and other people, in a range from 0 (no risk) to 8 (total risk). If the risk was not acceptable, we have found new design solutions to reduce it and additional protections adopted when these solutions were not enough. Then we have established which tests should be run in order to verify some characteristics of the system, both imposed by the normative or studied ad hoc. Then we have calculated the so-called residual risk, that's the risk which is still present in the system after the adopted measures have been taken. For this risk, that's now reduced to acceptable levels, we have then indicated the prescriptions of use (descriptive safety), that can be in the form of guide lines to furnish to the user through the manual of use, through labelling of the device or through training, and they point out dangerous zones, incorrect uses to avoid or other, according to the case.

7 CONCLUSIONS

The development of robotical systems for surgical applications has been very fast and substantial in the last years. The level of improvement achieved for these systems can already guarantee some advantages for the patients compared to traditional techniques. Most recent tendencies lead to always more "active" and flexible systems, that would justify the production of devices which are still very expensive. One of the principal "bottle-necks" that slow down the development of such systems to a commercial level, is probably the almost total absence of specific technical norms, that forces the builders to find by themselves the technical solutions to make their commodities safe, with an evident waste of resources and contrasts among the various sectors involved. The development of guide-lines with strong international agreement will allow the proliferation of robotical systems in the next years also in surgery, and the attainment of notable advantages as previously stated. In particular, the system will soon begin the clinical tests in the Niguarda Hospital of Milan, and then the procedures for the examination by an Advised Certification Organisation and the attainment of the European Conformity Certification.

REFERENCES

1 **Davies, B.** The application of robotics to surgery. World Medical Technology Update 1996/7, pp. 144-146.

2 **Rovetta, A., Sala, R., Bejczy, A.K.** and **AA. VV.** (1994) A transatlantic telerobotic experiment for laparoscopic surgery, Robotics and Manufacturing vol. 5, ASME Press Series, pp.261-265. New York, USA.

3 **Gentili, E., Rovetta, A., Reschigg, S.,** and **Valentini, G.** (1997) Safety and surgical robots, Proc. 32nd MATADOR Conf., pp. 559-563. Manchester, UK.

4 **Louhisalmi, Y.** and **Leinonen, T.** (1995) Development of a prototype surgical robot, Proc. 9th World Cong. IFToMM, vol. 3, pp. 2166-2170. Milan, Italy.

5 **Sala, R.** (1995) Construction of a new automatic telemeter for medical applications and robotic telesurgery, Proc. 9th World Cong. IFToMM, vol. 3, pp. 2171-2174. Milan, Italy.

6 **Davies, B.L.** (1993) Safety of medical robots, safety critical systems, Book Chapman Hall, part 4, ch. 15, pp. 193-201.

A003/021/2003 © IMechE 2003

An active walking aid system for the aged

T HONDA, J OKAMOTO, Y KIKUCHI, and **Y KUSHIHASHI**
Department of Computer Science, Tokyo University of Agriculture and Technology, Japan

SYNOPSIS

A walking aid device for the aged persons has been developed and tested. The purpose of the study is to develop an wheeled frame which can provide the mobility and balance/support of a human body, including rehabilitation of walking function for the aged persons. In these times of increase of aged persons, we have to hold the ability of self-mobility, that is, the ability to walk by ourselves as long as possible in our aged period. The system is assumed to apply to the persons who can walk a little but with difficulty. It supports the subject's body at the armrests, and the sensors detect the distribution of the forces on the frame to control locomotion of the device and its user. For forwarding, the system recognizes the walking intension of the person based on those data, and the two actively controlled wheels proceed the body so as to move the non-supporting leg forward to take a step. Another control to keep support of the person' body is also performed. The implementation and experiments have shown the effectiveness of the system.

1 INTRODUCTION

Aged persons (the aged are usually called for the persons who are over sixty five years old)are apt to have troubles in their walking. A walking aid device for the aged persons has been constructed and tested. The purpose of the study is to develop an wheeled frame which can provide the mobility and balance/support of a human body, including rehabilitation of those functions. In Japan, the ratio to the total population of the people over sixty five years old, so-called the aged, became 16.7 % in 1999 A.D., and is estimated to be 25 % in 2015. In the world, the aged occupied 6.9 % in 2000, and will have the ratio 16.4 % in 2050. In these times, we have to hold the ability of self-mobility, that is, the ability to walk by ourselves as long as possible in our aged period. The system is assumed to apply to the persons who can walk by themselves but need to have some assistance. For such assistance, conventional walking assistance goods are there. Augmenting actively driving function to those assistance tools may extends the human activities. The device proposed here supports the subject's body at the armrests, and the sensors detect the distribution of the forces along X,Y and Z axes on the mechanical structure to locomote the device body and the user. For forwarding, the system recognizes the walking intension of the person based on those data, and the two actively controlled wheels proceed the body so as to move the non-supporting leg forward to take a step. Another control to keep support of the person' body at staggering is also performed. The implementation and experiments have shown the effectiveness of the system. The prototype system is called as "Pegasus-1" as its nickname.

2 DESIGN CONCEPT

The design purposes of the system have been listed as follows:

1. To aid or assist smoothly walking of the persons who need some mechanical support during walking
2. To give a means for rehabilitation of walking to persons who temporarily have some troubles at their feet or legs

As the conditions, it has been assumed that the walking is limited to the one along linear direction, and the user's upper half of body is healthy.

For those purposes, the design policies below has been set:

a. The mechanical structure is both to support the human body and to assist the locomotion of him.
b. The mechanical body is open to the forward direction to walk, that is, nothing is there in front of the human body not to make the subject feel presence of occlusion or blockaid.
c. The system actively locomote the subject, namely, the mechanical body is actively driven at need.
d. Intention of walking to walk and stop is transmitted to the system by the subject's action itself, that is, there is no need of control, operation, handling or manipulation.
e. The system to support the human body against the sudden staggering, too.
f. Nothing including sensors is attached or mounted to the subject's body not to disturb the user physically and mentally.

    A003/046/2003 © IMechE 2003

3 SYSTEM CONFIGURATION

3.1 Total structure

To fulfil the design concept shown in chapter 2, the elements required to the total system are mainly a mechanical body, a control system to actively drive the mechanical body and sensors to recognize the intention of the subject for walking including the speed and stopping. As the system is of course to move together with the user, those elements described above are to be put together to one body. That is, the control system and the sensors are to be mounted to the mechanical body.

3.2 Mechanical body

A mechanical body shown in Figure1. as the structure has been designed and constructed. The structure looks like con____ional walking aid tools without active driving function in the external appearance.

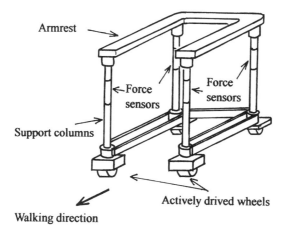

Figure 1. Mechanical structure of body

3.3 Sensors

According to the item f. of the design policies shown in chapter 2., the sensors are to be mounted on the mechanical body. To detect the subject's intention to walk and stop, furthermore to detect sudden unintentional staggering, detection of strains in the mechanical structure caused by forces delivered by a human body are essential. From the reason, force sensors using strain gages have been mounted on the four support columns of the mechanical body so as to detect the three directional components of forces. The axes for force components are defined as ,
    X axis: along forward direction,
    Y axis: along lateral direction,
    Z axis: along downward direction.

According to the axes, the forces are noted as,

F^X_L : Force along X axis of the left support column,

F^X_R : Force along X axis of the right support column,

F^Y_L : Force along Y axis of the left support column,

F^Y_R : Force along Y axis of the right support column,

F^Z_L : Force along Z axis of the left support column,

F^Z_R : Force along Z axis of the right support column.

3.4 Controller

For the availability and portabiliy, a note type PC has been used as the main controller and datalogger. In order to reduce the PC's load, a PIC(Peripheral Interface Controller, a micro controller with CPU and some interfaces) is utilized for A/D conversion of sensor signals, PWM(Pulth Width Moduration) drive of actuators, etc. The configuration of the system is shown in Fig.2

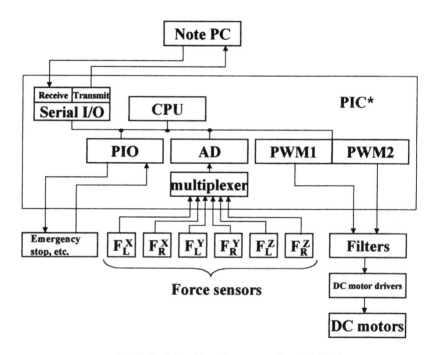

* PIC: Peripheral interface controller with CPU.

Figure 2. Configuration of control system

     A003/046/2003 © IMechE 2003

4 IMPLEM ENTATION AND EXPERIMENTAL RESULTS

4.1 Obsevations of walking signals

The prototype of the system Pegasus-1 has been constructed as shown in Figure 3., and the preliminary experiments have been performed to observe the behavior especially of F^Y_L, F^Y_R, F^Z_L and F^Z_R during walking. As human body sways from right to left or vice versa for every stepping, it is expected that the phase difference between F^Z_L and F^Z_R is near to π, and between F^Y_L and F^Y_R to 0.

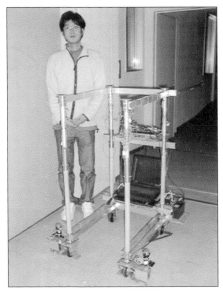

Figure 3. Whole external view of Pegasus-1

In Figure 4., two examples of the detected force signals are shown, (a) is in a good condition and (b) an ill one in contrast. Sliding walking with mall strides is preferable to getting good conditioned signals. On the contrary, walking by lifting legs high makes the force signals distorted.

From the experiments, the following results have been got:

(1) The phase differences described above have been roughly shown, but not a few scattering have appeared.
(2) Double peak have offen appeared around the maximum or minimum force value especially at the wide stride. It makes the recognition of phase difference more difficult.

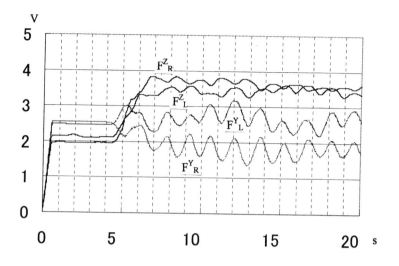

(a) Force signals in good condition

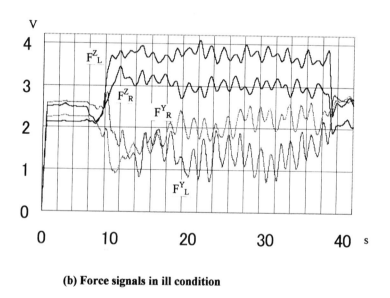

(b) Force signals in ill condition

Figure 4. Examples of detected force signals during human walking

A003/046/2003 © IMechE 2003

4.2 Control algorithm and experimental results

Based on these views, the final algorithm has been determined F^X_L and F^X_R being added for

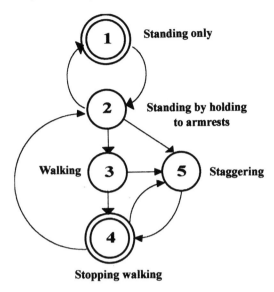

Figure 5. State transition diagram to control Pegasus-1

the speed control factor at walking and the recognition condition at human stopping and staggering. The total algorithm is shown in Figure 5. as a state transition diagram. In the Figure 5., the state numbers corresponds to the ones in Table 1. Table 1. shows the detailed control algorithm corresponding to the motions of human body. Figure 6. is a scene of experiment.

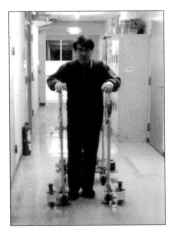

Figure 6. A scene of experiment

Table 1. Corresponding actions between human action, sensor's response and Pegasus-1

State	Human action	Forces to be detected	Expected action of Pegasus
1	Standing only without touching armrests	All $F^i_j \fallingdotseq 0$ $\cdots (1)$	Stop
2	Standing by holding to armrests	$F^Z_L \geqq F^Z_{Lth2}$ $F^Z_R \geqq F^Z_{Rth2}$ $F^Y_L \geqq F^Y_{Lth2}$ $F^Y_R \geqq F^Y_{Rth2}$ Others $\fallingdotseq 0$ $\cdots (2)$	Stop
3	Walking	$\Delta F^Z_L \leqq 0$ $\Delta F^Z_R \leqq 0$ $\Delta F^Y_L \leqq 0$ $\Delta F^Y_R \geqq 0$ $\cdots (3)$	Go with $V \propto F^X_L, F^X_R$
4	Stopping walking	$F^X_L \leqq F^X_{Lth4}$ $F^X_R \leqq F^X_{Rth4}$ $\cdots (4)$	Stop
5	Staggering	$F^X_L \geqq F^X_{Lth5}$ $F^X_R \geqq F^X_{Rth5}$ $F^Y_L \geqq F^Y_{Lth5}$ $F^Y_R \geqq F^Y_{Rth5}$ $\cdots (5)$	Stop

Notes: (1) V in the state 3 is the running speed of Pegasus.
   (2) Notation "th" in the subscript of F means the threshold value.

Figure 7. shows a result from experiments. In the figure, the subject almost only was standing without holding the armrests during the first 5 seconds. Then forces along some directions were put, and at 10 seconds later the conditions of the phases and the values being satisfied, Pegasus-1 began to go forward. After about 16 seconds gasus stopped by subject's staggering. Around 20 seconds, unpreferable chattering happened.

   A003/046/2003 © IMechE 2003

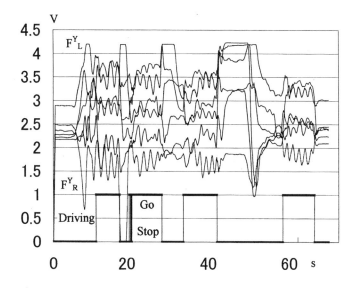

(a) Walking including staggerings

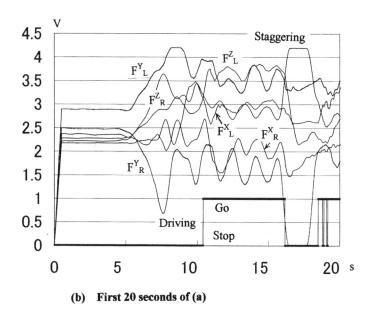

(b) First 20 seconds of (a)

Figure 7. A result of experiments : all of forces and control signals

5 CONCLUSIONS

A prototype of an actively drived walking aid system has been developed, and through the experiments the following results have been got.

(1) A walking aid system has been devised using actively driving mean through a convenient human interface.

(2) As a new interface, a possibility of a method to use the force information put from the user's body to the structure of the device has been shown.

(3) Accuracy of the recognition of human action and the control performance depends on the pattern of walking. To adapt miscellaneous walking pattern is the problem to cope with.

REFERENCES

(1)**Hiratsuka, M.,** and **Asada, H. Harry** (2000) Detection of Human Mistakes and Misperception for Human Perspective Augmentation: Behavior Monitoring Using Hybrid Hidden Markov Mmodels. Proc. of the 2000 IEEE Int. Conf. on Robotics & Automation, pp.577-582.

(2)**Wasson, G., Gunderson, J., Graves, S.** and **Felder,R.** (2000) Effective Shared Control in Cooperative Mobility Aids, American Association for Artificial Intelligence (www.aaai.org).

(3)**Bertani, A., Cappello, A., Benedetti, M. G., Simoncini, L.** and **Catani, F.** (1999) Flat foot functional evaluation using pattern recognition of grand reaction data, Clinical Biomechanics, 14, pp.484-493.

A003/046/2003 © IMechE 2003

Computer-aided Orthopaedic Surgery

Design and production of ceramics implants using virtual environment

T ITO and **T SATO**
Department of Mechanical Engineering, The University of Tokushima, Japan

SYNOPSIS

Custom-made design on hard-tissue implants are required by the market. However, appropriate design for custom-made production needs preliminary operation, which is not easily accepted. In addition to this design issue, material issue of implants is also under discussion. Porous ceramics are regarded as a suitable material for hard tissue implants, because they are compatible with bone and hard-tissue goes into the pore, which makes the bone formation with ceramics in the end. But their fragile features are not suitable for mechanical processing to create the complicated shape to fit the damaged portion of bone. We propose a new solution to these two issues, using a virtual environment for design and production on ceramics implants. The paper presents the overview of the idea behind the approach and describes some of the key technologies in our study.

1 INTRODUCTION

As one of the most aging countries in the world, the number of plastic surgery operation in Japan is increasing at the annual rate of 8 ~ 10 %. Implant products used in those operations covers wide range of hard tissues, including artificial hip joint, artificial knee joint, and other artificial bone materials, and their total market size reaches several billions of Japanese Yen, which is a very big market.

Although a variety of product ranges is provided, available hard tissue implants are based on prefabricated mass-production. For custom design of implants, pre-operation has to be made before the main operation. However, multi-stage operation is too much hard for the patients and is not usually conducted. Therefore, one of the most appropriate shapes of implants is selected from the wide range of products, and is applied to the damaged portion of the bone. Since it is almost impossible to reshape the implants to fit the damaged portion of the bone at the time of operation, normal portion of the bone in operation must be reluctantly cut off to apply the products.

In addition to this design issue, material issue of implants is also under discussion. As for the material of implants, titanium and ceramics are very often used. Although porous ceramics are regarded as a suitable one for hard tissue implants which grow into the tissue to become compatible after a certain period of time, their fragile features are not suitable for mechanical processing.

The objective of the study is to make custom-made design of implants without pre-operation, and to produce the designed implants with porous ceramics. The paper describes the needs for custom-made design on hard tissue implants. Then, the paper proposes a virtual manufacturing approach to design and production on hard tissue implants, including digital bone modeling, implant shape design, and implant evaluation. The paper also covers the mechanical processing on porous ceramics to prepare any complicated shape to fit the damaged portion of bone. Concluding remarks will follow.

2 NEEDS FOR CUSTOM-MADE DESIGN ON HARD TISSUE IMPLANTS

Plastic surgery operation using implants has to be completed, satisfying various requirements within shortest time as possible. If it is dental operation, dental doctor can take times to design and preparation, or can conduct multistage operation if necessary. However, in plastic surgery operation, it is not usually acceptable to conduct pre-operation to design the shape of implants which should be exactly fit the damaged portion of bone. Pre-operation means a very hard burden on patients, which should normally be avoided. Therefore, one of the most appropriate shapes of implants is selected from the wide range of products, and is applied to the damaged portion of the bone. It may be possible to reshape implants at the time of operation to fit the damaged portion, which is normally very complicated shape, if time permits. However, to minimize the operation time, time consuming task should be avoided by any means. Therefore, it is suitable to reshape the bone to fit the simple geometric shape of implants as shown in Figure 1.

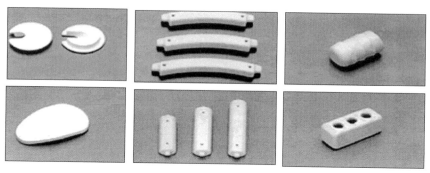

Figure 1. Popular shapes of ceramics implants

A003/015/2003 © IMechE 2003

Figure 2 shows two typical cases of damages portion of bone. In the left case, or the damages portion is in the middle of the bone, it would not cause any critical damage even if the bone is reshaped to fit the prefabricated implants. After the operation, bone tissue will grow into the pore of implants, which will become a part of the bone after a certain period of time. In the right case, however, the damaged portion is close to the shoulder joint, which is a very critical part of the bone. If the portion is reshaped to fit the prefabricated implant as shown in Figure 3, then the implant may be applied to the portion just like the previous case. However, the strength of the portion cannot be maintained, and it may cause the loss of whole joint portion. If the portion has not to be cut off in this way, the risk to lose the whole joint can be drastically reduced.

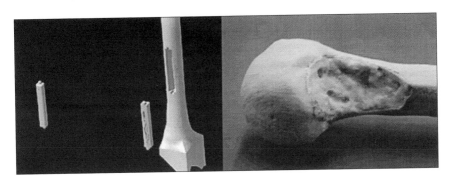

Figure 2. Examples of damaged portion of bone

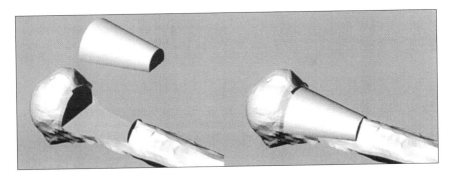

Figure 3. Plastic surgery using implant

To do so, the key solution would be associated with the custom-made design on implants which should be exactly fit the damaged shape of the bone, which is normally a very complicated shape. If the custom-made shape design is available, its production can be studied independently.

3 VIRTUAL MANUFACTURING APPROACH TO DESIGN AND PRODUCTION

To achieve the goal of custom-design on implants without pre-operation, or in other words, to evaluate the design, to produce the product in the virtual environment, and to provide the design data to the actual production site, we have been developing a virtual design and manufacturing environment.

From design to production, several stages are considered and modeled to implement the environment. The model includes: digital bone modeling, implant shape design, rapid prototyping and evaluation, and production.

3.1 Digital bone modeling

For implant design, digital model of the bone should be created precisely. We use medical data for the modeling, such as CT scan data, and creates digital bone model. Although the digital bone model is used for implant design, preparation of physical model is also available as shown in Figure 4. The upper bone in Figure 4 is a physically rapid-prototyped model based on the original mode shonw in the lower part of the figure. Direct scanning method is also applied to prepare digital bone model and its physical prototype model, which are used to evaluate the accuracy of modeling. Although digital modeling is an effective approach for evaluation, physical modeling is sometimes easier for evaluation. Physical modeling is always available if its digital model is prepared.

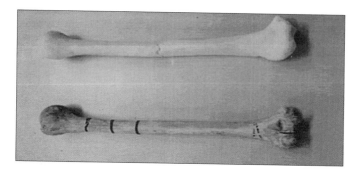

Figure 4. Physical model of digital bone and its original

3.2 Implant shape design

Implant shape is designed based on the digital model prepared in the previous stage. Figure 5 shows a sample design of implant, which is designed for the insert implant of damaged bone as shown in the right example of Figure 2. Compared to the prefabricated implant shown in Figure 3, the shape of the designed implant is exactly fit to the damaged portion, which means that cutting-off of the normal portion can be minimized. Once the digital design is made, its physical rapid prototype is also available if neccesary. Evaluation on both digital and physical model along with the bone model including the damaged portion makes it for sure to use the implant in the plastic surgery.

A003/015/2003 © IMechE 2003

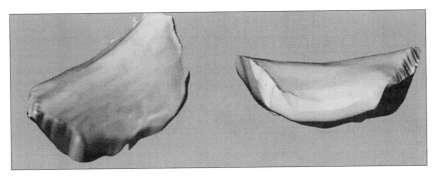

Figure 5. An exmple of designed implant

3.3 Implant evaluation

Designed implant is evaluated in the virtual environment, which can also be accessed via a network by participating person. Manupulating the implant and the bone, fitness of them is studied, which can also be used for giving information on the scheduled plastic surgery to the patient as an informed consent. As seen from the Figure 6, the designed implant is much more suitable for the damaged portion, which means that cutting-off of the normal portion can be minimized. For the use of cuscom-made implants, the risk to lose the joint function can be as minimal as possible. The design being provided to production site, the appropriate i
mplant can be manufactured.

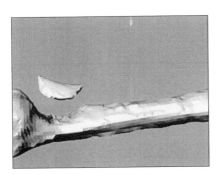 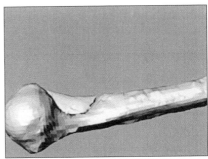

Figure 6. Implant evaluation in virtual environment

4 PRODUCTION OF IMPLANTS WITH POROUS CERAMICS

With the design of implant prepared in the virtual environment, its production can be available, if it is made from easily process-able materials, such as metal or hard ceramics. However, it is completely different in the case of porous ceramics.

The potential advantage offered by a porous ceramics implant is its inertness combined with the mechanical stability of the highly convoluted interface developed when bone grows into the pores of the ceramics. Mechanical requirements of the prostheses, however, severely restrict to the use of low-strength porous ceramics to low-load- or non-load- bearing applications. Studies show that, when load bearing is not a primary requirement, nearly inert porous ceramics can provide functional implants. When pore sizes exceed 100micro m, bone will grow within the interconnecting pore channels near the surface and maintain its vascularity and long-term viability. In this manner, the implant serves as a structural bridge and model or scaffold of bone formation. The microstructure of certain corals makes an almost ideal investment material for the casting of structure which highly controlled pore sizes.

To prepare the hard tissue implants using the porous ceramics based on the STL data, we apply a new method using wax combination. Porous ceramics material comprising a special wax is thermally processed to stabilize in its physical shape, and prepared as a starting material for hard tissue implant. On the contrary to normal porous ceramics, which are too fragile to be processed, mechanical processing can be available on this waxed ceramics. After the processing, the unnecessary wax is to be removed by second thermal processing. The picture on the right side of Figure 7 shows an example of sculptured ceramics after the second thermal processing, which can hardly be processed by a normal method due to fragile features of the porous ceramics.

Figure 7. Mechanical processing on porous ceramics material with special structure

The picture of the left side of Figure 7 shows a starting ceramic bar with special inner structure so that bone tissue can much easily grows into the ceramics. Although the structure is much more fragile than that of normal porous ceramics, mechanical processing to produce any complicated shape can also be possible with our method. This special structure will accelerate the bone formation, however, fragility of the structure may raise another issue which we will report as a separate study.

A003/015/2003 © IMechE 2003

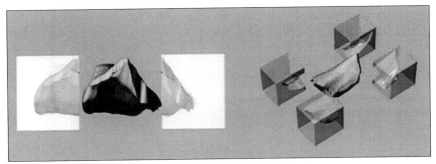
Figure 8. Virtual production of implant

5 CONCLUDING REMARKS

The paper described the current status of plastic surgery regarding hard tissue implants, and pointed out the critical issue for requirement on custom-made design and its difficulty in preparation for hard tissue implants. As a new approach to hard tissue implant design and production, we proposed a custom-made design approach using virtual manufacturing environment. Our study focuses on precision process for porous ceramics for hard tissue, and development of a concurrent design system for hard tissue implants. The paper covered the basic idea and picked up some key technologies in the system.

In the final stage of the project, medical doctor collaborates with engineers to design the most appropriate shape of hard tissue implants, evaluates it on a virtual environment, and prepares it as a rapid prototyping product. After the evaluation, the implant design will be transferred to the manufacturing site through the network, produce the implant, and deliver it to the doctor in a very timely manner. During the preparation, the doctor can present the patient what sort of material is going to be used for the operation under the virtual environment, and even show the rapid prototype product.

Collaborative implant design using virtual manufacturing approach of our study provides a new approach to plastic surgery operation using implant products, and a solution to ease the pain of patients.

REFERENCES

1 **Backstrom, M, et al**, 1998, CSCW tools for teleworkers and members of distributed teams, Euromedia'98, pp.167-176.

2 **Chiroff, R.T., White, E.W., Webber, J.N., and Roy, D.M.**, 1975, Tissue ingrowth of replamineform implants, J. Biomed. Mater. Res. Symp. 6:29-45.

3 **Cunha, G.G., Lopes, M.C.S., Lnadau, L., and Ebecken, N.F.F.**, 2001, SPAR - a realistic virtual environment offshore floating platform simulation system, European Simulation Symposium, pp.135-139.

4 **Fukuda, S. Matsuura, Y. and Dzbor, M.**, 1998, Internet-based motion still learning, Euromedia'98, pp.218-224.

5 **Godehardt, E., Koch, J.A., Pieper, P., Bergstedt, S. Kesper, B., and Moeller, D.P.F.**, 2000, From the skull to the face: inverse 3D-simulation in archaeology and forensic medicine, pp.377-381.

6 **Hartkopf, V, Shankavaram, J, and Loftness, V.**, 2000, European Concurrent Engineering Conference, Apr.17-19, pp. 5-12.

7 **Hench, L.L. and Ethridge, E.C.**, 1982, Biomaterials: An interfacial approach, Academic Press, New York.

8 **Hench, L.L.**, 1991, Bioceramics: From concept to clinic, J. Am. Ceram. Soc., 74/7:1487-1510.

9 **Holmes, R.E., Mooney, R.W., Buchloz, R.W., and Tencer, A.F.**, 1984, A coralline hydroxyapatite bone graft substitute, Clin. Orthop. Relat. Res., 188:282-292.

10 **Hulbert, S.F., Bokros, J.C., Hench, L.L., Wilson, J., and Heimke, G.**, 1987, Ceramics in clinical applications: past, present, and future, in Hight Tech Ceramics, Edited by P. Vincenzini, Elsevier, Amsterdam, Netherland, pp.189-213.

11 **Ito, T.**, 2002, Collaborative implant design using virtual manufacturing approach, International Seminar on Digital Enterprise Technology, Durham, United Kingdom, pp.87-90.

12 **Ito, T. and Sato, T.**, 2002, Virtual environment for for collaborative design on hard tissue implants, European Simulation Symposium, Dresden, Germany, pp.153-157.

13 **Kesper, B. and Moeller, D.P.F.**, 2000, Temporal database concept for virtual reality reconstruction, 2000, European Simulation Symposium, pp.369-376.

14 **Dos Santos, C.L.N., Cunha, G.G., and Mello e Luiz Landau, L.F.N.**, 2000, Simulations at petroleum industry using virtual reality techniques, The international workshop on harbour, maritime & multimodal logistics modeling and simulation, pp.10-15.

15 **White, E.W., Webber, J.N., Roy, D.M., Owen, E.L., Chiroff, R.T., and White R.A.**, 1975, Replamineform porous biomaterials for hard tissue implant applications, J. Biomed. Mater. Res. Symp., 6:23-27.

16 **Zini, A. and Rocca, A.**, 2000, Building a virtual ship, The international workshop on harbour, maritime & multimodal logistics modeling and simulation, pp.35-40.

A003/015/2003 © IMechE 2003

On agile manufacturing of custom hip stems

G-C VOSNIAKOS, A KRIMPENIS, P ZAMPAKOS, D PANTELIS, C PROVATIDIS, and B LEKOU
National Technical University of Athens, Greece
K KAROUZOS
Technological Education Institute of Athens, Greece

SYNOPSIS

Custom hip stems are preferred to standard shapes for total hip replacement when the patient's anatomy is outside common limits and / or cement-less systems are necessary. An integrated approach for developing a custom hip stem manufacturing system with flexibility and fast response to 'orders' is presented in this paper. The general framework is identified first. Work conducted so far on material selection involves experiments using Ti-6Al-4V or alumina on the stem and UMWPE on the cup in order to identify wear evolution with time and friction coefficient. Work on CAM consists of parametric modelling of the uncut block regions for fast rough cutting, optimisation concepts for z-heights of cut slices and design of experiments-based cutting condition and roughing strategy selection.

1 INTRODUCTION

Custom-made implants, as opposed to those produced in standard series, are required usually in facial or skull operations. In total hip replacement operations they can solve problems such as : loosening associated with cemented implant fitting by using press-fit techniques, implant shape peculiarities due to patient anatomy and material peculiarities due to bio-compatibility, strength and erosion characteristics in association with geometric parameters of the design (1,2). In order to widen use of custom made hip implants, however, special techniques must be employed in order to design their form and select the materials involved according to the patient's parameters and in order to make them quickly and cost-effectively.

As far as form design is concerned, this could be based on MRI scans or even radiographic images from which a 3D model of the implant may be constructed. Interactive software that can help surgeons define implant geometry, thus minimising design time, is presented in (3). The parameters of hip implant geometry have constituted the object of clinical studies (4) and can increase the level of automation of parametric design software.

As far as material selection is concerned, there have been numerous experimental studies. Metallic materials such as stainless steels, titanium alloys, superalloys and even composites have been tried for the load-bearing functionality part (substrates), and various coatings such as zirconia, alumina etc have been applied to the substrates in order to ensure low wear and biocompatibility (2). Such studies may be of particular importance to the custom-made implant area because of the peculiar connections of shape design parameters and material wear, e.g. a particular femoral stem head size is associated with higher wear rate and larger wear particle size distributions, depending on material used (5). Tribological characteristics of coatings fitted on particular substrates may be studied systematically in order to decide which combinations perform best in particular circumstances or custom-designed implants (6).

As far as manufacturing of custom-made implants is concerned, there are no standardized methods, not even for series manufacturing (2). Forging and casting in dies are associated with series production. Casting in conjunction with rapid prototyping could be adopted for custom manufacturing, whereas cutting from a solid billet or bar is normally the typical method for imparting the primary shape to the implant, but in order to make it economical CAD/CAM should be employed and cutting should make use of multi-axis high-speed machines (7). Such techniques should make provision for tight tolerances and good surface finish. Coating and surface modification processes should also be prescribed and conducted according to current practice (PVD, CVD, plasma etc.)

According to the concepts sketched above, a complete system addressing custom design and manufacture of hip implants could consist of the following parts :
1.Prosthetic shape acquisition, which should construct a geometric model from MRI or even radiographic data. There are already commercially available software programs addressing this need and, currently, their functionality is deemed satisfactory.
2.Prosthesis design, where a parametric CAD model of the implant is constructed incorporating the most important shape variations. The basic form parameters (head diameter, support for cup etc.) should conform to patient details and to results from tests and clinical evidence.
3.Selection of material for the stem and cup by matching details of the particular case to experimental material tests, simulation tests including FEA, and experience from clinical evidence.
3.Design of tool paths for CNC machining and selection of cutting conditions with a view to reduce time and increase quality.

Whereas the second area above is concerned with design of the implants and falls outside the scope of this paper, work on the last two areas is presented next and is subject to continuous research effort.

2 MATERIAL SELECTION STUDIES

Polymeric materials are suggested for the fabrication of cup in ball-socket arthroplasty, while the counterface (artificial femoral head) is made of ceramic (Al_2O_3-based or ZrO_2-based) or a corrosion resistant alloy (usually titanium or stainless steel). However, polyethylene wear debris, in some cases, lead to osteolysis, loosening and premature failure of artificial joints. Alumina is a bioceramic with low friction coefficient, low wear rate and outstanding chemical inertness. Titanium alloys and especially Ti6Al4V possess high biocompatibility, mechanical

A003/024/2003 © IMechE 2003

strength, fatigue and corrosion resistance. Surface treatments (thermal oxidation, oxygen diffusion, PVD-DLC, ion-implantation) can improve wear resistance of titanium alloy-polyethylene pairs.

Tribological properties and associated wear mechanisms of the UHMWPE/Alumina and UHMWPE/Ti6Al4V pairs under lubricating sliding conditions (using serum), were determined experimentally under loads of 70, 85 and 95 N, simulating realistic conditions using pin-on-disk tribotesting device, according to ISO 3696 standard. Wear is expressed as the average wear rate $(mm^3.mm^{-2}.m^{-1})$ calculated by the volumetric wear (mm^3) divided by the total sliding distance (m) and the contact surface (mm^2). The wear tracks were examined by a stereomicroscope and an optical microscope while wear debris were analyzed by SEM (8).

There are three points of view under which the two pairs of materials can be compared : coefficient of friction, wear rate and wear mechanisms. The UHMWPE/Alumina combination possesses a relatively constant friction coefficient (0.037 ± 0.002) as compared to that of the UHMWPE/Ti6Al4V pair, for which the friction coefficient increases abruptly from 0.01 to 0.058 and then decreases to 0.053 when normal load varies from 70 to 85 and then to 95 N. As for wear rate, for the combination of UHMWPE/Alumina an increase is observed with the applied normal load (10.5, 11.2 , 23 $* 10^{-2}$ $mm^3/m/mm^2$ for 70,85 and 95 N load). For the combination UHMWPE/Ti6Al4V, negligible wear rate values were observed (<0.78 $* 10^{-2}$ $mm^3/m/mm^2$).

According to SEM study of wear marks and signs, the UHMWPE/Ti6Al4V pair exhibits a progressive pitting, delamination and adhesive wear mechanism as the applied load increases. For the UHMWPE/Alumina pair, a transition of the wear mechanism from ploughing based to cutting wear mode is observed when applied load is increased. A difference in morphology of the wear debris is observed between the two material combinations. The UHMWPE/Ti6Al4V combination generates fine spherical/flaky shaped debris while UHMWPE/Alumina generates a coarser fibrous debris form. The change of Ra for the UHMWPE/Alumina pair is greater, see Fig. 1, verifying that wear mechanisms for this pair are more severe than in UHMWPE/Ti6Al4V, which is probably due to the marked difference in microhardness (Alumina/Ti6Al4V microhardness is 1900/250 HV respectively).

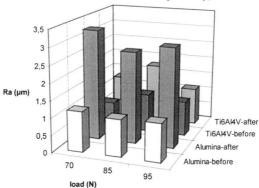

Figure 1. Variation in mean surface roughness.

3 PARAMETRIC COMPUTER AIDED MANUFACTURING TECHNIQUES

The implementation of the proposed methodology involves the optimal use of a CAM system. For this purpose, Delcam's PowerMill was chosen.

3.1 CAM system functionality

Delcam's PowerMill is CAM software that produces collision free toolpaths for CNC machine tools from CAD models. PowerMill reads both surface and solid models from IGES, VDA-FS, ACIS (SAT), STL and triangulated DUCT files. It offers a variety of roughing and finishing strategies and animates the produced toolpaths as most CAM software does. A post processor can convert the resulting toolpaths in G-code of an appropriate CNC machine tool, so that the custom implant can be cut.

Although CAM systems can deliver toolpaths for any kind of surface, when segmentation of a complex surface is needed, the user must define the different patches and apply the appropriate toolpath in each one. Moreover, the resulting toolpaths are not optimised in any sense but that of collision avoidance. Keeping this in mind, user-developed algorithms customise the regions for roughing and finishing of the applicable surface. This can be achieved either through the native macro language or through external programming. In Powermill '.mac' files are text editor files that batch-process commands. Macro language programming results are presented next.

3.2 Roughing

3.2.1 Block definition for roughing by region

A simplified hip implant can be as seen in Fig. 2. This can result from 3D model reconstruction of 2D tomography scans at predefined distances depending on the desired tolerance.

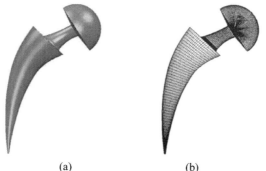

(a) (b)

Figure 2. Simplified model of hip implant (a) solid model (b) meshed model

A CAM system's primary objective is to create CNC toolpaths and to simulate CNC cutting. The simulation process needs the definition of the used tool, the cutting – roughing and finishing – strategies, the original rough block to be machined etc. A basic step in the whole process is to define the exact rough block to be machined. When a surface is quite complex the user breaks it down into smaller parts, each of which can be machined with one tool and a certain strategy in order to achieve optimisation of criteria such as minimum cutting time, maximum material removal etc. Otherwise, a single tool and strategy for the whole part would lead to lengthy toolpaths and/or too much remaining material on the part, see Fig. 3. To that purpose, appropriate macro code was written, namely the Block.mac.

Figure 3. Sub-optimal cutting of hip implant model.

This macro results in the creation of .pic files, each of them defining the boundaries of a cutting region, i.e. a block. The block files are : 'whole.pic', 'body.pic', 'neck.pic', 'head.pic', and 'hard.pic'. Dividing the original surface into regions makes it possible to apply a different roughing and finishing strategy in each region Fig. 4(a) depicts the macro's resulting blocks and Fig. 4(b) the point map used in the code.

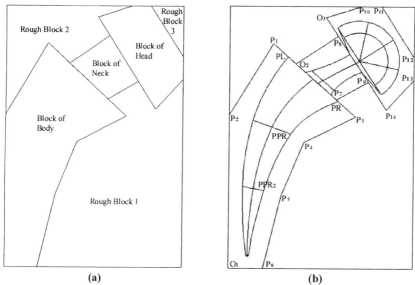

(a)   (b)

Figure 4. (a) Area map of resulting blocks (b) Point map for the Block.mac

As seen in Fig. 4(a), six areas are defined : Body, Neck, Head, Rough 1, Rough 2 and Rough 3. In fact, Rough 1, Rough 2 and Rough 3 are considered as one block, not physically, but as far as the applying cutting strategy is concerned. The limits of the blocks are calculated through the geometric characteristics (points, tangents and normals, magnitudes etc.) of their respective surfaces using equations in a parametric fashion. In some cases, the criterion of minimum distance from the surface was used, for instance for the points between Body and Rough 1 blocks, see Fig. 4.

The division of the primary block into parts is followed by tool selection and application of the different rough cutting strategies in each of them. Fig. 5 shows three successive stages of rough cutting the body, neck and head blocks.

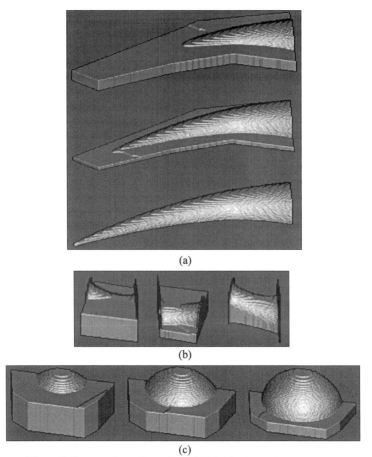

(a)

(b)

(c)

Figure 5. Stages of rough cutting of (a) body (b) neck and (c) head

3.2.2 Scallop height optimisation

In all machining processes a very important factor is surface roughness, so much so in the case of implants. Surface quality and machining time are two parameters that oppose each other. A need for very high surface quality results in the manufacturing time rising exponentially. This is why an appropriate optimisation criterion must be found in order to achieve the desired accuracy in reasonable machining time. The surface roughness can be indirectly evaluated via the concept of remaining material volume upon the desired surface. Machining time is related to feed and depth of cut.

A003/024/2003 © IMechE 2003

Most strategy parameters take real values, whilst tool geometric characteristics discrete values. The practical combination space is almost infinite. Artificial intelligence implementations help reduce the search space of these infinite possibilities. Off-the-shelf Genetic Algorithms (9) have been tried in case studies with good results.

A genetic algorithm is a stochastic optimisation procedure, which can solve complex problems by imitating Darwinian theories of evolution on a computer. The concept behind the creation of genetic algorithms is the global optimisation of an objective function in a complex multi-modal search space. Genetic operators, e.g. mutation, inversion, crossover, are applied to an initial population (solution) with an aim to improve its fitness.

A genetic algorithm was applied to the rough cutting optimisation of a hemisphere (10) as the first step of a broader objective. A variety of end-mill tools having different geometric characteristics were considered. Least machining time and least remaining volume were considered simultaneously. The value of the proposed method is that, for a typical toolpath, see Fig. 6(a), the optimal usage of the set of tools used is found and the height distribution of the created scallops is calculated, see Fig.6(b).

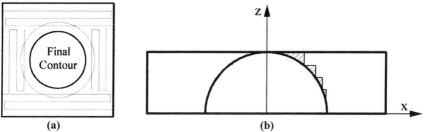

(a)   (b)

Figure 6. Roughing of a hemisphere (a) Toolpath and (b) Scallops.

Further generalisation involves different tool types (ball-end mills etc) and criteria for the correct choice of tool type in respect to definition of their 3D geometric effects on the cutting surface. Free-form surfaces such as those present in the implant model are much more difficult to deal with than a hemisphere. In that case, calculation of machining time and surface roughness need to make use of the CAM system API functions and the genetic algorithm must be tightly coupled with it. In addition, a narrowing-down exercise on the types of cutting strategy that must be considered for each type of surface should be conducted in order to keep the size of the problem manageable.

In order to recognise different types of surfaces and consequently assign them cutting strategies an obvious criterion is curvature, in particular constrained mean curvature as well as local minima / maxima of a region. In order to scan surfaces defined in different ways, e.g. explicitly as Bezier surfaces, implicitly through joining intermediate sections, etc. and consider curvature at different points, Powermill defines meshes. A macro named 'plegma.mac' was programmed in order to refine meshes in different parts of the model (viz. body, neck, head), see Fig. 2(b). An iterative refinement procedure was employed, that is controlled by a recursive subdivision index practically ranging from 1 to 4. The rationale is that surface properties are only available at mesh nodes, so enough of them should be created in order to draw conclusions on curvature changes along or across a surface.

3.3 Cutting strategy selection

All the possible strategies in the roughing and finishing domains include a relatively wide number of parameters that affect the machining time and the product's surface roughness. A major step in selecting the right one is to define influential parameters as well as the degree of influence. A systematic way to extract these influential parameters is Taguchi's Design of Experiments method. Design of Experiments (DoE) is a way to incorporate Taguchi's orthogonal arrays (OAs) to design and conduct fractional factorial experiments that collect all statistically significant data with the minimum possible number of repetitions. DoE dictates a series of steps to follow so as to yield improved understanding of product or process performance (11). Implementing DoE involves choosing the parameter levels (usually 2 or 3), parameter inter-dependency and the appropriate OA, as well as conducting the experiments accordingly.

In this case, a variety of roughing strategies were tested. For instance, in the raster roughing strategy the following parameters were studied : Raster Angle (orientation of the tool main direction), Step over distance (relative overlap between successive passes on the same plane), Tolerance (accuracy of curve approximation), Profiling, Lead in Moves (method of starting the cut), Z heights (number of horizontal parallel slices / vertical passes). Three levels were chosen for all parameters and the corresponding OA was L27, see Table 1, providing better resolution than L18 which was a candidate, too. The measured values were the machining time, the remaining volume and the remaining surface. Volume and surface were calculated using STLView software. The results were combined in a weighted function (respective weights were 0.4, 0.3 and 0.3). The best results correspond to Repetition 1 in Table 1.

Table 1. Design of experiments and optimisation results.

Rep. No.	Raster Angle	Step Over	Tolerance	Profiling	Lead in Moves	Z Heights	Machining Time (min)	Remaining Volume (mm³)	Remaining Surface (mm²)	Penalty Function
1	40	2	0,01	before	drill	25	11,4	102067	38940	19,3
2	40	2	0,05	during	ramp	27	23	101413	39078	26,8
3	40	2	0,1	after	plunge	30	33,73	100745	38957	31,1
4	40	3	0,01	during	ramp	30	23,68	101703	45223	27,1
5	40	3	0,05	after	plunge	25	24,05	101444	45471	27,2
6	40	3	0,1	before	drill	27	24,17	101543	45392	27,2
7	40	4	0,01	after	plunge	27	18,98	102631	48580	25,1
8	40	4	0,05	before	drill	30	19,12	102314	48824	25,3
9	40	4	0,1	during	ramp	25	18,65	102607	48774	25,0
10	60	2	0,01	during	plunge	27	33,65	100861	38280	30,9
11	60	2	0,05	after	drill	30	34,03	101079	38521	30,2
12	60	2	0,1	before	ramp	25	34,5	100825	38473	28,5
13	60	3	0,01	after	drill	25	24,25	101693	44885	26,2
14	60	3	0,05	before	ramp	27	24,72	101816	45126	26,5
15	60	3	0,1	during	plunge	30	23,9	101568	45046	26,1
16	60	4	0,01	before	ramp	30	19,88	102598	48207	24,5
17	60	4	0,05	during	plunge	25	19,05	102881	48422	24,3

 A003/024/2003 © IMechE 2003

Table 1. continued...

Rep. No.	Raster Angle	Step Over	Tolerance	Profiling	Lead in Moves	Z Heights	Machining Time (min)	Remaining Volume (mm³)	Remaining Surface (mm²)	Penalty Function
18	60	4	0,1	after	drill	27	19,43	102148	48355	24,4
19	auto	2	0,01	after	ramp	30	32,83	100901	38333	29,6
20	auto	2	0,05	before	plunge	25	32,8	101142	38634	29,6
21	auto	2	0,1	during	drill	27	32,43	101228	38582	27,5
22	auto	3	0,01	before	plunge	27	23,12	101879	45110	23,8
23	auto	3	0,05	during	drill	30	22,73	102299	45431	23,6
24	auto	3	0,1	after	ramp	25	23,12	102160	45341	23,7
25	auto	4	0,01	during	drill	25	17,7	102917	48641	21,8
26	auto	4	0,05	after	ramp	27	18,08	103303	48908	21,9
27	auto	4	0,1	before	plunge	30	18,08	103259	48811	21,8

4 CONCLUDING REMARKS

The work presented so far refers to two major processes pertaining to custom femoral implants, namely selection of the appropriate material and CNC cutting of the final shape.

In the first domain, the work so far is experimental in nature, trying to identify tribological characteristics of stem and cup material combinations. The work is being extended towards ultimately building a knowledge base that will advise the designer according to factors peculiar to patient anatomy, loads, type of activity and other relevant parameters that govern material choice. In order to achieve this more experiments need to be made, simulation mechanisms need to be materialised and finite element analysis needs to be taken on-board.

In the second domain, the work so far is computational in nature, trying to optimise CNC programs in terms of execution time and surface roughness. The work needs to be generalised in terms of surface types and tool types that can be tackled. In addition, more powerful multi-criteria optimisation techniques need to be connected on-line to the CAM software used, that will act as a cost function calculator.

A third domain which needs to be opened up is that of parametric design of implants according to data and knowledge coming from experience, tomography measurements and engineering calculations similar, if not the same, with those envisaged for the expansion of material selection activities.

Agility of manufacture would be achieved only when the above activities attain a sufficient level of completion so that it would, then, be fast and economical enough to produce custom implants instead of opting for standard ones.

ACKNOWLEDGMENT

This work is partly sponsored by General Secretariat of Research and Technology of the Greek Ministry of Development.

REFERENCES

1   **Schmalzried, T.P.,** and **Callaghan, J.J.** (1999) Current Concepts Review - Wear in Total Hip and Knee Replacements. Journal of Bone and Joint Surgery, 81(1), pp. 115 - 136.

2   **Dearnley, P.A.** (2000) A review of metallic, ceramic and surface treated metals used for bearing surfaces in human joint replacements. Proceedings of the Institution of Mechanical Engineers, Part H: Journal of Engineering in Medicine, Vol. 213, pp.107-135.

3   **Viceconti, M., Testi, D., Gori, R., Zannoni, C., Cappello, A.** and **De Lollis, A.** (2001) HIDE : a new hybrid environment for the design of custom-made hip prosthesis. Computer Methods and Programs in Biomedicine, Vol. 64, pp. 137 – 143.

4   **Kaneuji, A., Matsumoto, T., Nishino, M., Miura, T., Sugimori, T.** and **Tomita, K.** (2000) Three-dimensional morphological analysis of the proximal femoral canal, using computer-aided design system, in Japanese patients with osteoarthrosis of the hip. Journal of Orthopedic Science Vol. 5, pp. 361 – 368.

5   **Hirakawa, K., Bauer, T.W., Stulberg, B.N., Wilde, A.H. and Secic, M.** (1996) Characterization and comparison of wear debris from failed total hip implants of different types. The Journal of Bone and Joint Surgery, Vol. 78, pp. 1235-43.

6   **Liu, C., Bi, Q.** and **Matthews, A.** (2002) Tribological and electrochemical performance of PVD TiN coatings on the femoral head of Ti–6Al–4V artificial hip joints. Surface and Coatings Technology, (In Press).

7   **Werner, A., Lechniak, Z., Skalski, K.** and **Kedzior, K.** (2000) Design and manufacture of anatomical hip joint endoprostheses using CAD/CAM systems. Journal of Materials Processing Technology, Vol. 107, pp. 181 – 186.

8   **Pantelis, D.I., Pantazopoulos, G., Sarafoglou, Ch. and Lekou, B.** (2002) Wear Behaviour of UHMWPE-Al_2O_3 and UHMWPE-Ti6Al4V pairs for hip joint arthroplasty (submitted for publication).

9   **Giannakoglou, K. C., Giotis, A. P. and Karakasis, M. K.** (2001) Low-cost genetic optimization based on inexact pre-evaluations and the sensitivity analysis of design parameters. Inverse Problems in Engineering, Overseas Publishers Association, Vol. 9, pp. 389-412.

10 **Krimpenis, A.** and **Vosniakos, G.** (2002) Optimisation of Multiple Tool CNC Rough Machining of a Hemisphere as a Genetic Algorithm Paradigm Application. International Journal of Advanced Manufacturing Technology, (In Press).

11 **Montgomery, D.C.** Design and analysis of experiments. 3rd ed., John Wiley, 1997.

Index